ESO ASTROPHYSICS SYMPOSIA
European Southern Observatory

Series Editor: Philippe Crane

Springer-Verlag Berlin Heidelberg GmbH

Francesco Paresce (Ed.)

Science with the VLT Interferometer

Proceedings of the ESO Workshop
Held at Garching, Germany,
18–21 June 1996

Springer

Volume Editor

Francesco Paresce
European Southern Observatory
Karl-Schwarzschild-Strasse 2
D-85748 Garching, Germany
email: fparesce@eso.org

Series Editor

Philippe Crane
European Southern Observatory
Karl-Schwarzschild-Strasse 2
D-85748 Garching, Germany

Cataloging-in-Publication data applied for

Die Deutsche Bibliothek - CIP-Einheitsaufnahme

Science with the VLT interferometer : proceedings of the ESO workshop, held at Garching, Germany, 18 - 21 June December 1996 / Francesco Paresce (ed.). - Berlin ; Heidelberg ; New York ; Barcelona ; Budapest ; Hong Kong ; London ; Milan ; Paris ; Santa Clara ; Singapore ; Tokyo : Springer, 1997
(ESO astrophysics symposia)

ISBN 978-3-662-22399-4 ISBN 978-3-540-69398-7 (eBook)
DOI 10.1007/978-3-540-69398-7

Originally published by Springer-Verlag Berlin Heidelberg New York in 1997.
Softcover reprint of the hardcover 1st edition 1997

Typesetting: Camera ready by authors/editors
Cover design: *design & production* GmbH, Heidelberg
SPIN: 10552350 55/3144-543210 - Printed on acid-free paper

Preface

It has always been ESO's aim to operate the VLT in an interferometric mode (VLTI) which allows the coherent combination of stellar light beams collected by the four 8-m telescopes and by several smaller auxiliary telescopes. In December 1993, in response to financial difficulties, the ESO Council decided to postpone implementation of the VLTI, Coudé trains and associated adaptive optics for all the UTs but included provisions for continuing technological and development programmes devoted to the aim of reintroducing these capabilities at the earliest possible date.

The desirability of carrying out the full VLTI programme as originally envisaged at the earliest possible moment has not, however, diminished, especially in view of VLTI's exceptional capabilities and resulting potential for new and exciting discoveries. In recent years, interferometric projects have begun to play a central role in ground-based high-resolution astronomy, and numerous instruments have been completed or are in the process of construction. Several large-aperture interferometers will probably come on-line near the turn of the century. The impending presence of these new instruments represents an important incentive both for clarifying the scientific cases for various VLTI implementation plans and for ensuring VLTI's competitiveness in the international context over the next 10–20 years.

The primary scientific issues that the VLTI seeks to address are well defined and they can be used to drive the design of the facility and development of an implementation plan that will optimally exploit the various technological stages of the project and ensure their compatibility with a vigorous, yet realistic and timely, astrophysical programme. In order to study these issues and to establish a clear set of guiding principles for the development of VLTI, an Interferometry Science Advisory Committee (ISAC) was established in April 1995. This committee reviewed the technical status of VLTI, its scientific rationale as elaborated by past advisory panels, and the present recovery plans. The committee has defined and prioritised the key science drivers for the program and the technical specifications that flow from them. A summary of these recommendations were published in the March 1996 issue of the ESO Messenger, p.14. The last step in this process was to present these science drivers and technical recommendations to the wider community and to entertain new scientific ideas for its utilization and new technical concepts for its implementation.

To provide a forum for just such a discussion, ESO hosted a workshop on "Science with the VLTI", in Garching during June 18–21, 1996 with ISAC as the Science Organizing Committee. More than 100 scientists from all over the world participated in the meeting, assuring its unqualified success. Particularly encouraging was the presence of a substantial number of very young researchers whose interest and motivation are critically important to the continued progress of this emerging technique. This book collects most of the talks presented at the meeting together with the excellent technical tutorials given on the first day of the meeting for the benefit of the still significant number of scientists not yet conversant with all the intricacies and complexities of the science of interferometry. The organizers hope this book will serve well the purpose of preparing the community to carry out compelling scientific investigations with the VLTI and to make sure the VLTI satisfies its considerable expectations.

The workshop could not have taken place without the expert assistance from Christina Stoffer and Georgette Hubert in all practical aspects of its organization and smooth management. These proceeedings, on the other hand, would not have seen the light of day without the able collaboration of Pam Bristow. To all three, the organizers owe a heartfelt debt of gratitude for their efforts.

Garching, May 1997 *Francesco Paresce*

Contents

1. INTERFEROMETRY TUTORIALS (All You Wanted to Know About Interferometry But Were Afraid to Ask)

Introduction to Interferometry
T.R. Bedding 3

An Introduction to Interferometry with the ESO Very Large Telescope
O. von der Lühe 13

Interferometry in Practice
A. Quirrenbach 35

Astrometry with O/IR Interferometers
M. Shao 48

2. SOLAR SYSTEM, EXTRASOLAR PLANETS, LOW MASS STARS AND BROWN DWARFS

Solar-System Observations with the VLTI
T. Encrenaz 57

Giant Planet Formation and the Masses of Extrasolar Planets
G. Wuchterl 64

Theory of Very Low Mass Stars, Brown Dwarfs and Extra-Solar Giant Planets
I. Baraffe, G. Chabrier 72

On the Detection of Exo-Zodiacal Light by Nulling Interferometry with the Magellan Telescopes
W.A. Traub, N.P. Carleton, J.R.P. Angel 80

Direct Observation of Extrasolar Planets with an Infrared Interferometer
V. Coudé du Foresto, J.-M. Mariotti, G. Perrin 86

3. AGN AND STARBURST GALAXIES

VLTI Observations of Radio-Loud Active Galaxies
G.K. Miley 91
Observations of Extragalactic Sources Near Bright Stars
B. Théodore, P. Petitjean 100
AGN Models and Fringe Visibilities
F. Malbet, P. Petitjean, G. Henri 102
AGNs at VLTI Scales
G.M. Voit 104
The AGN Dance: Astrometric Variability as the Result of Supernova Explosions in Nuclear Starbursts
J. Melnick, I. Aretxaga, R. Terlevich 112
Adaptive Optics Observations of Starburst Galaxies
O. Lai, D. Rouan, E. Gendron, F. Rigaut 119
Cosmology with the Very Large Telescope Interferometer: Using a Space-Based Astrometric Reference Frame
D. Tytler 121

4. THE POTENTIAL OF VLTI AND THE CONNECTION WITH RELATED TECHNIQUES

Tailoring an Interferometer to Its Science and Vice Versa
T.A. ten Brummelaar, H.A. McAlister, W.G. Bagnuolo, W.I. Hartkopf, S.T. Ridgway, N. Turner 133
Some Aspects of the VLTI's Potential for Binary Star Research
J. Davis 143
Combining Mid-Infrared Lunar Occultations and VLTI Measurements
B. Stecklum, H.-U. Käufl, A. Richichi 153

5. HOT STARS

Interferometric Observations of Be Stars
A. Quirrenbach ... 163
Spectroscopic and Photometric Support for Interferometric Observations of Be Stars
P. Harmanec ... 171
Imaging of Wolf–Rayet Stars with the VLTI
J. Davis, A.F.J. Moffat, L.F. Smith ... 179

6. CIRCUMSTELLAR ENVIRONMENTS AND MASS LOSS

AGB and Post-AGB Stars: An Overview
H. Habing ... 183
Winds of Red Giants and Supergiants: Basic Physical Processes, Structure and Appearance
E. Sedlmayr, J.M. Winters, A.J. Fleischer ... 192
Study of Late-Type Stars with the Infrared Spatial Interferometer of the University of California at Berkeley
B. Lopez, W.C. Danchi, P. Cruzalèbes, M. Bester, C.H. Townes ... 200
Interferometric Studies of Late Phases of Stellar Evolution
G. Weigelt, Y. Balega, K.-H. Hofmann, N. Langer, R. Osterbart ... 206
Interferometric Imaging of Late-Type Stars with the VLTI
C.A. Haniff ... 212
Geometry of Post-AGB Shells
C. Waelkens, R. Waters, H. Van Winckel ... 220
New Insights in Close Circumstellar Physics Using Interferometry
J.-P.J. Lafon, N. Berruyer ... 227

7. STAR FORMATION AND EARLY STELLAR EVOLUTION

Circumstellar Disks in the Early Phases of Star Formation
A. Natta 237
Outflows from Young Stars
J. Eislöffel, C. Dougados 240
Disks Around Young Stars
F. Ménard, F. Malbet, J.-L. Monin, J.-P. Berger, S. Guilloteau, A. Dutrey, G. Duvert 248

8. STELLAR POPULATIONS AND GALACTIC CENTER

Observing the Galactic Center with the VLTI
A. Eckart, R. Genzel, R. Hofmann, S. Drapatz, R. Katterloher, A. Quirrenbach, L. Tacconi-Garman 259
The VLTI and the Galactic Center He I Cluster
F. Najarro, A. Krabbe, R. Genzel, D. Lutz, R.P. Kudritzki 267
Globular Cluster Proper Motions
G. Meylan 274

9. STELLAR DIAMETERS AND SURFACE STRUCTURE

Indirect Imaging of Stellar Surface Structure
A.P. Hatzes 293
Stellar Surface Observations with the VLTI
O. von der Lühe 303
On the Scale of Photospheric Convection
B. Freytag, H. Holweger, M. Steffen, H.-G. Ludwig 316
Observing Stellar Surfaces with a High-Precision Infrared Interferometer
G. Perrin, V. Coudé du Foresto, S.T. Ridgway, J.-M. Mariotti, N.P. Carleton, W.T. Traub 318

Studying the Coolest Giant Stars at Milliarcsecond Resolution
A. Richichi 326
The Temperature Scale of the M Dwarfs
A.J. Booth, J. Davis, M. Bessell 332
Observations of Cepheids with the VLTI
D. Mourard 334

CONFERENCE SUMMARY

The VLTI and the Universe: Conference Summary
A. Quirrenbach, J.-M. Mariotti 339

POSTER PAPERS

Simulations of VLTI/VISA Imaging Observations of Young Stellar Objects at 2.2μm
N. Ageorges, O. von der Lühe 349
Speckle Observations of Post-Hipparcos Late-Type Double Stars
E. Aristidi, B. Lopez, M. Carbillet, P. Cruzalèbes, F. Mignard, Y. Bresson, G. Helmer 351
Wavefront Characterization for High Angular Resolution Techniques
R. Avila, A. Ziad, J. Borgnino, F. Martin, A. Agabi, A. Tokovinin 353
Interfero Coronography at O.C.A.
P. Baudoz, Y. Rabbia, J. Gay 355
Simulation of Speckle Images for a Field Wider Than the Isoplanatism Angle
H. Beaumont, C. Aime, L.J. Sánchez, H. Lantéri 357
Visibility Losses Due to Atmospherical Spectral Decorrelation in Michelson Interferometry
P. Berio, D. Mourard, F. Vakili 359

High-Resolution Measurements of Close Binary Stars and the Probability Imaging Technique
M. Carbillet, C. Aime, E. Aristidi, G. Ricort 361
Late-Type Stars at High Angular Resolution: Observations and Modelling
P. Cruzalèbes, B. Lopez, M. Bester 363
Interferometric Observations of Mira Stars
K.-H. Hofmann, Y. Balega, M. Scholz, G. Weigelt 367
Results of Studies on the Environmental Factors Affecting the VLTI Performance
B. Koehler, S. Lévêque, O. von der Lühe 369
Internal Optical Path Calibration System for VLTI
S. Lévêque, B. Koehler, O. von der Lühe 375
Pupil Mask Interferometry of M8E-IR
M. Löwe, B. Stecklum, O. von der Lühe, A. Quirrenbach 379
Observations of the Close Binary System β Lyrae with the GI2T Interferometer
F. Morand, D. Bonneau, D. Mourard, P. Harmanec 381
Stellar Winds of T Tauri Stars
G. Paatz, M. Camenzind 383
Operating the VLTI in the Thermal Infrared
A. Quirrenbach 385
Multi-speckle Interferometric Imaging with the VLTI and the LBT at Optical Wavelengths
Th. Reinheimer, K.-H. Hofmann, M. Schöller, G. Weigelt 387
Infrared Interferometric Imaging with the VLTI in the Multi-speckle Mode with a Combination of the 8 m UTs and the 2 m ATs
Th. Reinheimer, G. Weigelt 389
Thermal IR Imaging with MAX: Pushing the Limit of Single-Dish Ground-Based Observations
M. Robberto, T.M. Herbst, S.V.W. Beckwith, C. Birk, P. Bizenberger 391
Differential Interferometry and the Non-axisymmetric Envelope of γ Cas
L.J. Sánchez, R.G. Petrov, V. Vasyuk, S. Lagarde, A. Blazit, S. Cuevas, R. Foy 393

HST Imaging of GM Aurigae: A Circumstellar Disk Seen in Scattered Light
K. Stapelfeldt, and the WFPC2 Science Team 395

Software Concept for Four-Dimensional Bispectral Analysis of Stellar Interferograms
H. Weghorn 397

Real-Time Image Compression Method for Astronomical Interferograms
H. Weghorn, J. Colberg, T. Bauer 399

Synthetic Images from Dynamic Model Atmospheres of C-Rich Long-Period Variables
W. Windsteig, E.A. Dorfi, S. Höfner, J. Hron, F. Kerschbaum . 401

Brightness Profiles and Spatial Spectra of Dynamical Models for C-Rich Circumstellar Dust Shells Around Long-Period Variables
J.M. Winters, A.J. Fleischer, E. Sedlmayr 403

Author Index 405

List of Participants

Name	Institution
AGEORGES, Nancy	ESO, Garching nageorge@eso.org
ALBERDI, Antxon	Laboratorio de Astrofisica Espacial y Fisica Fundamental, Madrid alberdi@laeff.esa.es
ANDERSEN, Johannes	Niels Bohr Institute for Astronomy, Copenhagen ja@astro.ku.dk
ARISTIDI, Eric	Université de Nice aris@procyon.unice.fr
AVILA, Remy	Université de Nice remy@elisa.unice.fr
BACKMAN, Dana	Caltech, IPAC dbackman@ipac.caltech.edu
BARAFFE, Isabelle	Ecole Normale Supérieure de Lyon ibaraffe@cral.ens-lyon.fr
BEAUMONT, Hubert	Université de Nice hubert@castor.unice.fr
BECKWITH, Steven	MPI für Astronomie, Heidelberg svwb@mpia-hd.mpg.de
BEDDING, Tim	University of Sydney, Chatterton Astronomy Dept. bedding@physics.usyd.edu.au
BERGERON, Jacqueline	ESO, Garching jbergero@eso.org
BERIO, Philippe	Observatoire de Calern berio@altair.obs-azur.fr
BERRUYER, Nicole	Observatoire de la Côte d'Azur, Nice nicole@obs-nice.fr
BEUZIT, Jean-Luc	Observatoire de Grenoble, Lab. d'Astrophysique beuzit@gag.observ-gr.fr

BOOTH, Andrew	University of Sydney, School of Physics booth@physics.usyd.edu.au
CARBILLET, Marcel	Université de Nice marcel@procyon.unice.fr
CASSÉ, Martin	ESO, Garching mcasse@eso.org
CIOLEK, Glenn	University of Chicago, Dept. of Astronomy & Astrophysics ciolek@jets.uchicago.edu
CLAES, Peter	ESA - VILSPA pclaes@iso.vilspa.esa.es
COLBERG, Jörg	MPI für Astrophysik, Garching jgc@mpa-garching.mpg.de
COUDÉ DU FORESTO, Vincent	MPI für Astronomie, Heidelberg foresto@mpia-hd.mpg.de
CRUZALÈBES, Pierre	Observatoire de la Côte d'Azur, Grasse cruzalebes@obs-azur.fr
DANZIGER, John	Osservatorio Astronomico di Trieste danziger@oat.ts.astro.it
DAVIS, John	University of Sydney, Chatterton Astronomy Dept. davis@physics.usyd.edu.au
DE MARCHI, Guido	ESO, Garching demarchi@eso.org
DERIE, Frédéric	ESO, Garching fderie@eso.org
DI BENEDETTO, Paolo	Istituto di Fisica Cosmica, Milano pdibene@ifctr.mi.cnr.it
DOUGADOS, Catherine	Observatoire de Grenoble, Lab. d'Astrophysique dougados@gag.observ-gr.fr
DUSCHL, Wolfgang	Institut für Theoretische Astrophysik, Heidelberg wjd@ita.uni-heidelberg.de

ECKART, Andreas	MPI für Extraterrestrische Physik, Garching eckart%mpeis1 @mpe.mpe-garching.mpg.de
EIROA, Carlos	Universidad Autonoma de Madrid carlos@astro1.ft.uam.es
EISLÖFFEL, Jochen	Thüringer Landessternwarte, Tautenburg eisloeffel@tls-tautenburg.de
ENCRENAZ, Thérèse	Observatoire de Paris, Meudon Therese.Encrenaz@obspm.fr
FELDT, Markus	Max-Planck-Gesellschaft AG, Jena mfeldt@astro.uni-jena.de
FLEISCHER, Axel	Technische Universität Berlin, Inst. für Astronomie & Astrophysik Fleischer@physik.TU-Berlin.DE
FREYTAG, Bernd	Universität Kiel, Institut für Astronomie und Astrophysik supas024@astrophysik.uni-kiel.d400.de
FRIDLUND, Malcolm	ESA/ESTEC mfridlun@astro.estec.esa.nl
GLINDEMANN, Andreas	MPI für Astronomie, Heidelberg glindem@mpia-hd.mpg.de
GROENEWEGEN, Martin	MPI für Astrophysik, Garching groen@mpa-garching.mpg.de
HABING, Harm	Sterrewacht, Huygens Lab., Leiden habing@strw.leidenuniv.nl
HANIFF, Chris	Mullard Radio Astronomy Observatory, Cavendish Lab., Cambridge cah@mrao.cam.ac.uk
HARMANEC, Petr	Astronomical Institute, Academy of Sciences of the Czech Republic, Ondrejov hec@sunstel.asu.cas.cz hec@mbox.cesnet.cz
HATZES, Artie	The University of Texas at Austin artie@astro.as.utexas.edu

HOFMANN, Karl-Heinz	MPI für Radioastronomie, Bonn khh@specklec.mpifr-bonn.mpg.de
JASNIEWICZ, Gérard	Observatoire de Strasbourg gerard@simbad.u-strasbg.fr
KÄUFL, Hans-Ulrich	ESO, Garching hukaufl@eso.org
KOEHLER, Bertrand	ESO, Garching bkoehler@eso.org
KOUBSKY, Pavel	Astronomical Institute, Ondrejov Observatory koubsky@sunstel.asu.cas.cz
KRAUTTER, Joachim	Landessternwarte Heidelberg jkrautte@lsw.uni-heidelberg.de
KULKARNI, Shri	California Institute of Technology, Pasadena srk@astro.caltech.edu
LAFON, Jean-Pierre	Observatoire de Paris, DASGAL, Meudon JPJ.Lafon@obspm.fr
LAI, Olivier	Observatoire de Paris, Meudon lai@hplyot.obspm.fr
LE POOLE, Rudolf	Sterrewacht Leiden lepoole@strw.LeidenUniv.nl
LÉGER, Alain	Institut d'Astrophysique Spatiale, Université de Paris XI leger@iaslab.ias.fr
LEIBUNDGUT, Bruno	ESO, Garching bleibund@eso.org
LÉNA, Pierre	Université & Observatoire de Paris lena@megasx.obspm.fr
LEVEQUE, Samuel	ESO, Garching sleveque@eso.org
LÖWE, Mathias	Max-Planck-Gesellschaft AG, Jena loewe@astro.uni-jena.de
LOPEZ, Bruno	Observatoire de la Côte d'Azur, Nice lopez@rameau.obs-nice.fr
MARIOTTI, Jean-Marie	Observatoire de Paris, Meudon mariotti@obspm.fr

MATHIAS, Philippe	Observatoire de la Côte d'Azur, Nice mathias@obs-nice.fr
MAYOR, Michel	Observatoire de Genève mayor@scsun.unige.ch
MELNICK, Jorge	ESO-Chile jmelnick@eso.org
MENARD, François	Observatoire de Grenoble, Lab. d'Astrophysique menard@gag.observ-gr.fr
MENDEZ, René	ESO, Garching rmendez@eso.org
MEYLAN, Georges	ESO, Garching gmeylan@eso.org
MILEY, George	Sterrewacht Leiden miley@strw.LeidenUniv.nl
MONNET, Guy	ESO, Garching gmonnet@eso.org
MOORWOOD, Alan	ESO, Garching amoor@eso.org
MORAND, Frédéric	Observatoire de la Côte d'Azur morand@ocar01.obs-azur.fr
MOURARD, Denis	Observatoire de la Côte d'Azur mourard@obs-azur.fr
MUNDT, Reinhard	MPI für Astronomie, Heidelberg mundt@mpia-hd.mpg.de
NAJARRO, Francisco	Universitäts-Sternwarte München paco@usm.uni-muenchen.de
NATTA, Antonella	Osservatorio Astrofisico di Arcetri natta@arcetri.astro.it
OSTERLOH, Martin	Max-Planck-Gesellschaft AG, Jena osterloh@fred.astro.uni-jena.de
PAATZ, Gernot	Landessternwarte Heidelberg gpaatz@mail.lsw.uni-heidelberg.de
PADGETT, Deborah	IPAC / JPL dlp@ipac.caltech.edu
PARESCE, Francesco	ESO, Garching fparesce@eso.org

PENNY, Alan	Rutherford & Appleton Laboratories alan.penny@ast.star.rl.ac.uk
PERRIN, Guy	Observatoire de Paris, Meudon – DESPA perrin@hplyot.obspm.fr
PETITJEAN, Patrick	Institut d'Astrophysique, Paris petitjean@iap.fr
PETROV, Romain	Université de Nice romain@ayalga.unice.fr
PORETTI, Ennio	Osservatorio Astronomico di Brera poretti@merate.mi.astro.it
QUIRRENBACH, Andreas	MPI für Extraterrestrische Physik, Garching quirrenbach@mpe-garching.mpg.de
REINHEIMER, Thorsten	MPI für Radioastronomie, Bonn tr@specklec.mpifr-bonn.mpg.de
RENZINI, Alvio	ESO, Garching arenzini@eso.org
RICHICHI, Andrea	Osservatorio Astrofisico di Arcetri richichi@arcetri.astro.it
ROBBERTO, Massimo	MPI für Astronomie, Heidelberg robberto@mpia-hd.mpg.de
SANCHEZ, Leonardo	Université de Nice leonardo@elisa.unice.fr
SCHALINSKI, Cornelius	Institute of Space Sensor Technology (DLR), Berlin schalinski@dlr.de
SCHOBER, Hans Josef	Karl-Franzens-Universität Graz, Institut für Astronomie schober@bkfug.kfunigraz.ac.at
SEDLMAYR, Erwin	Technische Universität Berlin, Inst. für Astronomie & Astrophysik Sedlmayr@physik.TU-Berlin.de
SHAO, Michael	Jet Propulsion Lab., Pasadena mshao@huey.jpl.nasa.gov
SHAVER, Peter	ESO, Garching pshaver@eso.org

SHU, Frank	University of California, Dept. of Astronomy, Berkeley shu@bkyast.berkeley.edu
SPYROMILIO, Jason	ESO, Garching jspyromieso.org
STAPELFELDT, Karl	NASA Jet Propulsion Lab., and MPI für Astronomie, Heidelberg krs@wfpc2-mail.jpl.nasa.gov
STECKLUM, Bringfried	Thüringer Landessternwarte, Tautenburg stecklum@tls-tautenburg.de
TANGO, William	University of Sydney, Chatterton Astronomy Dept. tango@physics.usyd.edu.au
TARENGHI, Massimo	ESO, Garching mtarengh@eso.org
TEN BRUMMELAAR, Theo	Georgia State University, CHARA theo@chara.gsu.edu
THEODORE, Bertrand	CNRS, Verrières-le-Buisson theodore@aerov.jussieu.fr
TRAUB, Wesley	Harvard Smithsonian Center for Astrophysics traub@cfaft1.harvard.edu
TYTLER, David	University of California, San Diego tytler@cass155.ucsd.edu dtytler@ucsd.edu
ULRICH-DEMOULIN, Marie-Helene	ESO, Garching mhulrich@eso.org
VAKILI, Farrokh	Observatoire de la Côte d'Azur vakili@obs-azur.fr
VAN LOON, Jacco	ESO, Garching jvloon@eso.org
VOIT, Mark	STScI, Baltimore voit@stsci.edu
VOLONTÉ, Serge	ESA, Paris SVOLONTE%ESA.bitnet @listserv.gmd.de

VON DER LÜHE, Oskar	ESO, Garching ovdluhe@eso.org
WAELKENS, Christoffel	Katholieke Universiteit Leuven, Instituut voor Sterrenkunde christoffel@ster.kuleuven.ac.be
WALSH, Jeremy	ST-ECF, Garching jwalsh@eso.org
WARD, Martin	University of Leicester mjw@star.le.ac.uk
WATERS, Rens	Astronomical Institute, University of Amsterdam rensw@astro.uva.nl
WEGHORN, Hans	Sternwarte der Universität Bonn weg45@amc.de hweghorn@astro.uni-bonn.de
WEIGELT, Gerd	MPI für Radioastronomie, Bonn weigelt@mpifr-bonn.mpg.de
WIEDEMANN, Günter	ESO, Garching gwiedema@eso.org
WINDSTEIG, Walter	Universität Wien, Institut für Astronomie windsteig@auro.ast.univie.ac.at
WINTERS, Jan Martin	Technische Universität Berlin, Inst. für Astronomie & Astrophysik winters@physik.TU-Berlin.de
WUCHTERL, Günther	Inst. für Astronomie der Universität Wien & Inst. für Kernphysik der TU Wien wuchterl@amok.ast.univie.ac.at
ZIAD, Aziz	Université de Nice aziz@procyon.unice.fr
ZINNECKER, Hans	Astrophysikalisches Institut Potsdam hzinnecker@aip.de

Part 1

INTERFEROMETRY TUTORIALS
(All You Wanted to Know
About Interferometry
But Were Afraid to Ask)

Introduction to Interferometry

Timothy R. Bedding

School of Physics, University of Sydney 2006, Australia

Abstract. This tutorial gives a general introduction to optical and infrared interferometry. The observables measured by an interferometer are fringe visibility and closure phase, which can then be fitted by a model or used to reconstruct an image. In order to measure fringes, the path lengths from the source to the detector via the different individual apertures must be equalized. This requires some form of fringe tracking, which constrains the type of observations that are feasible and has important implications for the complexity and size of the resulting image.

1 Basic Principles

The purpose of these tutorials is address two questions: "Can I use VLTI to observe my favourite object?" and "What will it tell me?" Technical details of interferometry will only be discussed in so far as they affect the scientific capabilities. For more details, the reader is referred to reviews by Woolf (1982), Roddier (1988) and Shao & Colavita (1992).

1.1 Angular Resolution

We begin with the fact that big telescopes produce sharper images. Neglecting atmospheric effects, the angular resolution (in radians) of a telescope of diameter D is given by

$$\theta = \lambda / D. \tag{1}$$

This omits a factor of 1.22 appropriate to a circular aperture, which we dismiss as a technical detail, and neglects any image degradation from imperfections in the optics. The following table shows typical angular resolutions in milliarcseconds (mas) for three wavelength regimes:

	Visible 0.4–1 μm	Near IR 1–5 μm	Mid-IR 10–20 μm
D=8 m	18 mas	77 mas	400 mas
D=100 m	1.4 mas	6 mas	30 mas

Interferometry allows us to achieve high angular resolution without building a telescope of enormous size (Figure 1). Instead, we synthesize a large aperture by combining light beams from several small telescopes. When we observe a distant object in this way, we see interference fringes. These fringes arise because of the wave nature of light and they contain information about the object being observed.

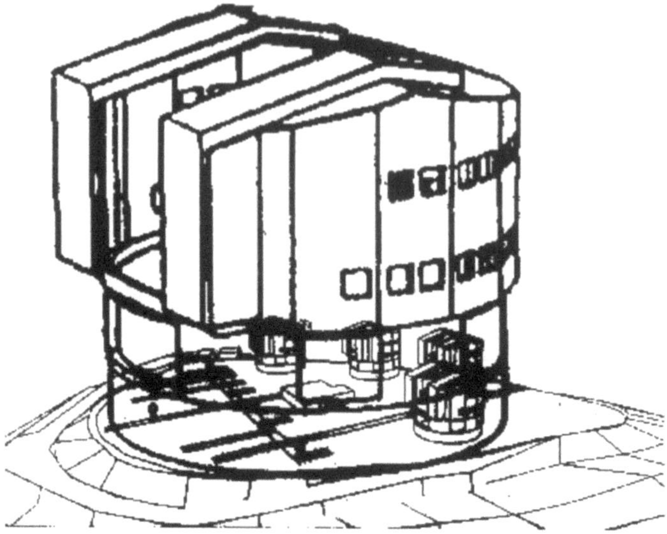

Fig. 1. An expensive way to achieve high angular resolution from Paranal.

1.2 Fringe Visibility

The visibility of fringes is a number between zero and one which measures the fringe contrast (see Figure 2). It is defined as $V = (I_{\rm max} - I_{\rm min})/(I_{\rm max} + I_{\rm min})$. If the dark regions in the fringe pattern go down to zero, the fringes have a visibility of one and we say the object is unresolved. If $V = 0$ there are no fringes and the object is completely resolved.

It is important to remember that interferometers record fringes, not images (unlike adaptive optics systems, which *do* record images). So the question "Can I use VLTI to observe my favourite object?" becomes "Can I measure fringes from my object?" If the object is very large and diffuse, the answer may be no.

1.3 Interpreting Fringe Visibility

In practice, fringes from an interferometer will never have perfect contrast ($V = 1$), even for a point source, because of atmospheric and instrumental effects. One must therefore accompany each measurement of the target by a similar measurement of an unresolved calibrator source. The 'true' visibility of the object is then the observed visibility of the object divided by the observed visibility of the calibrator.

As an example, consider a two-telescope interferometer which is measuring the angular size of a star. The procedure is to measure V at different baselines, one at a time, by moving the telescopes. From the sequence of measurements we find that the fringe visibility goes down as the baseline increases (see Figure 2). The star is becoming more resolved on longer baselines. Fitting a curve to the measured visibilities allows you to determine the angular diameter of the star.

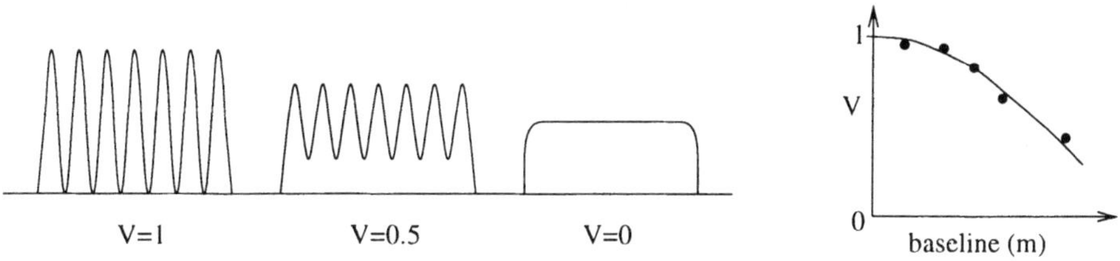

Fig. 2. Left: examples of fringes with visibilities of 1, 0.5 and 0. Right: visibility as a function of baseline for a resolved star.

This visibility curve has a special meaning: it is the Fourier transform of the object's brightness distribution. Thus, to make an *image* of the sky we need to reverse this process, which means performing an inverse Fourier transform of the measured visibilities. To do this accurately for anything but the most simple object requires that you sample the visibility at many different baselines. We talk about needing good coverage of the (u, v) plane, a phrase borrowed from radio astronomers. Having more telescopes in your interferometer saves time, since it allows you to measure visibilities on several baselines simultaneously (the number of baselines from an array of M telescopes is $M(M-1)/2$). Allowing the sky to rotate during the night also improves (u, v) coverage, and this technique is called earth rotation synthesis.

1.4 Closure Phase

As stated above, the fringe visibility measured using a particular baseline tells you one component of the object's Fourier transform. However, a Fourier transform is a complex quantity, having both an amplitude and phase. While the amplitude is given by the fringe visibility, the phase is given by the position of the fringes.

Unfortunately, the fringes from an interferometer continually move backwards and forwards because of atmospheric turbulence, which changes the path length in front of each aperture. Therefore, it seems that the position of the fringes tells you nothing about the object. However, consider the case of three telescopes. We can observe three sets of fringes simultaneously, all moving, but they don't move independently. Their movement is due to the atmosphere, but their *relative* positions contain information about the object. Thus, with three telescopes you can measure an additional piece of information, called the closure phase, which contains some (but not all) of the phase information. For example, the closure phase tells you about the asymmetry of the object (but not its absolute position on the sky). An array of more than three telescopes yields a closure phase for each triangle, although not all these measurements are independent. Combining these closure phases with the visibility measurements allows you to reconstruct an image of the object.

2 Interferometry in Practice

2.1 Atmospheric Effects

An interferometer measures fringes to obtain visibilities and closure phases. Why is this so difficult? A plane wavefront from a distant star travels through the atmosphere and is distorted by fluctuations in refractive index. The wavefront that reaches our ground-based telescope is no longer flat. The angle between different parts of the wavefront is about one arcsecond, which is why the image seen by a single large telescope is a fuzzy blob about one arcsecond in size. A useful analogy is to represent the distorted wavefront by a crumpled piece of paper. An adaptive optics system tries to flatten the paper using a large number of fingers (actuators). On the other hand, an interferometer takes small pieces of the wavefront and combines them.

Note that the shape of the wavefront is approximately the same at all wavelengths: it has the same physical size in microns (because the refractive index of air varies only weakly with wavelength). That is why the seeing in arcseconds is approximately the same at all wavelengths. Why then is interferometry easier in the infrared? The important thing is the size of the wavefront measured not in microns, but in wavelengths. As the wavelength increases, the distortions become relatively less important.

We define the atmospheric coherence length r_0 to be the size of a 'flat' patch on the wavefront, where this is decided relative to the wavelength. More precisely, r_0 is defined so that the rms variation in the wavefront across a patch of that diameter is $\frac{\lambda}{2\pi}$ (one radian of phase). In typical seeing r_0 is 10–20 cm at visible wavelengths and becomes larger at longer wavelengths. The increase is not exactly linear, with r_0 depending on wavelength as $\lambda^{6/5}$.

What about time variation? The shape of the wavefront changes with time, but to good approximation the dominant effect is the wind blowing the whole pattern past the telescope. This model is known as frozen turbulence. We can then define an atmospheric coherence time τ_0, which is roughly the time taken for one r_0-sized patch to move past. In the visible this is a few tens of milliseconds, again increasing with wavelength as $\lambda^{6/5}$.

2.2 How Does an Interferometer Work?

An interferometer combines two or more separate parts of the wavefront to produce fringes. This requires two corrections to the wavefront, as shown in Figure 3. Firstly, although the wavefront across each individual telescope is approximately flat, it will have a tilt, so we require tip-tilt correction to make the beams overlap.

The second correction is to make the path lengths equal, without which no fringes will be visible. The crude part of this correction is to allow observations of objects all over the sky, not just those which are directly overhead. This is achieved by delaying one beam (by bouncing it back and forth along a tunnel), which allows you to steer the interferometer. Much harder is to make the small correction for the atmosphere, a few tens of microns, which varies rapidly and randomly. This explains the need for fringe tracking.

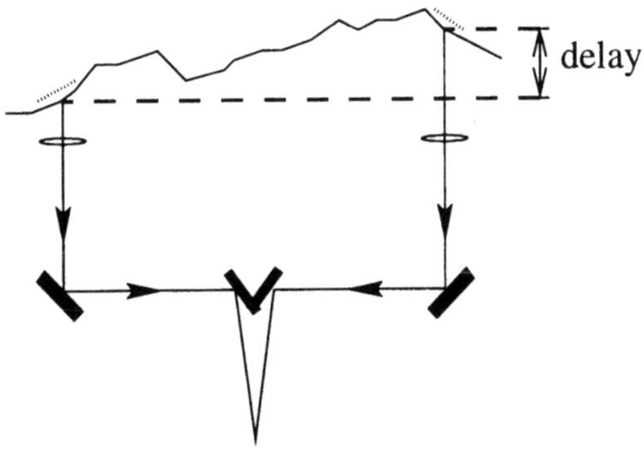

Fig. 3. The wavefront above a two-element interferometer, showing the corrections required for tip-tilt and delay

2.3 Fringe Tracking

To measure fringes, we must maintain equal optical paths for a reasonable period (minutes at a time). There are three ways to achieve this:

1. track fringes on the scientific target (which requires the target to have a bright compact component)
2. track fringes on a nearby reference star (note that a laser beacon is *not* compact enough)
3. blind tracking (which requires a very stable and well-calibrated system; this may be possible in the mid-IR)

In addition, there are two *levels* of fringe tracking. The more difficult to achieve is 'cophasing,' which means maintaining equal optical paths to within a fraction of a wavelength. The fringes will then be steady, allowing integration for many times τ_0. This is clearly useful in the case of a bright reference star and a faint and/or resolved target: fringes can be tracked to high accuracy on the reference star, allowing long integrations on the target.

Less difficult is to track to within the coherence length of the light ($\lambda^2/\Delta\lambda$), so that the fringes are still detectable but are continually moving. In this case, called 'coherent' tracking, you must measure the fringes in many short exposures (τ_0).

There are some tricks which help with fringe tracking. One is bootstrapping, which involves tracking fringes on a series of short baselines (see Figure 4). A possible problem is the propagation of errors, perhaps resulting in an unacceptably large error on the long baselines. Of course, it also requires at least four telescopes to make the gain worthwhile and, in terms of (u, v) coverage, these must be deployed rather wastefully.

Another trick is to track at one wavelength and measure fringe visibilities at another. For example, at longer wavelengths the object should be less well resolved (Equation 1), so you can track at a long wavelength and observe at

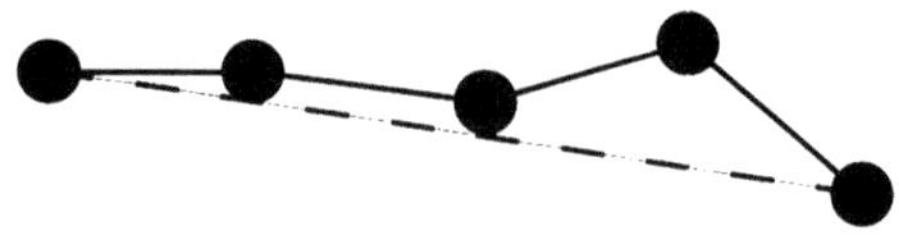

Fig. 4. Fringe tracking by bootstrapping. The dots are the individual telescopes. On the short baselines (solid lines) the target is only partially resolved and fringe tracking should therefore be easy. By tracking on each of these short segments, the fringes on the longest baseline (dashed line) will be stable and can be measured, despite their low visibility.

a shorter one. For some objects (e.g., Be stars) you can track on a compact continuum source and observe extended structure in an emission line.

Nevertheless the ability to track fringes is a difficult requirement and will usually set the limiting magnitude of the interferometer.

3 What Can We Expect?

3.1 Image Complexity

We should not expect to produce complex images like those from the VLA and similar radio interferometers. The best analogy between radio and optical interferometry is VLBI, which use six or seven apertures over Intercontinental baselines. An array of 27 radio telescopes gives $27 \times 26/2 = 351$ simultaneous baselines, allowing very efficient coverage of the (u, v) plane. This works well in the radio because you can amplify the signal from each telescope before combining them. Unfortunately, in the visible and near infrared this cannot be done without introducing noise that swamps the signal. Every extra telescope adds noise, and there comes a point where it is detrimental to add more telescopes. For fuller discussion of these points see Nityananda (1994) and Prasad (1994).

Of course, extra telescopes could be used in a completely separate array, by combining their light on a second detector. But then the number of baselines being measured simultaneously will be much less than $M(M-1)/2$. For this reason, no optical/infrared interferometer is being built with more than six elements. It is therefore unlikely that image complexity will match that of the VLA. Roughly speaking, the number of non-zero resolution elements in the final image equals the number of independent measurements in the (u, v) plane. For VLTI this will be a few tens to a few hundreds. Finally, for on-source fringe tracking there is an extra constraint on image complexity, namely that the object have a compact component.

3.2 Field of View

How big will the final image be? One restriction comes directly from the preceding discussion on image complexity. The *number* of non-zero pixels is limited by (u, v) coverage while their *size* is set by the resolution of the interferometer. Since this resolution is typically reckoned in milliarcseconds, the images of a complex object (one with many non-zero pixels) cannot be very large.

There is another restriction on the field of view. Equal path lengths, necessary to detect fringes in the first place, must be maintained over the whole field. Fringe tracking will (hopefully) ensure that paths are equal at the centre of the field. But looking off-axis by an angle θ introduces an extra path length of $D\theta$, where D is the length of the baseline. This extra path difference will only be unimportant if it is small compared to the coherence length of the light $(\lambda^2/\Delta\lambda)$. Thus, the limit on the field of view is

$$\theta \lesssim \frac{\lambda}{D}\frac{\lambda}{\Delta\lambda}.$$

Writing it in this way, we see that the field of view is set by the product of the angular resolution and the spectral resolution. This limits the size of the region over which fringes can be detected. For example, if the field width is to be 100 resolution elements, which would be about 0.6 arcsec for VLTI in the near IR, the spectral resolution must be better than 100.

In principle this limit can be overcome by building a Fizeau interferometer (also called homothetic mapping in ESO publications), which requires very complicated extra optics to ensure path equality over the whole field. Another approach is to build a separate delay compensator for each part of the field. This is done for a dual-feed system, in which one (small) field is centred on the target and the other on the reference star. In this case, the maximum separation between object and reference is set by isoplanatism.

3.3 Dynamic Range

This is the ratio of the brightest part of the image to the faintest part. Images with 100:1 are already possible with optical interferometers, while 1000:1 will be difficult. This must be kept in mind when deciding what to expect from VLTI observations of your favourite object.

4 Conclusion

Interferometers give improvement in angular resolution of several orders of magnitude over current ground- and space-based telescopes, but at a price. They cannot produce beautiful images of complex objects like those from HST or even the VLA. Despite the difficulties and limitations, the enormous potential gains in resolution have spurred more than a dozen major interferometry projects around the world. The latest information on these can be found at `http://huey.jpl.nasa.gov/~shaklan/olbin/` (Optical Long Baseline Interferometry News).

Acknowledgements

I have benefited greatly from discussions with many colleagues over the years, and would particularly like to acknowledge Gordon Robertson, Ralph Marson and Oskar von der Lühe. I also thank Gordon Robertson for comments on this paper. For financial assistance I am grateful to the University of Sydney Research Grants Scheme and the Australian Research Council.

References

Nityananda, R. (1994): Comparing optical and radio – quantum issues, in *IAU Symposium 158: Very High Angular Resolution Imaging,* edited by J. G. Robertson and W. J. Tango, p. 11, Kluwer: Dordrecht.

Prasad, S. (1994): Implications of light amplification for astronomical imaging, *J. Opt. Soc. Am. A,* **11,** 2799.

Roddier, F. (1988): Interferometric imaging in optical astronomy, *Phys. Rep.,* **170,** 97.

Shao, M. and Colavita, M. M. (1992): Long-baseline optical and infrared stellar interferometry, *Ann. Rev. Astr. & Ap.,* **30,** 457.

Woolf, N. J. (1982): High resolution imaging from the ground, *Ann. Rev. Astr. & Ap.,* **20,** 367.

DISCUSSION

Theo ten Brummelaar: I think you glossed over the calibration problem just a little bit too much. At the longest baselines, finding unresolved objects will be difficult.

Tim Bedding: The object you use as a calibrator doesn't have to be unresolved, it just has to be known and preferably smaller than the target object. But you are right, at the longest baselines almost everything you look at will be resolved. You have to bootstrap, calibrating on partially resolved stars. This is more of problem in the visible than in the near and mid infrared, where the resolution is less.

Andreas Quirrenbach: Another comment on calibration. You are quite right that this can be a big problem and not so much because the calibrator is resolved but because of variations in time and over the sky. If you want to have an image with a dynamic range of 100:1 or 1000:1, you also require that precision in calibration of visibility measurements. You have to measure fringe contrast to a precision of 1% or 0.1% through the changing atmosphere. It's basically the same problem as calibrating the point-spread function in adaptive optics observations. The changes on the timescale of ten minutes are very worrisome and you have to change between your target object and the calibrator as fast as you can.

Theo ten Brummelaar: If you are cophasing three telescopes, aren't you servoing out one of the very things you are trying to measure?

Oskar von der Lühe: If you have three telescopes, it is sufficient to stabilize fringes between two pairs. The phase on the third baseline will be determined by the object. So, no, you are not destroying a good observable when you fringe track.

Martin Ward: How good could the astrometry be in principle?

Mike Shao: Tens of microarcseconds

George Miley: You mentioned that adding baselines wouldn't improve things. But of course the dynamic range is limited by the number of (u, v) points, which is limited by the number of telescopes. Adding baselines helps dynamic range, at least for bright objects, even though you do add noise.

Tim Bedding: As far as image making is concerned I agree, the more telescopes the better. I'm talking here about signal detection: the need to detect fringes and track them. If you build an optical array with too many telescopes, you won't be able to measure fringes on any baseline, because you are adding photons from all of those telescopes at once. [The observed fringe visibility on a given baseline will decrease because the fringes are seen against the background light from all the other telescopes.]

Chris Haniff: If you're doing regular astronomy, most astronomers agree that if you have bigger telescopes you get better data. If you are building an interferometer and someone gives you extra money, what would you do? Should you build more telescopes or bigger telescopes?

Tim Bedding: Making the telescopes bigger only helps up to a point. Once the telescope diameter reaches a few times r_0, you don't gain nearly as quickly. For example, the VLTI Auxiliary Telescopes (ATs) are designed to be 1.8 m because that is a few times r_0 in the near infrared. You wouldn't try to operate a visible interferometer with 2 m telescopes. So the answer depends on the wavelength. If you want to operate at 10 μm then make them 8 metres, since at this wavelength an 8 m aperture is almost diffraction limited (Section 1.1). The wavefront across those Unit Telescopes on Paranal is almost flat in the mid-IR, which is why it's such a great interferometer. So if you want to operate in the mid-IR you would build one or two more UTs. However, if you want to operate in the near-IR you want more ATs, five or six of them.

Chris Haniff: But not twenty?

Tim Bedding: No.

Wesley Traub: If you had the money, another possibility is to invest in making a laser guide star and adaptive optics for each telescope.

Tim Bedding: Yes, it would be great to have adaptive optics for the Unit Telescopes in the near infrared. The other thing you could buy is one of these superconducting tunnel junction detectors for fringe tracking.

Francesco Paresce: Do you think that in five minutes you could tell us about phase closure?

Tim Bedding: The figure shows three telescopes all looking at the same star. Above each telescope there is an unknown and continually changing thickness of atmosphere. We measure the phase of fringes between each pair of telescopes. By phase, I mean the position of the fringes. The phases are continually changing and tell you nothing about the object.

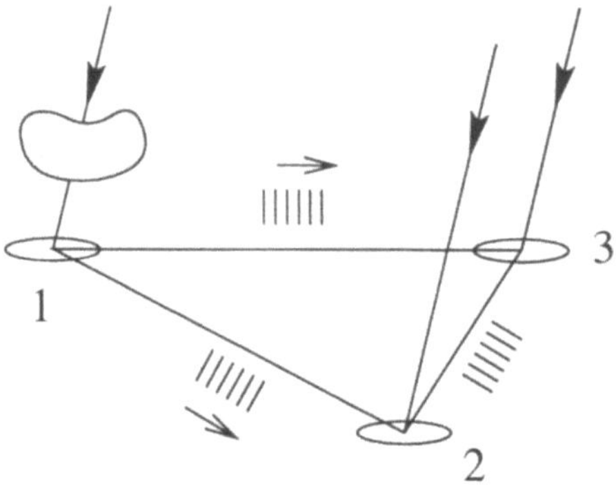

Now imagine adding an extra piece of atmosphere in front of Telescope 1. The fringes on baselines 1-2 and 1-3 will both shift in the direction shown by the arrows, while the fringes on baseline 2-3 will be unaffected. What I measure is the algebraic sum of the three fringe shifts, defining positive as clockwise round the triangle. Then the extra atmosphere in front of Telescope 1 has shifted one set clockwise, one set anticlockwise, and one set not at all. The algebraic sum $\phi_{12} + \phi_{23} + \phi_{31}$ is called the closure phase and, from the preceding argument, it is completely unaffected by atmospheric fluctuations. If the atmosphere disappeared entirely we would get the same number. This number is telling you something about the object.

Francesco Paresce: If you have four telescopes?

Tim Bedding: From four telescopes you can make four triangles, but only three of these are independent. A nice thing is that, unlike with visibility amplitudes, you don't need to calibrate closure phase. If you look at a point source, the closure phase is always zero. It is much more resistant to variations in seeing. And it's also sensitive to object structure, telling you about asymmetries even below the diffraction limit of the interferometer.

Jim Beletic: You could add that laser guide stars won't help you phase the three apertures because the light from the laser spot to each telescope is coming through a different part of the atmosphere. You can only use the laser to flatten the wavefront across each telescope.

Tim Bedding: Yes.

An Introduction to Interferometry with the ESO Very Large Telescope

Oskar von der Lühe

European Southern Observatory
Karl-Schwarzschild-Str. 2, D – 85748 Garching

1 Introduction

The interferometric mode of the ESO Very Large Telescope (VLT) permits coherent combination of stellar light beams collected by four Unit Telescopes (UTs) with 8m diameter and by several Auxiliary Telescopes (ATs) of the 2m class. While the position of the UTs is fixed, ATs can be moved on rails and can operate from 30 stations distributed on the top of the observatory site for efficient UV coverage. Coherent beam combination can be achieved with the UTs alone, with the ATs alone, or with any combination of up to eight telescopes. A distinct feature of the interferometric mode is the high sensitivity due to the 8m pupil of the UTs with the potential for adaptive optics compensation in the near-infrared spectral regime.

The VLT Interferometer is conceived as an evolutionary programme where a significant fraction of the interferometer's functionality is initially funded, and more capability may be added later while experience is gained and further funding becomes available. The scientific program is now defined by a team which consists of a VLTI Scientist at ESO and fifteen astronomers from the VLT community. ESO has recently decided to resume the construction of the VLTI which was delayed in December 1993, in order to achieve first interferometric fringes with two of the 8m telescopes around the year 2000, and routine operation with ATs from 2003 onwards.

This paper presents an overview of the history of the VLT Interferometer and its basic capabilities. The synthesis capability and the sensitivity of VLTI are addressed in more detail. An extensive bibliography on VLTI–relates papers is also provided.

2 History of VLTI

A coherent combined mode has been part of the Very Large Telescope program since its beginning in the early 80's (Enard, 1984, Merkle, 1986, 1988a and 1988b, Léna, 1987 and 1988). Combining the collecting power of 8m diameter telescopes with the resolution of a baseline of about 100m seemed to be an opportunity too good to miss. The main driver for including the interferometric mode were making images of celestial sources at a resolution of the milli-as. At that time, interferometry at radio wavelengths was becoming recognized as a very powerful tool, which could provide much of the data analysis techniques needed for the optial spectral ranges. Speckle interferometry and related techniques were intensely researched and well on the way to their current "toolbox" state. Exploiting the diffraction limit with adaptive optics appeared as a more distant possibility. There was only litte experience with distributed interferometers at optical wavelength, essentially concentrated in Europe.

The main thrust for interferometry with large aperture telescopes was for observations in the near infrared domain (Roddier and Léna, 1984). It was recognised early on that a fixed configuration of large telescopes would prove too inflexible to make faithful images of sources with even moderate complexity. Therefore, the VLT includes several smaller Auxiliary Telescopes of the 2m class which can be relocated to easily change baselines which complement those of the fixed large telescopes. Also, more baselines could be provided at a given time which would result in a better and faster covered UV plane. It was not until later that the power of the small telescopes for providing an independent interferometric array which could be used all of the time was recognised.

The early definition of conceptual choices for the VLTI was performed by an ESO Working Group (VLT Report No. 49, 1986). The development of a detailed technical concept commenced, after the ESO Council approved the VLT Programme in 1987, in 1988 with a team led by Jacques Beckers for more than three years (Beckers et al., 1990a, Beckers, 1990b, Beckers, 1990c, Beckers, 1990d, Beckers, 1991a, Beckers, 1991b). An Interferometry Panel set up from experts in high angular resolution observations and related science fields made many contributions until end of 1992. As a result of the work of this body, the first implementation plan for VLTI was published in 1989 (Beckers et al., VLT Report 59 1989). In hindsight, it is interesting to see how many of the current science targets were already addressed in this report. However, astrometric capability was not considered feasible, as the potential of narrow-angle astrometry was not yet recognised. A second report (Beckers et al., VLT Report 65 1992) addressed the concepts for instrumentation for combined focus (Claret et al., 1991, Reinheimer et al., 1991), and still serves as a reference for the variety of possible techniques to combine beams.

In the meantime, the location for the observatory on Cerro Paranal was decided and the work on the site progressed. The layout of the VLTI – a subject of debates since years – was completed after some fulminant weeks early 1992 and remained unchanged ever since (von der Lühe et al., 1991). The four 8m Unit Telescopes (UTs) are placed in some sort of a semi-circle. A more accentuated circle would have been more desirable but was not possible for reasons of wind obstruction and the quality of the rock in certain locations. The locations of the delay line tunnel and 30 stations for the ATs telescopes were identified and are now cast in concrete. The goal was to optimise the array layout for imaging and to leave the door open for a variety of phase recovery methods.

At this time, VLTI consisted of the coudé optical trains on all UTs, two movable ATs (Plathner et al., 1991), two delay lines with a 60m stroke each

(Jörck et al., 1991), and a beam combination telescope in the interferometry laboratory with an 8 as continuous field (Beckers et al., 1990a, Beckers et al., 1991c). The large field requirement came from the desire to use sources in the field for phase referencing (the equivalent to adaptive optics with a refernce star) but resulted in very large dimensions for the delay lines and the transfer optics. An extra budget was foreseen for the development of the fringe detection instrumentation within the community.

The development of VLTI had its all time high when the French Centre National de la Recherche Scientifique (CNRS), the Max-Planck Society (MPG) and ESO signed an agreement on the enhancement of the interferometric mode of the VLT in December 1992 (The Messenger, 1993a). Under this agreement, a third AT and a third delay line could be developed togther with the first two, thereby substantially enhancing the synthesis capability of the array by adding phase closure capability. VLTI could be used all of the time with the ATs in phase closure modes, removing much of the pressure from the UTs.

An all time low was reached just a year later, when the ESO Council determined in December 1993 that the development of VLTI should be put on hold for reasons of cash flow levelling (The Messenger, 1993b). The development continued in the form of a "research and development programme" while means to overcome the budgetary situation were sought (von der Lühe et al., 1994 and 1995a, Koehler et al., 1994, Ferrari et al., 1994a and 1994b, Gay et al., 1994a and 1994b). Also, a fresh look at the concept and its scientific justification were taken when the successor to the Interferometry Panel, the Interferometry Science Advisory Committee (ISAC) was assembled in 1995. A summary of a science programme and the resulting conceptual consequences is described in the ESO Messenger (Paresce et al., 1996). One proposal was to limit the operational wavelength for the early phases with Auxiliary telescopes to the near infra-red and with UTs to the thermal infra-red spectral ranges. The VLTI would then not require higher order adaptive optics systems which considerably reduces system complexity and cost. Atmospheric and instrumental control can be limited to tip-tilt compensation only.

Although the efforts to acquire additional funding were not succesful, ESO decided in December 1995 to go ahead with the development of a reduced VLTI. A "new plan" which accounts for the recommendations of the ISAC was developed during the first half of 1996 (von der Lühe et al., 1996) in coordination with the partners of the VLTI agreement, in order to secure their continued support. Further details of this plan are described in this paper. The detailed defintion of the VLTI is continuing (Koehler et al., 1996, Lévêque et al., 1995a, 1995b and 1996). This ESO workshop is another milestone on the way towards a new and improved VLTI.

The implementation of the VLTI is now distributed over two phases. The first phase (Phase A) includes the coudé optical trains on two UTs, two delay lines, two auxiliary telescopes and the beam combination instrumentation for the near-infrared (ATs) and the thermal infrared (UTs). This phase is funded and will carry the program through the year 2003. A second phase (Phase B) includes a third AT, the remaining two coudé trains on the UTs and two more delay lines will require additional funding. Phase B can be concluded by 2005 if funding is possible from 2000 onwards.

3 General Capabilities

3.1 Layout of the VLT Interferometer

Fig. 1 shows the layout of the VLTI on Paranal. The North direction is approximately 20 degrees clockwise from topwards. The four UTs are seen on the upper half of the picture. The delay line tunnel transverses the mountain in the middle. The lower part of the mountain is filled with Auxiliary telescope stations and the tracks on which they will be transported. Two tracks extend across the tunnel to the northern part of the site, the western one passes close by the central facilities where a station is located which will be used for servicing the coudé insets.

Starlight from telescopes is directed underground towards the tunnel in the form of collimated beams. Mirrors direct the light beams along the tunnel towards the opposite sides where the cat's eyes optics of the delay line retro-reflect the beams. Another set of mirrors redirects the light near the center of the tunnel towards the beam combination equipment in the laboratory. Beam combination and detection of interferometric fringes occur in the laboratory.

3.2 Availability

Whenever the UTs will be used for the interferometric mode, their time allocation will go through ESO's ordinary review process, which means competition with single telescope programs. Only observations which require high sensitivity and / or high spectral resolution will usually require UTs. We expect that the use of them will be limited to a small percentage of the observing time initially, and will evolve with experience.

In contrast, the Auxiliary telescopes will be available all of the time for interferometry, as these specialised telescopes will not support any other observing mode. They will therefore support obseving programs which need a lot of observing time, but less sensitivity, and for synoptic programs.

3.3 Sky Coverage and Access

The sky coverage is limited by the sky coverage of the telescopes and the optical delay that the delay lines can produce. All telescopes support a coverage of 60 deg from Zenith with full performance. The delay lines support a coverage of 60 deg from Zenith for baselines up to 130m (i.e., all UT baselines). For AT configurations with baselines larger than 120m, the coverage is reduced and depends on the exact configuration. A good value to use here is 30 deg. Ultimately, the sky coverage can be increased to 60 deg on all baselines.

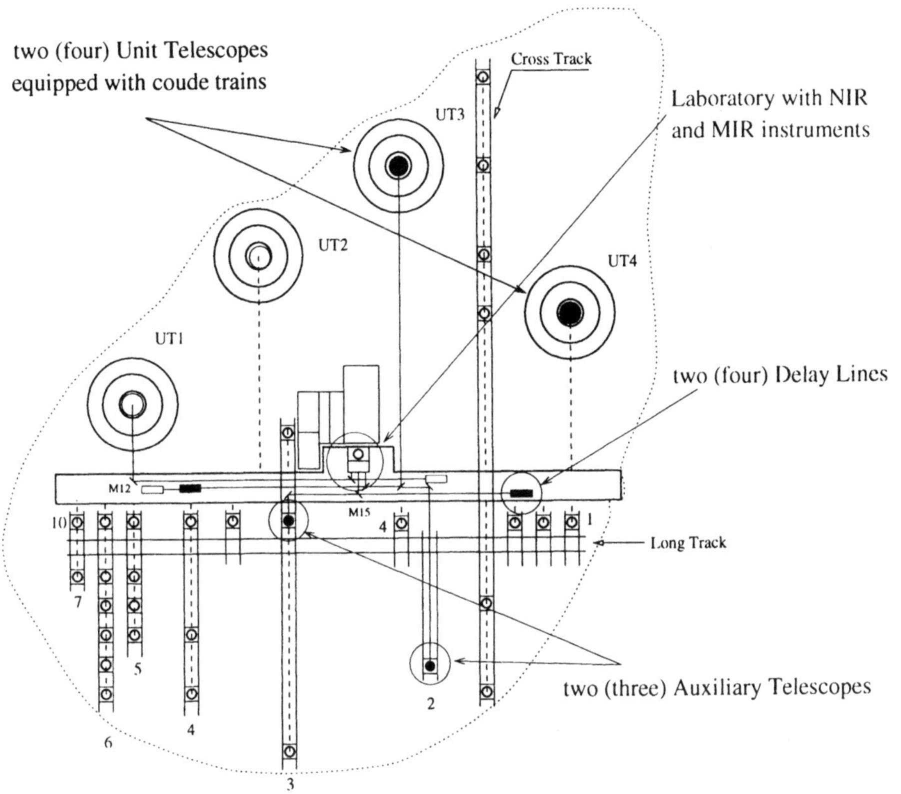

Fig. 1. Configuration of the VLT on Cerro Paranal.

VLTI supports rapid slewing between sources to enable efficient observation of mutiple sources and calibrator sources. This is essential for maximizing on–source time. The target time for acquiring fringes on near–by sources (within 1 degree) is of the order of 30 sec. The telescopes can acquire a source anywhere in the sky within 3 min.

3.4 Field of View and Dual Feed

The desire for including narrow-angle 1992 astrometric capability into VLTI with Auxiliary telescopes became apparent in 1994 (Léger et al., 1995, Quirrenbach 1995). The original idea to accommodate this capability was to augment the optics for the Auxiliary telescopes (von der Lühe et al. 1995b). In view of the current changes in the concept we intend to pursue a different approach.

The coude foci of Unit and Auxiliary telescopes will be equipped with dual feeds which support two field angles, one of which is on the axis, to be selected and fed into the interferometer. There will be a 2 as field centered at each of the two field positions. One field position will be used to transfer the light from the source under investigation, the other will be used for phase referencing (UTs and ATs) or astrometric referencing (ATs only). The range of the selection will be $5 \cdots 60$ as from the axis, this accomodates the complete field at the coudé foci. The two beamlines share the delay lines optics.

The concept behind the phase reference is similar to a reference star for adaptice optics; it will serve for image and fringe tracking should the science target be too faint. Atmospheric effects introduce differential phase errors between the two field positions of the dual feed which result in fringe contrast loss. For typical conditions at Paranal, a reduction of fringe contrast by a factor 0.7 occurs for field separations of 10 as at a 2.2 μm wavelength, and of 40 as at 10 μm.

The requirements and engineering implications of narrow angle astrometry are, on the other hand, very severe and need thorough studies. The atmospheric limits to astrometric precision depend linearly on field angle and amount to about 10 μas with a 100 m baseline and 30 min of integration. Achieving this precision in practice requires the knowledge of the vector baselines to 50 μm accuracy and of the instrumental differential delay between the primary and secondary beam to 5 nm accuracy. At this time, the technical limits to determine the differential delay are not known.

The concept for the dual feed is still under development. Although designs for dual feeds exist elsewhere, the requirements for VLTI are not trivial. For example, the requirement for spectral coverage and astrometric precision essentially exclude the use of transmissive optics.

3.5 Fringe–Tracked and Blind Operation

The term fringe–tracked operation refers to the generation of a fringe position error signal near the point of beam combination with a suitable sensor. The Observatoire de Cote d'Azur currently develops a fringe sensor prototype under contract with ESO (Gay et al., 1994b). The sensor operates independently from any other beam combination instrument in the H band (1.6 μm) on the science source or a reference source. The error signal is fed to the control system of the delay lines to adjust their position. The high frequency part of the fringe position error operates on a piezoelectric element which controls the piston of the cat's eye field mirror.

No fringe sensor is used when operating "blindly". The delay lines track the expected geometric delay, which is calculated using a model for the instrumental delays and telescope positions, similar to a telescope pointing model. Short exposure interferograms which "freeze" the disturbance effects of the atmosphere are taken. Those will show fringes only if the deviation from zero path difference is small (within ≈ 3 μm in the V band, 10 μm in the K band, and 50 μm in the N band). Atmospheric piston will produce path differences of $10 \cdots 30$ μm *rms*. This mode requires accurate knowledge of the instrumental characteristics of VLTI.

3.6 Beam Combination and Instruments

The beam combination laboratory area will accommodate instruments, at least for the near and thermal IR, along with a number of other subsystems. Figure 2 shows an overview of the layout. Stellar light beams enter the laboratory from below. A switchyard which consists of dichroic and fully reflective mirrors direct the light towards the instruments and to the fringe and image sensors. A calibration unit which provides various sources also feeds both instruments and sensors and assures intercalibration. Laser metrology beams feed the fringe sensor, as well as the instruments, for internal calibration. Other metrology beams travel back to the telescopes and determine the delay line zero position for the internal path. Primary and secondary beams emerging from the delay line carriages (hatched boxes in the tunnel) are directed towards the laboratory in the upper half. The switchyard is near the center of the layout. Intruments are on the top and right. Fringe and image sensors are to the left. The structure to the lower right is the metrology beam launching system.

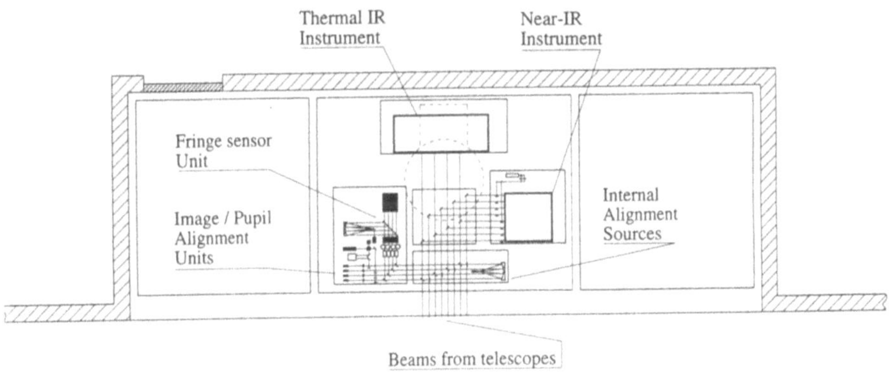

Fig. 2. Layout of the beamcombination Laboratory.

The initial operation of VLTI will concentrate on the near IR with ATs and on the thermal IR with the UTs. Top-level capabilities of beam combination instruments as recommended by the Science Advisory Committee are given below.

- ISAC recommendations for the Near-IR
  - support four beam lines

- spectral range $0.7 \cdots 4 \mu m$ with a spectral resolution up to 10,000
- calibration through single mode operation ("spatial filter")
- internal fringe tracking capability
- support dual feed

– ISAC recommendations for Mid-IR
- support two to four beam lines
- spectral range N+Q bands (extension to M), resolution up to 1,000
- calibration through single mode operation
- internal fringe tracking capability
- fringe detection by temporal modulation
- may not deliver "zero spacings"

3.7 Operational Capabilities

The development of VLTI's capabilities in Phases A and B is shown in table 3.7. The concept of VLTI is prepared for evolution beyond the capabilities of phases A and B, aan indication for this evolution is given in the last column.

<table>
<tr><th></th><th>Phase A
(2000 +)</th><th>Phase B
(2003 +)</th><th>Ultimate</th></tr>
<tr><td>Number of UTs</td><td>2</td><td>4</td><td>4</td></tr>
<tr><td>Number of ATs</td><td>2</td><td>3</td><td>3+</td></tr>
<tr><td>Number of delay lines (maximum number of elements)</td><td>2 (3 with fixed cat's eye and reduced sky coverage)</td><td>4 (5 with fixed cat's eye and reduced sky coverage)</td><td>8</td></tr>
<tr><td>Adaptive Optics</td><td>tip-tilt only</td><td>tip-tilt only</td><td>NIR AO with UT</td></tr>
<tr><td>Diff.–limited wavelength range</td><td colspan="2">5 - 20 μm with UTs, 1 - 20 μm with ATs</td><td>1 - 20 μm</td></tr>
<tr><td>Relocation of ATs</td><td>during the day</td><td>during the day</td><td>within 1h</td></tr>
</table>

Table 1. Growth of VLTI capabilities with time.

4 Array Configuration and UV Coverage

Table 2 shows the coordinates of the UTs and the Auxiliary telescope stations. The notation of the array conficuration is also indicated in fig. 1. Both latitude and longitude, as determined by a GPS measurement of the location of UT4, and the north and east relative coordinates in meters, relative to the position of UT1, are shown. The nominal height above sea level of the intersection of altitude and azimuth axes is 2676 m for UTs and 2669 m for Auxiliary telescopes. The delay constant is the distance along the optical path from the coude area at each telescope or station to the center of the delay line tunnel, this distance must be taken into account when the position of the cat's eyes is determined.

Station	Latitude	Longitude	N_{rel} [m]	E_{rel} [m]	Delay constant [m]
UT1	-24.37.33.117	70.24.11.642	0.00	0.00	88
UT2	-24.37.31.465	70.24.10.855	51.08	24.31	88
UT3	-24.37.30.300	70.24.09.896	87.08	53.97	96
UT4	-24:37:31.000	70:24:08.000	65.45	112.59	104
Long track 1	-24:37:32.707	70:24:07.394	12.68	131.33	72
Long track 2	-24:37:32.793	70:24:07.638	10.00	123.79	64
Long track 3	-24:37:32.880	70:24:07.882	7.32	116.25	56
Long track 4	-24:37:33.226	70:24:08.857	-3.38	86.10	24
Long track 5	-24:37:33.573	70:24:09.832	-14.09	55.94	32
Long track 6	-24:37:33.746	70:24:10.320	-19.45	40.86	48
Long track 7	-24:37:33.919	70:24:10.808	-24.80	25.79	64
Long track 8	-24:37:34.092	70:24:11.295	-30.16	10.71	80
Long track 9	-24:37:34.179	70:24:11.539	-32.83	3.17	88
Long track 10	-24:37:34.265	70:24:11.783	-35.51	-4.37	96
Cross track 1	-24:37:31.260	70:24:08.73	57.42	89.97	80
Cross track 2	-24:37:30.772	70:24:08.905	72.50	84.62	96
Cross track 3	-24:37:30.041	70:24:09.164	95.11	76.59	120
Cross track 4	-24:37:29.309	70:24:09.424	117.73	68.56	144
Cross track 5	-24:37:33.698	70:24:07.866	-17.97	116.74	72
Cross track 6	-24:37:34.430	70:24:07.606	-33.05	122.10	88
Track 2	-24:37:34.359	70:24:08.180	-38.40	107.02	72
Track 3-1	-24:37:35.523	70:24:09.140	-74.40	77.36	96
Track 3-2	-24:37:32.841	70:24:10.092	8.52	47.91	96
Track 4-1	-24:37:34.894	70:24:10.461	-54.96	36.49	96
Track 4-2	-24:37:35.382	70:24:10.288	-70.04	41.84	96
Track 5-1	-24:37:34.580	70:24:11.122	-45.23	16.06	96
Track 5-2	-24:37:34.824	70:24:11.036	-52.77	18.74	104
Track 5-3	-24:37:35.068	70:24:10.949	-60.31	21.42	112
Track 6-1	-24:37:34.667	70:24:11.366	-47.91	8.52	104
Track 6-2	-24:37:34.910	70:24:11.279	-55.45	11.20	112
Track 6-3	-24:37:35.154	70:24:11.193	-62.99	13.88	120
Track 6-4	-24:37:35.398	70:24:11.106	-70.53	16.55	128
Track 6-5	-24:37:35.642	70:24:11.020	-78.07	19.23	136
Track 7-1	-24:37:34.753	70:24:11.610	-50.59	0.98	112

Table 2. Locations of UTs and AT stations.

The information presented here can be used to calculate the UV response of any configuration of telescopes in the VLTI and the corresponding positions of the delay line carriages.

Baselines between UTs range from some 30m (UT2 and UT3) to some 130m (UT1 and UT4). Baselines are shorter (and the UV coverage is thinner) in the NW–SE direction. Baselines between ATs range from 8m in the SW corner to 200m in the NS direction. The station layout is very flexible, virtually any desired configuration of three or four ATs can be obtained. Some stations have been placed in such a way that the limited UV coverage of the UTs can be complemented with baselines between a UT and an AT ("hybrid mode"), to result in an even and circular coverage. The centers of all telescopes and stations are located on an 8m by 8m grid.

Figure 3 shows the UV coverage of the array which consists of only the four UTs. This array is referred to as the "VLT Interferometer Main Array" (VIMA). The source declinations cover the range which is accessible with the VLT. It is evident that the UV coverage of the four main telescopes is less than perfect. There are substantial holes and the NW-SE corner is not well covered.

Some locations of AT stations were selected such as to optimally complement the VIMA array in terms of UV coverage with baselines between a UT and an AT. Figure 4 presents the UV coverage of the array which consists of all four UTs and three ATs located at the stations *long track 4* and *long track 7* and *cross track 2*, as an example. This configuration closes many of the holes which are left in the VIMA coverage, but there is still some reduced coverage in the NW direction. This can be improved by including an AT station near the south-eastern end of the site, e. g., *cross track 6*.

In all cases, a substantial lack of UV coverage will occur for relatively short baselines. This "central hole" is a common problem which is due to the mutual obstruction of UTs by their enclosures, so it cannot be solved with UTs. Only baselines between ATs can be made short enough (less than 8m) to fill the central 25m of baseline space. The cluster of short spacings south of UT1 has been specifically laid out to provide these short spacings.

5 Sensitivity and Limiting Magnitudes

The sensitivity of an interferometer depends critically on the context of the measurement and cannot be comprehensively covered here. We will however give some guidelines which can be used to assess the feasibility of most experiments. An important sensitivity measure is the signal-to-noise ratio of the fringe's contrast magnitude, which is relevant to measurements of visibility magnitude and of referenced phases. Closure phase measurements involve triple product statistics which are significantly less sensitive than visibility magnitude measurements. The visibility magnitude signal-to-noise ratio (SNR) depends mostly on the following factors.

1. *Detected source flux:* the detected source flux determines directly the strength of the measured signal. Depending on source magnitude, spectral bandwidth and integration time the detected source flux may vary enormously. The associated photon noise grows with the square root of the signal strength. There are two regimes which are distinguished by the significance of the photon noise due to the source. In the *photon rich* regime, the photon noise of the source dominates all other noise sources, and the signal-to-noise ratio grows with the square root of the flux (one decade of SNR per 5 magnitudes). In the *photon starved* regime, the photon noise of the source flux is less significant that the noise from other sources. The SNR is now directly proportional to the flux (one decade of SNR per 2.5 magnitudes).

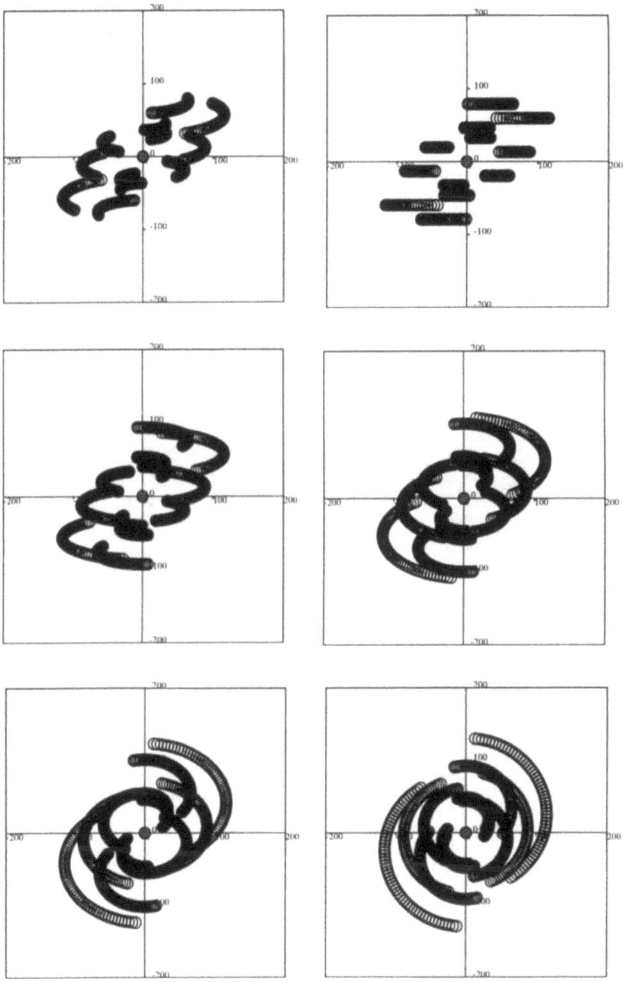

Fig. 3. UV coverages of the VLT Main Array. Declinations: top left +15°, top right 0°, middle left −15°, middle right −30°, bottom left −45°, bottom right −65°. Units are meters

2. *Optical quality:* any differential aberration of the wavefronts brought to interference results in a loss of fringe signal. The optical quality can be expressed in terms of the *Strehl ratio* (the ratio of the peak intensities of an point source image with aberrations and an aberration-free point source image), a number which is practically always less than unity. For many methods of beam combination, the Strehl ratio determines directly the fraction of source flux which contributes to the fringe signal. Some beam combination techniques

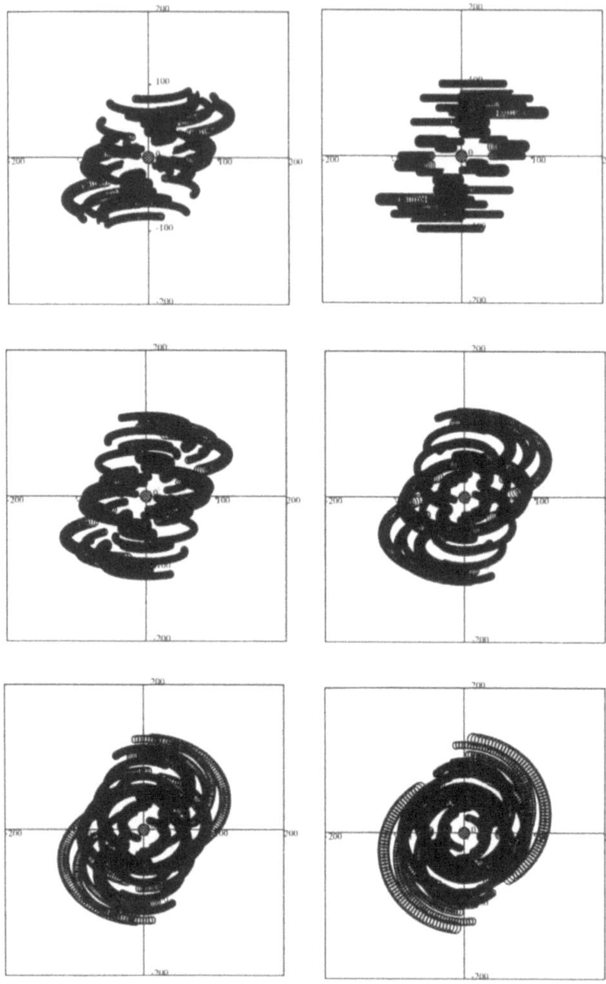

Fig. 4. UV coverages of the VLT Main Array. Declinations: top left +15°, top right 0°, middle left −15°, middle right −30°, bottom left −45°, bottom right −65°. Units are meters

are able to reject the excess flux which does not contribute to the fringe signal, those have better photon noise characteristics. A common source of aberrations is atmospheric seeing which is incompletely compensated by an adaptive optics system, or not compensated at all.

3. *Thermal background:* measurements in the near and thermal infrared are mostly limited by the emission of the atmosphere and the optics. The thermal background can be modelled by the spectrum of a blackbody at a observa-

tory temperature. It creates signal independent photon noise which increases with the square root of the exposure. The detected background flux is given by the product of the emission coefficient of the atmosphere and the optics, the optical etendue of the detection system, and the integral of the Planck function weighted by the spectral response of the detection system. For an interferometer which has lots of optics, the emission coefficient can be assumed as $(1 - \tau_{opt}$ where τ_{opt} is the overall transmission of the optics. A beamcombiner which limits the detected field to the Airy disk of an array element has an entendue of $\overline{\lambda}^2$ where $\overline{\lambda}$ is the mean detected wavelength.

4. *Detector noise:* many detectors generate thermal or readout noise which degrades the signal and needs to be taken into account. This type of noise may have different characteristics with respect to integration time. Thermal noise can be reduced by increasing integration time while readout noise occurs for every readout of a detector, irrespective of how long an integration takes.
5. *Intrinsic visibility:* the proper ("intrinsic") visibility of the source at the observed baseline determines directly the strength of the fringe signal. Most interesting sources have low intrinsic visibility, a useful range is values from $10^{-3} \cdots 10^{0}$.
6. *Coherent and incoherent integration times:* the term "coherent" integration time refers to a single detector readout of a stabilised fringe. Depending on the experiment, the coherent integration time may be milli-seconds (for a measurement limited by the atmospheric correlation time) to minutes (for a measurement with fringe stabilisation which is only limited by the time dependence of the baseline). Several coherent integrations of the fringe signal may be combined to an "incoherent" integration by, e. g., averaging the squared moduli of the fringe amplitude measurements. This kind of integration is relevant when, e. g., each coherent integration would be too noisy to reliably detect the position of a fringe. The SNR of one incoherent integration increases with the square root of the number of accumulated coherent integrations.
7. *Number of elements:* the number of array elements which are *simultaneously* detected may influence the measured signal on a given baseline. The fringe signal of a non-redundant baseline is given by the intrinsic visibility of the source divided by the number of array elements which contribute to the detection. For example, with beam combination in the image plane where the light of all array elements is superimposed to a fringed source image, all array elements contribute to the detection. Another option is to split the light of each element several ways and to beamcombine pair-wise, then the number of contributing elements is two, but the loss of flux due to the multi-way beamcombination needs to be accounted for.

For a measurement where the detected flux is limited to the fraction of energy corresponding to the Strehl ratio, the signal-to-noise ratio for a measurement of visibility magnitude is given with

$$\mathrm{SNR}_V = \frac{SN_p}{\sqrt{1+N_b+SN_p+N_d^2}}\frac{V}{N_{Tel}}\sqrt{K}\ . \tag{1}$$

For a measurement where all of the flux is detected, the signal–to–noise ratio is given with

$$\mathrm{SNR}_V = \frac{SN_p}{\sqrt{1+N_b+N_p+N_d^2}}\frac{V}{N_{Tel}}\sqrt{K}\ . \tag{2}$$

S is the Strehl ratio, N_p is the number of detected photons of the source, N_b is the number of detected photons of the background, N_d is the *rms* detector noise expressed in photon counts, V is the intrinsic visibility of the source (i. e., the quantity of interest), N_{Tel} is the number of array elements which contribute to the fringe measurement, and K is the number of coherent integrations averaged to form the (incoherent) measurement.

Figures 5 to 5 show the dependence of the signal–to–noise ratio with source magnitude under a variety of conditions which apply to the combination of two elements of the VLT Interferometer. Each figure refers to a different photometric band and shows two sets of four panels. The four upper panels refer to combination of two UTs, the lower four refer to combination of two ATs. Each panel has four plots which represent, from top to bottom, intrinsic source visibilities of 10^0, 10^{-1}, 10^{-2}, and 10^{-3}.

In each set of four panels, the top left panel refers to the sensitivity of a fringe tracker with a coherent integration time adapted to the atmospheric correlation time and using the full width of the photometric band. The top right panel refers to coherent integration of 900 sec with the exception of the N band where an incoherent integration of 50 msec exposures, covering 900 sec, is shown. The spectral bandwidth is 200 mÅ and represents a high spectral resolution mode. The bottom two panels show the SNR for incoherent averages of many short exposures with a spectral bandwidth of 100 Å, totalling 900 sec integration time. All panels except for the bottom right ones assume the operation of an adaptive optics compensation system which improves the Strehl ratio. The adaptive optics is optimised for the near–infrared (4.5 degrees of freedom per m^2 area in the entrance aperture and a closed loop bandwidth of 50 Hz) and the improvement in the V band is marginal. The bottom right panel represents the "blind tracking" mode without the use of adaptive optics and fringe control loops. All cases assume a single pixel detector which is photon counting for the V band (no detector noise). The assumed overall optical throughput including detector sensitivity is 0.1.

These plots are intended to provide an indication of the sensitivity of the VLTI, and require some comments to avoid misunderstandings.

1. All initial observations with the VLTI will be done with a single baseline, so the SNR shown was calculated for $N_{Tel} = 2$. Adding more baselines either reduces the source flux by a factor $N_{Tel} - 1$ if an N_{Tel}–way pairwise beamcombiner is used, or N_{Tel} in eqns. 1 and 2 needs to be adjusted.

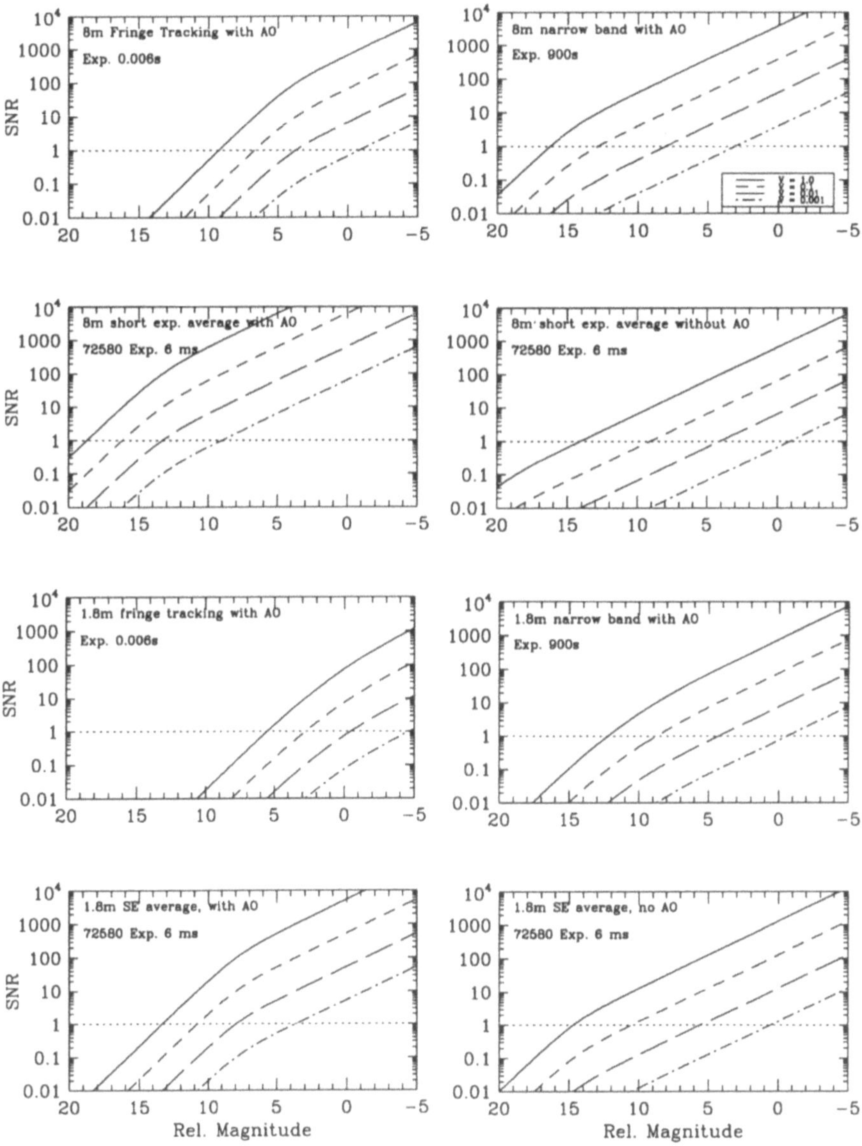

Fig. 5. SNR of the visibility magnitude for the V band.

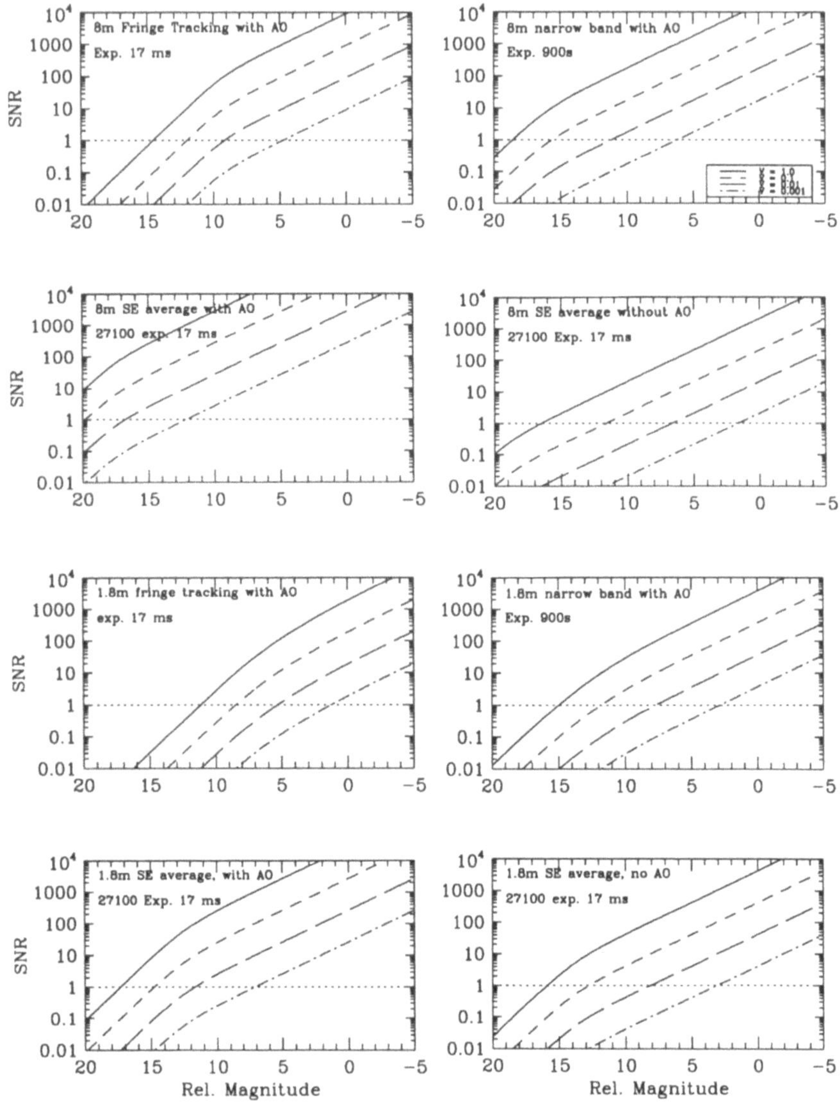

Fig. 6. SNR of the visibility magnitude for the J band.

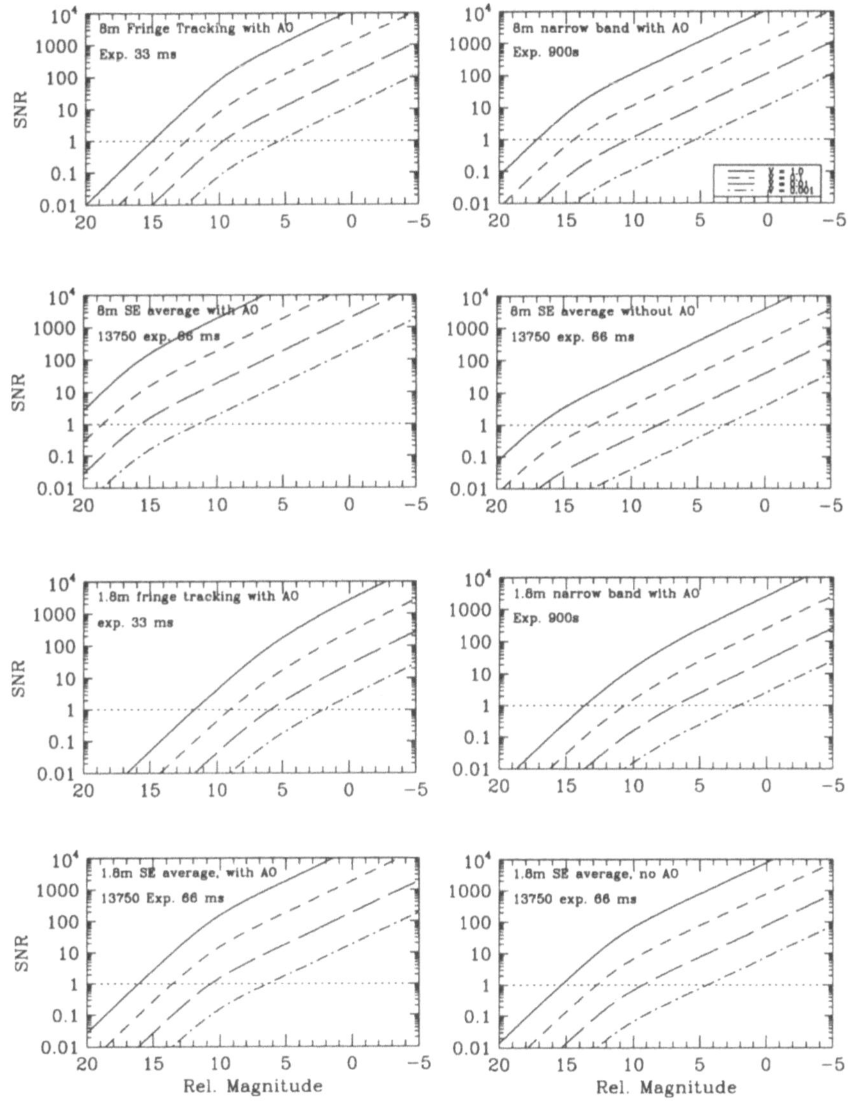

Fig. 7. SNR of the visibility magnitude for the K band.

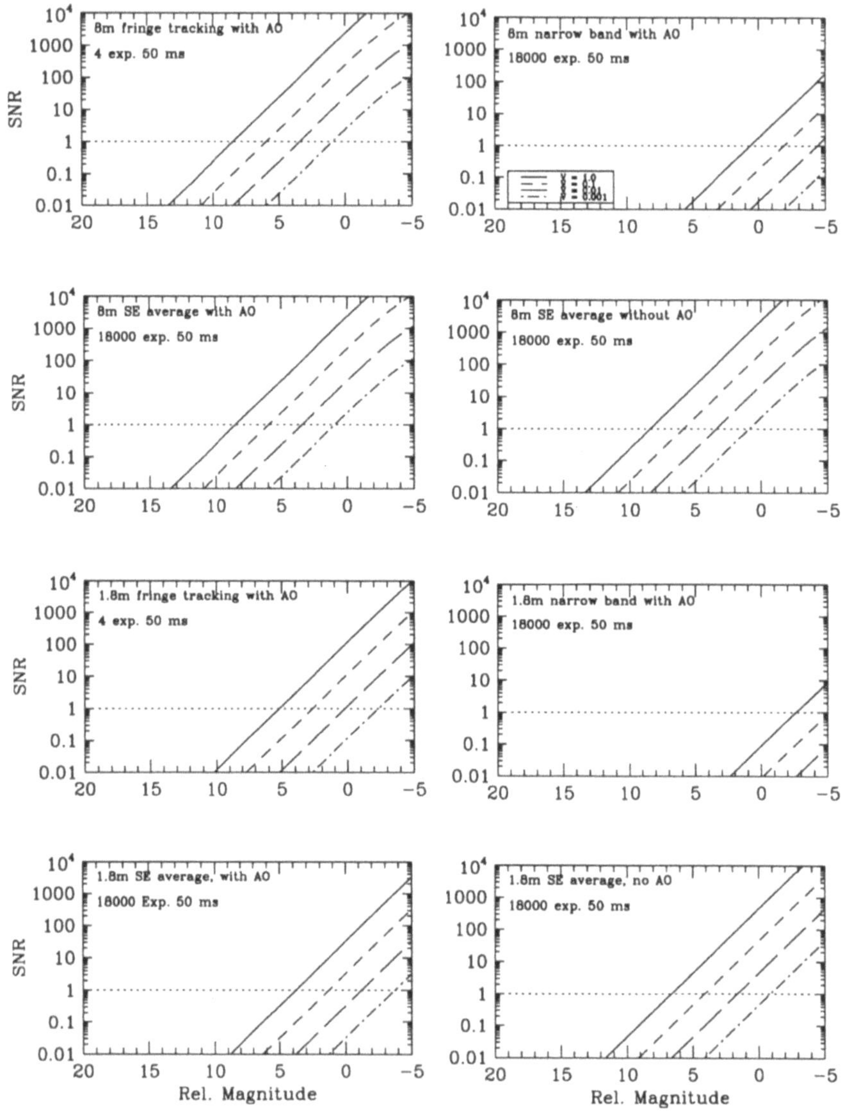

Fig. 8. SNR of the visibility magnitude for the N band.

2. The fringe tracking case appears to be less sensitive than the long integration case which presupposes fringe tracking. With the VLTI, fringe tracking will be performed in the H band, so one should take the J band curves which are similar to the H band performance. The plots are representative for the ATs where "adaptive optics" will be realized in terms of fast tip–tilt guiding, rendering the ATs diffraction–limited in the near–infrared. The plots are representative for the UTs only with an adaptive optics system, which is currently not part of Phases A or B. Therefore the sensitivity *for fringe tracking* with UTs without AO systems will be somewhat lower than indicated here, about at the level of the sensitivity of ATs.
3. In case that more than two telescopes are used, fringe tracking may be performed on shorter baselines which tend to sample higher intrinsic visibilities, while long term integration is performed on the longer baselines with lower visibility. This potential may unveil the real power of the fringe tracking mode.
4. The long exposure cases all assume a spectral bandpass of 200 mÅ which is relevant to shorter wavelengths, but definitely too narrow for the K and N bands. The curves shown are therefore rather conservative. Also, the SNR *per spectral channel* is shown. A beamcombiner with a dispersive element may permit many channels to be detected, where the signal could be combined to increase the SNR even further.

6 Conclusions

We have described the current status of the interferometric mode of the ESO Very Large Telescope. The development plan for the VLTI has been adapted to the difficult financial situation by substantial modification of the concept and by implementig capability gradually in two phases. We feel that the modifications simplfy the project in many areas while even strengthening the scientific potential. However, introducing the astrometric capability is a new requirement which will certainly be difficult. When Phases A and B are completed, a large subset of the initial goals for VLTI will have been put into existence. At this time, the VLTI will be far from complete, and one should prepare for further upgrades.

An impotant capability will be the near infrared capability of Unit telescopes through adaptive optics. Those can be systems which are optimised for interferometric use and therefore may be simpler than stadalone adaptive optics systems developed for other instruments. Introducing adaptive optics will tremendously increase the sensitivity with the UTs, as indicated by the SNR plots. Another important capability would be extending the spectral regime towards the visible, opening a wide range of spectral diagnostics and even higher resolution. Similarly important would be adding a fourth Auxiliary telescope and more delay lines to bring the number of beamlines up to a number of eight. VLTI will then unveil its full potential as an optical aperture synthesis array.

References

J. M. Beckers, R. Braun, G. P. di Benedetto, R. Foy, R. Genzel, L. Koechlin, F. Merkle, G. Weigelt, *The VLT Interferometer Implementation Plan,* Report by the ESO/VLT Interferometry Panel, ESO VLT Report No. 59, October 1989

J. M. Beckers, D. Enard, M. Faucherre, F. Merkle, G. P. di Benedetto, R. Braun, R. Foy, R. Genzel, L. Koechlin, G. Weigelt, "The VLT Interferometer: 1. Proposed implementation," Proc SPIE Vol. 1236, pp. 108ff, 1990a.

J. M. Beckers, "The VLT Interferometer: 2. Factors affecting on-axis operation," Proc SPIE Vol. 1236, pp. 364-371, 1990b.

J. M. Beckers, "The VLT Interferometer: 3. Factors affecting wide field-of-view operation," Proc SPIE Vol. 1236, pp. 379-389, 1990c.

J. M. Beckers, "The VLT Interferometer: 4. The utility of partial adaptive optics," Proc SPIE Vol. 1236, pp. 379-389, 1990d.

J. M. Beckers, "Interferometric Imaging with the Very Large Telescope," J. Optics (Paris) 22,2 pp. 73-83, 1991a.

J. M. Beckers, "Blind operation of optical astronomical interferometers: options and predicted performance," Exp. Astronomy 2, pp. 57-71, 1991b.

J. M. Beckers, M. Faucherre, B. Koehler, O. von der Lühe, "A progress report on the implementation of the VLT Interferometer," ESO Conference on *High Resolution Imaging by Interferometry II,* J. M. Beckers, F. Merkle (Eds.), ESO Garching, pp. 775-789, 1991c.

J. M. Beckers, R. Braun, G. P. di Benedetto, A. Eckard, M. Faucherre, R. Foy, R. Genzel, P. Léna, O. von der Lühe, L. Koechlin, F. Merkle, G. Weigelt, *Coherent Combined Instrumentation for the VLT Interferometer,* Report by the ESO/VLT Interferometry Panel, ESO VLT Report No. 65, April 1992.

T. R. Bedding, J. M. Beckers, M. Faucherre, N. Hubin, B. Koehler, O. von der Lühe, F. Merkle, Zhu N., "The VLT Interferometer," in J. G. Robertson and W. J. Tango (Eds.), *Very High Angular Resolution Imaging,* IAU Symposium 158, pp. 143-150, 1993.

O. Citterio, D. Downes, A. Labeyrie, P. Léna, J. E. Noordam, F. Roddier, J. Wijnbergen, G. Weigelt, L. Weliachev, D. Enard, F. Merkle, M. Sarazin, R. Wilson, *Interferometric Imaging with the Very Large Telescope,* Final Report presented by the ESO/VLT Working Group on Interferometry, ESO VLT Report No. 49, June 1986.

A. Claret, J.-M. Mariotti, P. Léna, "Multi-spectral diffraction-limited imaging with the double-Fourier method: simulations for the VLTI," ESO Conference on *High Resolution Imaging by Interferometry II,* J. M. Beckers, F. Merkle (Eds.), ESO Garching, pp. 835-844, 1991.

D. Enard, "ESO 16 metre Very Large Telescope: The linear array concept," Proc. of the IAU Colloquium No 79, *Very Large Telescopes, their Instrumentation and Programs,* M.-H. Ulrich and K. Kjär (Eds.), Garching, pp. 767-787, April 1984.

M. Faucherre, R. Maurer, "On metrology systems for delay lines," ESO Conference on *High Resolution Imaging by Interferometry II,* J. M. Beckers, F. Merkle (Eds.), ESO Garching, pp. 1205-1214, 1991.

M. Faucherre, B. Koehler, O. von der Lühe, F. Merkle, Zhu N., "The VLT Interferometer: current status and expectations for the next 20 years," Proc. of the ESA Colloquium on Targets for Space-based Interferometry, ESA SP-354, Beaulieu, pp. 91-98, 1992.

M. Ferrari, G. Lemaître, "Variable curvature mirrors for the ESO VLTI," in Proc. SPIE 2201 *Adaptive Optics in Astronomy,* 1994a.

M. Ferrari, Lemaître, G., Mazzanti, S., von der Lühe, O., di Biagio, B., Montiel, P., Revest, D., Joulié, P., Carré, J.-F., "Highly variable curvature mirrors for the Very Large Telescope Interferometer," in M. A. Ealey and F. Merkle (eds.), *Adaptive Optics in Astronomy,* SPIE Proceedings Vol. 2201, 1994b.

J. Gay, Y. Rabbia, "LAMP: a concept for the ESO VLTI fringe sensor," in Proc. SPIE Proc. Vol. 2200 *Amplitude and Intensity Spatial Interferometry II,* J. Breckinridge (Ed.), 1994a.

J. Gay, Y. Rabbia, P. M. Bourlon, P. Antonelli, M. Dugue, J. Marchal, S. Menardi, M. Faucherre, N. Hubin, "Prototype for the ESO VLTI fringe sensor," in Proc. SPIE Proc. Vol. 2200 *Amplitude and Intensity Spatial Interferometry II,* J. Breckinridge (Ed.), 1994b.

H. Jörck, R. Maurer, J. Käse, M. Faucherre, J. M. Beckers, G. Kühn, H. Hupe, "The design of the Delay Lines for the VLT Interferometer," ESO Conference on *High Resolution Imaging by Interferometry II,* J. M. Beckers, F. Merkle (Eds.), ESO Garching, pp. 1157-1168, 1991

B. Koehler, O. von der Lühe, "Engineering aspects of environmental factors in VLTI," in J. Breckinridge (ed.), *Amplitude and Intensity Spatial Interferometry II,* SPIE Proceedings Vol. 2200, 1994.

B. Koehler, S. Lévêque, O. von der Lühe, "Results of studies on the environmental factors affecting the VLTI performance," in *Optical Telescopes of Today and Tomorrow,* SPIE Proc., June 1996.

A. Leger, Mariotti, J.-M., Rouan, D., Schneider, "How to search for extra-solar planets with the VLT/VISA?" in *Science with the VLT,* Proceedings of the ESO Workshop 28 June - 4 July 1994, J. R. Walsh and I. J. Danziger (eds.), Springer 1995, pp. 21-32, 1995.

P. Léna, F. Merkle, "Interferometry with the European Very Large Telescope," ESO-NOAO Conference on *Interferometric Imaging in Astronomy,* J. W. Goad (Ed.), NOAO Tucson, pp. 169-170, 1987.

P. Léna, "The interferometric mode of the European Very Large Telescope," NOAO-ESO Conference on *High Resolution Imaging by Interferometry,* F. Merkle (Ed.), ESO Garching, pp. 899-908, March 1988.

S. Lévêque, B. Koehler, O. von der Lühe, "Measurement of Optical Path Fluctuations due to internal Seeing for the VLT", in *Advanced Imaging Technology and Commercial Applications,* SPIE Proc. 2566, J. Gonglewski (Ed.), 156-163, 1995a.

S. Lévêque, B. Koehler, O. von der Lühe, "Longitudinal dispersion compensation for the Very large Telescope Interferometer," subm. to Experimental Astronomy, 1995b.

S. Lévêque, B. Koehler, O. von der Lühe, "Optical path difference model for the Very large Telescope Interferometer", in *Optical Telescopes of Today and Tomorrow,* SPIE Proc., June 1996.

F. Merkle, P. Léna, "Spatial interferometry with the European VLT," Proc. SPIE Vol. 628 *Advanced Technology Optical Telescopes III,* pp. 261-272, 1986.

F. Merkle, "Synthetic-aperture imaging with the European Very Large Telescope," Journ. Opt. Soc. Am. A 5,6, pp. 904-913, 1988a.

F. Merkle, "Current concept of the VLT interferometric mode," NOAO-ESO Conference on *High Resolution Imaging by Interferometry,* F. Merkle (Ed.), ESO Garching, pp. 909-919, 1988b.

"ESO, CNRS and MPG Sign Agreement on Enhancement of the VLT Interferometer," The Messenger 71, ESO, p. 1, March 1993a.

"News from Council," The Messenger 74, ESO, p. 1, December 1993b.

F. Paresce, and the Interferometry Science Advisory Committee, "A New Start for the VLTI", ESO Messenger 83, pp. 14-21, March 1996.

D. Plathner, J. M. Beckers, "The design of the Auxiliary Telescopes for the VLT Interferometer," ESO Conference on *High Resolution Imaging by Interferometry II,* J. M. Beckers, F. Merkle (Eds.), ESO Garching, pp. 939-948, 1991.

A. Quirrenbach, " Astrometric detection ond investigation of planetary systems with the VLT Interferometer," in *Science with the VLT,* Proceedings of the ESO Workshop 28 June - 4 July 1994, J. R. Walsh and I. J. Danziger (eds.), Springer 1995, pp. 33-37, 1995.

Th. Reinheimer, K.-H. Hofmann, G. Weigelt, "Computer simulations of interferometric imaging with the VLT interferometer," ESO Conference on *High Resolution Imaging by Interferometry II,* J. M. Beckers, F. Merkle (Eds.), ESO Garching, pp. 827-834, 1991.

F. Roddier, P. Léna, "Long baseline Michelson interferometry with large ground-based telescopes operating at optical wavelengths II. Interferometry at infrared wavelengths," J. Optics (Paris) 15,6 pp. 363-374, 1984.

Shao, M., M. M. Colavita, "Potential od long-baseline infrared interferometry for narrow-angle astrometry," Astron. Astrophys. 262, pp. 353-358, 1992.

O. von der Lühe, J. M. Beckers, R. Braun, "The configuration of the VLT Interferometer on the Paranal site," ESO Conference on *High Resolution Imaging by Interferometry II,* J. M. Beckers, F. Merkle (Eds.), ESO Garching, pp. 959-968, 1991.

O. von der Lühe, J. M. Beckers, M. Faucherre, B. Koehler, F. Merkle, Zhu N., "Interferometry with the Very Large Telescope," Proc. of an ESA Workshop on *Solar Physics and Astrohysics at Interferometric Resolution,* ESA-SP 344, Paris, pp. 247-254, 1992.

O. von der Lühe, D. Ferrand, B. Koehler, Zhu N., Th. Reinheimer, "The Interferometric Mode of the ESO Very Large Telescope", in Proc. SPIE Proc. Vol. 2200 *Amplitude and Intensity Spatial Interferometry II,* J. Breckinridge (Ed.), pp. 168-179, 1994.

O. von der Lühe, F. Derie, D. Ferrand, B. Koehler, S. Lévêque, F. Paresce, L. d'Arcio, "Interferometry with the ESO Very Large Telescope", in Proc. SPIE Proc. Vol. 2566 *Advanced Imaging Technology and Commercial Applications,* N. Clark, J. Gonglewski (Eds.), pp. 124-133, 1995a.

O. von der Lühe, A. Quirrenbach, and B. Koehler, "Narrow-angle Astrometry with the VLT Interferometer," in *Science with the VLT,* Proceedings of the ESO Workshop 28 June - 4 July 1994, J. R. Walsh and I. J. Danziger (eds.), Springer 1995, pp. 445-450, 1995b.

O. von der Lühe, A. Quirrenbach, "Calibration of the VLT Interferometer," in *Understanding and Calibrating HST and VLT Data,* ESO Conference, P. Benvenuti (Ed.), 173-179, 1995c.

O. von der Lühe, "VISA: the sub-array mode of the ESO Very Large Telescope Interferometer," in *Programmes Astrophysiques pour VISA,* Workshop by the French National Program for High Angular Resolution in Astronomy, J. Bouvier, Ed. 20-36, 1995d.

O. von der Lühe, F. Derie, B. Koehler, S. Lévêque, F. Paresce, M. Verola, "Interferometry with the ESO Very Large Telescope", in *Optical Telescopes of Today and Tomorrow,* SPIE Proc., June 1996.

Interferometry in Practice

Andreas Quirrenbach

Max-Planck-Institut für Extraterrestrische Physik, Postfach 1603, D-85740 Garching, Germany

Abstract. This tutorial contains a brief discussion of some topics that are important for the planning and succesful execution of observations with an optical interferometer. It touches upon sensitivity, calibration, fringe tracking, the field-of-view, synthesis imaging, and model fitting. Some examples and references to the more detailed literature are included.

1 Signal-to-Noise Ratio and Sensitivity in Interferometry

Sensitivity predictions for optical and near-infrared interferometry such as those given for the VLTI in ESO (1989) are usually based on signal-to-noise calculations, which assume that systematic errors can be controlled to the extent that unavoidable random noise dominates at the performance limit of the instrument. The sources of this random noise are essentially the same in an interferometer as in a single telescope. Photon noise is normally most important in the visible wavelength regime, where photon-counting detectors with low dark count rates can be used. Detector noise, i.e. read noise and dark current, will usually be most important in the near infrared. The photon noise associated with the thermal background from the telescope and from the sky dominates the noise budget at even longer wavelengths.

The interferometric signal, on the other hand, depends not only on the number of photons received (i.e. on source brightness, telescope diameter and throughput, spectral bandwith, and detection efficiency), but also on the fringe contrast (i.e. on source compactness, baseline length, atmospheric wavefront distortion, and loss of contrast within the instrument). In the photon-noise-limited case, for example, the signal-to-noise ratio (SNR) of the square[1] of the visibility amplitude, V^2, is given by

$$\mathrm{SNR}\,(V^2) = \frac{1}{4} \cdot \sqrt{M} \cdot \frac{NV^2}{\sqrt{1+\frac{1}{2}NV^2}} \ , \tag{1}$$

where N is the total number of photons detected during the integration time, and M is the number of "frames" or samples of V^2. A similar expression holds for the SNR of the visibility phase. Some very important conclusions can be drawn from Equation 1:

[1] It is usually more convenient to work with V^2 rather than V, since it is possible to construct an unbiased estimator for V^2 in the presence of photon noise, see Shao et al. (1988).

1. The SNR does not depend on N alone, but on NV^2. This means that the sensitivity is greatly reduced for extended sources compared to point sources. This effect can be very dramatic. For example, the sensitivity is reduced by 5 magnitudes for objects with $V = 0.1$. Since generally V drops with baseline length, the SNR will normally be worst on the longest baselines.
2. It is instructive to look at the asymptotic behavior of Equation 1 for $NV^2 \gg 1$ and for $NV^2 \ll 1$. These two cases are usually referred to as the "photon-rich" and "photon-starved" regimes, respectively (although this is actually a misnomer, in view of the previous point). In the photon-rich regime, SNR $\propto \sqrt{MNV^2}$, i.e. the signal-to-noise ratio is proportional to the square root of the total number of photons received during the on-source integration time. In the photon-starved regime, however, SNR $\propto \sqrt{M} \cdot NV^2$, i.e. the signal-to-noise ratio drops precipitously with diminishing brightness or increasing resolution when NV^2 gets much smaller than 1.
3. From the previous point it follows that in the photon-starved regime the coherent integration time τ (the "frame time") should be as long as possible. For a *fixed total on-source integration time*, we have $NV^2 \propto \tau$ and $M \propto \tau^{-1}$, and therefore SNR $\propto \sqrt{\tau}$. While τ is normally restricted by the coherence time of atmospheric phase fluctuations, it is sometimes possible to use a phase reference to extend τ beyond this limit, resulting in a substantial increase in sensitivity (see Quirrenbach et al. 1994).

Qualitatively similar arguments apply also in the detector- and background-noise-limited cases, and for other observables such as the closure phase. The relevant quantitative expressions can be found in Shao and Colavita (1992).

2 The Quality of Interferometer Data

As with any astronomical instrument, the quality of the data obtained with an interferometer depends on the signal-to-noise ratio, and on residual systematic errors left after calibration. However, there are a few complications specific to interferometry, which have to be taken into account for the design of the instrument, for the planning and execution of observations, and for the data reduction. These complications are due to the sensitivity of an interferometer to spatial and temporal phase variations, and to the peculiar shape of the entrance pupil, which is non-contiguous and highly diluted. Tracing the light from the source to the detector, the following sections briefly discuss the most important parameters which have to be considered for an evaluation of the performance of optical and infrared interferometers.

2.1 The Source

The most important source parameters affecting the quality of interferometric data are: brightness, compactness, complexity, and variability. As discussed above, the SNR depends not only on the source brightness, but also on the visibility. Therefore the reference to a "limiting magnitude" for an interferometer can be quite misleading. It would be more appropriate to use the term "point source sensitivity", and it must be kept in mind that even for sources subtending only a few resolution elements the sensitivity can be many magnitudes worse. This is usually the case for low-contrast objects such as stellar photospheres, which are essentially limb-darkened disks with little modulation from granulation and star spots. To illustrate this point, I should mention that several attempts to observe the *really* bright star α Ori with the Mk III interferometer failed; it is so big (~ 42 mas) that NV^2 is fairly low even on the shortest baselines (3.0 m and 4.2 m) of the Mk III.

While a filled aperture samples all spatial frequencies up to the diffraction limit, the very dilute apertures characteristic of long-baseline interferometers measure only selected Fourier components of the source brightness distribution. If the source has structure at spatial frequencies that are not sampled by the interferometer, the information obtained may be quite misleading, and one has to be aware of the potential pitfalls. For example, a companion may not be distinguishable from a stellar surface feature with a given set of data (see Di Benedetto and Bonneau 1990). Another problem for synthesis imaging is source variability. Since the instantaneous coverage of the *uv* plane provided by few-element arrays is rather poor, Earth rotation synthesis and / or observations using multiple configurations are frequently required to obtain a sufficient sampling of the *uv* plane. The data are therefore taken over time spans ranging anywhere from a few hours to several weeks. Any unaccounted variability of the target source, which includes not only photometric variations, but also rotation of a spotted stellar surface, or orbital motion of a binary, will give rise to artifacts in synthesis images, or even spoil them completely.

2.2 The Atmosphere

Atmospheric phase fluctuations distort the wavefront of light from astronomical sources and thus destroy the coherence in an interferometer. The two most important parameters characterizing the atmospheric turbulence are the length scale r_0, and the time scale τ_0. The most simplified model of the atmosphere is that of a screen consisting of turbulence cells with diameter r_0, which change on the time scale of τ_0. (This is sufficient for most heuristic arguments and rough calculations; a good summary of the more rigorous theory has been given e.g. by Roddier (1989).) As long as the telescope diameter $D < r_0$, and the coherent integration time $\tau < \tau_0$, the wavefront can be regarded as fully coherent. When $\tau > \tau_0$, the fringe phase changes within the integration time, which leads to a loss in fringe contrast and thus to a reduced SNR and calibration problems. If $D > r_0$, the fringe phase varies across the pupils of the interferometer elements, and multi-speckle data recording and data reduction techniques have to be used.

It is now useful to consider the "coherence volume" of the atmosphere, i.e. a cylinder with diameter r_0 and length $c \cdot \tau_0$, where c is the speed of light. We can now calculate the SNR from Equation 1, where now N is the number of photons within the coherence volume, and M is the number of these cells that arrive at the telescope pupil during the total integration time. It is then clear that the transition from the photon-rich to the photon-starved regime occurs for a point source when there is one photon in the coherence volume. (For an extended source the transition is at $NV^2 = 1$.) One could also expect this intuitively, since only those photons can interfer with each other that are contained within the same coherence volume; if the average number of these photons is less than one, most of them simply add noise without contributing to the signal. The atmosphere thus sets a fundamental limit to the sensitivity of an interferometer which is difficult to overcome even with large apertures. It is important to note, however, that both r_0 and τ_0 scale with $\lambda^{6/5}$, i.e. the coherence volume increases towards longer observing wavelengths with a $\lambda^{18/5}$ scaling. This is the reason why interferometry is very difficult in the blue spectral region, but comparatively easy in the infrared.

If an adaptive optics system is available, it can be used to flatten the wavefront. Note, however, that adaptive optics also needs a bright guide star, which emits a fair number of photons into each atmospheric coherence volume. Adaptive optics is therefore useful mainly for resolved sources (with $NV^2 \ll 1$, but $N \gg 1$), for observations with a narrow spectral bandwidth, and in conjunction with laser guide stars.

Optical and near-infrared interferometers normally use a fast guider (tip-tilt tracker) to maintain the alignment of the source images from the individual telescopes in the presence of tracking errors, vibrations, and atmospheric fluctuations. The removal of overall tip and tilt from the wavefront distortions effectively increases r_0 by a factor of nearly 3. For good seeing conditions at an excellent site like Paranal, a 1.8 m telescope will be fully coherent at a wavelength of 2.2 μm, and an 8 m telescope at 10 μm, using just tip-tilt correction. It will therefore be possible to exploit the full sensitivity of the VLTI main array (consisting of the 8 m telescopes) in the thermal IR, and of the auxiliary telescopes in the near-IR, without using high-order adaptive optics.

2.3 The Telescope and Delay Line System

The photon flux arriving at the beam combiner is directly proportional to the telescope size and to the throughput of the telescope / delay line system. For signal-to-noise calculations, additional factors have to be considered. The effect of the atmosphere for telescope sizes $D > r_0$ has already been discussed in the previous section, but "internal seeing" can also lead to a loss of coherence. It is therefore necessary to either evacuate the light path from the telescopes to the beam combiner, or at least to suppress air circulation and temperature gradients in the delay line tunnels. Aberrations and micro-roughness of the mirrors also introduce phase errors across the wavefront, which can add up to a substantial degradation after the many reflections necessary in an interferometer.

While these problems cannot be avoided completely, mistakes in the optical design can add more trouble in the form of polarization and field rotation effects. Every reflection at non-normal incidence introduces a phase shift, which is different for the two states of polarization (in the plane of the reflection and orthogonal to this plane). When two beams that have undergone different reflections are combined, no value of the delay can therefore be found where the two polarizations will both interfere additively, and the fringe contrast will be reduced. While this problem can be solved in principle by using only one polarization state, the addition of a polarizer increases the complexity and throws away half of the valuable photons. It is better to make the reflections in all interferometer arms the same, even if this means adding extra mirrors to some of them. *From the point of view of interferometry, it is a very bad idea to use different coatings on individual telescopes, to optimize them for specific wavelength regions.* Having identical reflections in all arms also avoids differential field rotation, which would reduce the fringe contrast on-axis, and completely destroy coherence off-axis.

In an interferometer with air delay lines, such as the VLTI, longitudinal dispersion (i.e. wavelength-dependent pathlength differences) is a serious problem. The light path through the atmosphere is equal for all telescopes, if all telescopes are at the same height above sea level and if the curvature of the Earth can be neglected, which is usually the case for baselines of less than 1 km. However, the length of the light path through the delay lines changes continually with the positions of the delay line carts. The resulting air mismatch must be corrected in broad band systems, e.g. with glass compensators.

2.4 The Beam Combiner and Detector

Imperfections in the beam combiner contribute to degradations of the wavefront in a similar way as those in the telescopes and delay lines. They can be lumped together and then be described by a single (wavelength-dependent) number, the system visibility V_{sys}. (Sometimes the atmospheric wavefront degradation is also included in V_{sys}.) This is the fringe contrast that is obtained in observations of an unresolved source. For SNR calculations (using e.g. Equation 1) the product of the source visibility and V_{sys} has to be used.

There are many different ways of combining the light from the individual telescopes. The choice of the beam combiner design depends on the operating wavelength, spectral resolution $R = \lambda/\Delta\lambda$, number of spectral channels, number of telescopes to be combined, telescope size, desired field-of-view, and available detector technology. Examples of beam combiner designs have been described by Buscher (1988), Mozurkewich (1994), and in the ESO report (1992). A particular point of interest for imaging applications is the question of whether the beam combiner introduces non-closing phase errors; some beam combiner designs do, others do not. Monitoring such errors with a laser metrology system is necessary if the closure phase is to be used. A sophisticated metrology system is also needed when two beam combiners are used simultaneously, which is the case, e.g., when the tasks of fringe tracking and taking of scientific data are separated.

In contrast to radio interferometry, where the signal can be amplified before the correlation, the SNR is always reduced when more than two telescopes are combined in an optical or near-IR interferometer. If the beams are combined pairwise, beam splitters must be used that reduce the signal; if all beams are combined on a single detector, the signal of each baseline has to be detected against the combined noise from all telescopes. The combination of telescopes of unequal size presents additional problems (Beckers 1989). Combining a 1.8 m telescope with an 8 m telescope gives the same sensitivity as the combination of two 1.8 m telescopes in the worst case, and that of two 3.8 m telescopes in the best case.

The requirements for the detector are intimately linked to the layout of the beam combiner. While a single pixel per baseline can in principle be sufficient, as in the Mk III (Shao et al. 1988), covering a "wide field" of 2″ with 0.5 mas pixels is not feasible with current IR array technology. Low noise and fast read-out are always important considerations; photon counting is very advantageous. Superconducting tunnel junction devices (Peacock et al. 1996) hold the promise of overcoming some of the limitations of the detectors presently in use; they may turn out to be particularly useful when moderate R and a limited number of spectral channels are required.

2.5 Observing Strategy and Data Reduction

Because of the loss of fringe contrast due to the atmosphere and instrumental imperfections, the measured raw visibilities have to be converted into "true" calibrated source visibilities using

$$V_{\rm cal} = V_{\rm raw} / V_{\rm sys} \quad . \tag{2}$$

Any errors in the determination of $V_{\rm sys}$ affect the final data directly; the uncertainty of $V_{\rm cal}$ is in many practical cases dominated by these calibration problems rather than by the formal SNR. Temporal variations of $V_{\rm sys}$ due to changes of the seeing are normally the dominant source of error. Fortunately it is possible to use internal seeing indicators such as delay fluctuations, which are monitored by the delay line metrology, in the calibration process (Mozurkewich et al. 1991). It is also useful to switch rapidly between the astronomical target source and visibility calibrators. For example, a typical observing night with the Mk III consisted of $\sim$ 200 individual observations. Since nearly half of them were scans of visibility calibrators, it was normally possible to obtain good fits of $V_{\rm sys}$ as a function of seeing, zenith angle, and time.

The sensitivity to seeing changes can be substantially reduced by spatial filtering, e.g. by coupling the light into a single-mode fiber (e.g., Coudé du Foresto 1994, Shaklan et al. 1992). At the other extreme, specific algorithms have been developed for the processing of data taken with large apertures in the multi-speckle regime (Mourard et al. 1994). A more complete account of the calibration of VLTI observations has been given by von der Lühe and Quirrenbach (1995).

3 On-Source and Off-Source Fringe Tracking

Interferometers can be operated in active and in passive mode. In a passive interferometer, the delay lines are commanded to track at the sidereal rate and the fringes are measured "blindly". This requires exact knowledge of the interferometer geometry and a coherence length longer than the delay variations caused by internal pathlength changes and the atmosphere; the coherent integration time is limited by τ_0. In an active interferometer, the fringe phase is measured in real time and the delay is corrected by a closed-loop fringe tracking system. Once the fringes are acquired, the system can follow atmospheric and instrumental delay fluctuations, which considerably relaxes the requirements on the stability of the interferometer. Moreover, if the fringe servo can be run with residual phase errors of less than 1 rad, it is possible to integrate coherently for arbitrarily long times[2]. It is therefore always desirable to operate an interferometer in the active fringe tracking mode.

Fringe tracking requires SNR $\gtrsim 10$ *within each coherence volume of the atmosphere.* If for the source of interest $NV^2 \gg 1$, it can be used easily as a reference for the fringe tracker. If the source is bright enough, but V^2 very small, baseline bootstrapping or wavelength bootstrapping may work. Baseline bootstrapping (see Figure 1) uses the signal on the short baselines of the array for the fringe-tracking servo, while data are taken on the long baselines, where V^2 can be much lower. The configuration of the Navy Prototype Optical Interferometer, which is optimized for observations of stellar surface structure, has been laid out specifically to make use of baseline bootstrapping (Mozurkewich and Armstrong 1992).

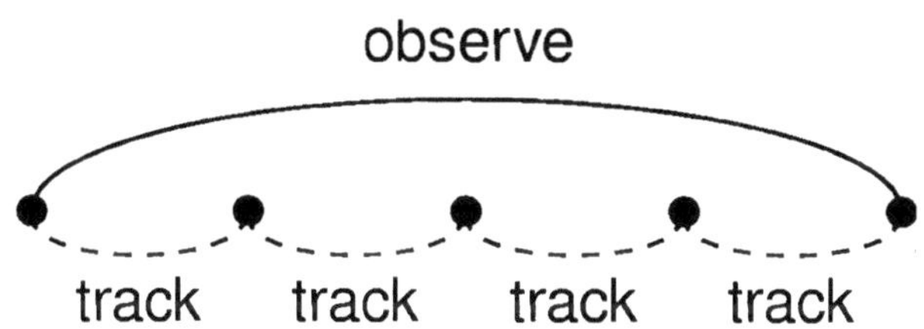

Fig. 1. Baseline bootstrapping. The fringes are tracked on the short baselines of the array (in this case a five-element linear configuration), where V^2 is high. The observations can then be done on the long baselines.

Wavelength bootstrapping (see Figure 2) uses the fact that on the same baseline V^2 may be high at a certain wavelength λ_1, but low at another wavelength λ_2. It is then possible to observe at λ_2, while the fringe tracker is working at λ_1. Wavelength bootstrapping is useful, e.g., for imaging of stellar photospheres, where λ_1 can be in the IR and λ_2 in the visible, and for imaging of circumstellar matter, where λ_1 can be in the continuum (where the small stellar photosphere dominates) and λ_2 in a line emitted by the extended material. For such dual-wavelength observations a number of effects have to be taken into account which introduce differential phase noise between λ_1 and λ_2 (Quirrenbach et al. 1994).

[2] Running a low-precision fringe servo is sometimes called "coherencing" of the interferometer, while running a servo with precision < 1 rad is called "cophasing".

It is possible, of course, to combine baseline bootstrapping and wavelength bootstrapping.

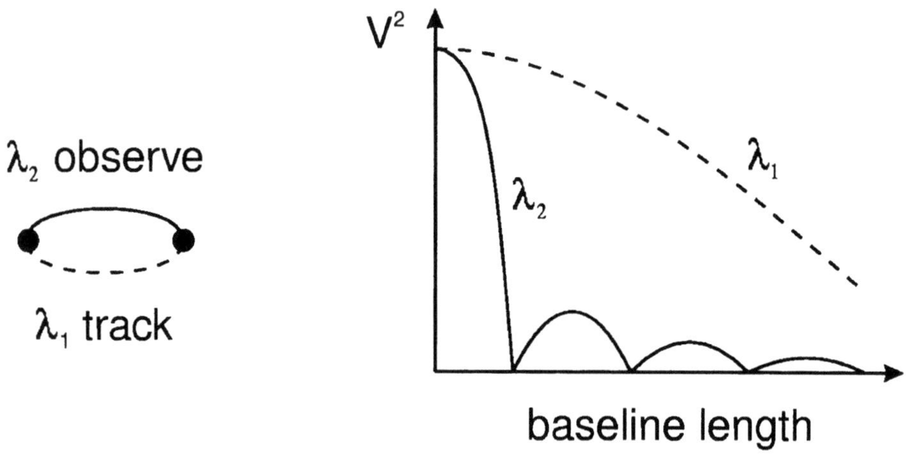

Fig. 2. Wavelength bootstrapping. The fringes are tracked at a long wavelength. The observations are done on the same baseline, but at a shorter wavelength, where the resolution is higher, but V^2 much smaller.

If $N < 1$, or if bootstrapping is not possible for other reasons, fringe tracking on a separate bright reference star may still be possible. However, the reference source must be within the isoplanatic angle θ_0 of the science target. (θ_0 is a few arcseconds in the visible, and scales with $\lambda^{6/5}$.) The reference source can also be used for adaptive optics wavefront sensing, if such a system is available. In this case the whole entrance pupil of the interferometer is made fully cophased and the sensitivity of the interferometer is essentially identical to the sensitivity of a single telescope with the same diameter. It is thus important to realize that *bright objects are needed to cophase an interferometer, but very faint sources can be observed in a limited field arount these reference sources.*

4 The Field-of-View of an Interferometer

The useful field-of-view of an interferometer is limited both by vignetting and by coherence effects. The unvignetted field-of-view through the delay lines of the VLTI is only a few arcseconds, which sets an upper limit to the useful field regardless of the way the beams from the individual telescopes are combined. In a Michelson stellar interferometer, however, the image formation process is more complicated than in a single telescope (or in a Fizeau interferometer) and imposes additional restrictions on the useful field-of-view. In general, the delay varies with field angle, which destroys the coherence for off-axis sources when a non-zero spectral bandwidth is used. Another way of saying this is that the position of the white-light fringe is not the same as the position of the image formed by the beam-combining telescope for off-axis objects. This effect is known as "bandwidth smearing" in radio astronomy and limits the field-of-view to a diameter of approximately $R = \lambda/\Delta\lambda$ spatial resolution elements.

The approach to beam combination and data processing can also have an influence on the interferometric field-of-view. For example, in the Mk III afocal beams are combined "on-axis" by 50% beam splitters. This means that all data correspond to exactly the same baseline vector (namely that between the centers of the two telescopes), no matter how big the individual telescopes are. In the GI2T, on the other hand, the beams are combined in an image plane, with the consequence that all baseline vectors occuring within the combined entrance pupil contribute. If in this case the visibility is estimated by an integration in the Fourier plane over all frequencies in the fringe peak, "aperture smearing" occurs, and the field-of-view is further restricted to the size of one airy disk of the individual telescopes (Tallon and Tallon-Bosc 1992).

If a wide field-of-view is desired, the delay must be made independent of field angle. This is the case when the output pupil of the interferometer is an exact scaled replica of the input pupil. This "homothetic beam combination" is extremely difficult to achieve, however, since the input pupil varies constantly due to the Earth's rotation, and because the tolerances on the beam positions, rotations, and magnifications are extremely tight. It is much easier to implement a dual-feed system that accepts two smaller fields and sends them through the delay lines separately. The delay difference between the two fields can then be taken out with an additional short-stroke differential delay line. If the two small fields can be selected independently from each other, the dual-feed system is as good as a single large field for almost all scientific purposes, since it is not possible to cover a field of several arcseconds diameter with detector pixels at the required resolution anyway.

5 Aperture Synthesis, Imaging, and Modelling

5.1 Interferometric Imaging

The primary observable in long-baseline interferometry is the complex visibility, which is related to the sky brightness distribution $B(\xi,\eta)$ through a Fourier relation:

$$V(u,v) = \int\!\!\int B(\xi,\eta)\,\mathrm{e}^{-2\pi i(u\xi+v\eta)}\,d\xi\,d\eta \quad . \tag{3}$$

In this relation, ξ and η are the coordinates in the tangent plane of the sky, and u and v are the two components of the baseline vector in the corresponding Fourier plane, measured in units of the observing wavelength. This means that every short observation measures one Fourier component of the sky brightness distribution. The fundamental task of interferometric imaging is the reconstruction of $B(\xi,\eta)$ from the visibility measurements. It is quite analogous to imaging in radio interferometry (see Thompson et al. 1986 and Perley et al. 1989), and software packages developed by radio astronomers are often used to reduce data from optical and infrared interferometry. Since an interferometer works in the Fourier plane, noise and calibration errors are not local in the image plane. They cause artifacts which are sometimes easy to confuse with real source structure. However, the two main problems in inverting Equation 3 are the uncomplete coverage of the *uv* plane, and the corruption of the phases through the atmosphere and instrumental imperfections.

Fourier inversion requires adequate sampling of the *uv* plane; for an array with few elements this means relying on Earth rotation synthesis and on the use of multiple configurations, if possible. The movable auxiliary telescopes of the VLTI provide a lot of flexibility, and the positions of their stations have been chosen to optimize the *uv* coverage. However, when only the 8 m telescopes can be used for sensitivity reasons, there will be considerable holes in the *uv* plane, which will lead to artifacts in the images. If the source structure depends on wavelength in a predictable way, or in a way that can be described by few parameters, the effective *uv* coverage can be improved by multi-frequency synthesis. This technique uses data at different wavelengths in a combined imaging process. One example where it works extremely well is the determination of binary orbits. It is known a priory that the relative position of the two stars does not depend on wavelength, and only one parameter per waveband is needed to describe the brightness ratio. It should also be possible to apply multi-frequency synthesis to the imaging of spotted stellar surfaces, which can be described by a brightness and a temperature at each point.

When the interferometer can be cophased on a point source, the relative phases of the visibility can be measured, and Equation 3 can be inverted directly. Otherwise the closure phase (Jennison 1958) has to be used, which means that the Fourier phases have to be found in an iterative process (see Perley et al. 1989). Because the point spread function of an interferometer is very complicated and has high side lobes, the "dirty image" obtained after the Fourier inversion has to be deconvolved. Algorithms that can be used for this task are CLEAN (Högbom 1974), Maximum Entropy (see Narayan & Nityananda 1986), and the Richardson-Lucy method (Richardson 1972, Lucy 1974). In the multi-speckle case, imaging methods similar to those used in single-telescope speckle interferometry can be used (Reinheimer et al. 1993).

5.2 The Dynamic Range of Images from Interferometry

Images obtained with the very dilute apertures characteristic of few-element interferometer arrays (i.e. with sparse sampling of the *uv* plane) always have a limited dynamic range, and are prone to artifacts. They can give misleading results if these limitations are not fully understood. The dynamic range of interferometric images is limited by four factors: the signal-to-noise ratio, calibration errors, source variability and gaps in the *uv* plane coverage. The total SNR in a map of a point source in an otherwise empty field can be calculated easily by multiplying the SNR of each individual visibility measurement with the square root of the number of these obervations. This same number can usually be used as a guideline in the more complicated cases of resolved or multiple sources. Systematic calibration errors do not average out with the number of observations; they produce artifacts that limit the dynamic range. Variations of the source brightness or structure have a similar effect as calibration errors. As a rule of thumb, a calibration error of $n\%$, or variability with a relative amplitude of $n\%$, can cause spurious structure in the image which contains up to $n\%$ of the total flux.

The effect of gaps in the uv plane coverage are more difficult to quantify. A very common problem in synthesis imaging is the lack of short-spacing data, which leads to the suppression of large-scale structure. In the case of the VLTI, measurements with the auxiliary telescopes on short baselines can be used to fill the hole that occurs in the center of the uv plane when only the 8 m telescopes are used.

5.3 The Power of Model Fitting

Because of the problems of Fourier imaging, the fitting of few-parameter models is often the best way to extract astronomical information from interferometric measurements. Model fitting gives quantitative results with error estimates, which are extremely difficult if not impossible to obtain from the analysis of synthesized images. Furthermore, a priori knowledge about the source structure (e.g. that it is a binary star which should be seen as two point sources by the interferometer) can easily be used. This makes it possible to use very limited data sets, which do not contain nearly enough information for Fourier inversion or image reconstruction. In particular, it is possible to use data from single-baseline instruments in a straightforward way, although they normally contain no useful phase information.

In fact, model fitting was the method of choice for a long time during the pioneering years of radio interferometry, and *all* astrophysical information from optical interferometry has so far been extracted by fitting very simple models to the visibilities. Examples of interesting astronomical conclusions that could be drawn from extremely small data sets, which contain only amplitude information, are measurements of the diameters of the envelope of γ Cas (Thom et al. 1986) and of Nova Cygni 1992 only ten days after maximum light (Quirrenbach et al. 1993). With slightly better uv coverage, but still without any phase data, it is possible to measure the overall geometry of astronomical sources; this has been demonstrated with the Mk III for the disks of Be stars (Quirrenbach et al. 1997). The most sophisticated application of model fitting to optical long-baseline interferometry so far is the fitting of orbital parameters of binary stars to visibilities with a global χ^2 minimization algorithm (Hummel et al. 1994).

6 Suggested Reading

The material collected in this volume should provide an up-to-date overview of the scientific topics that can be addressed with the VLTI, along with some technical information. More details about important issues like sensitivity, imaging performance, and concepts for the VLTI instrumentation can be found in the VLT reports No. 59 and 65 (ESO 1989, 1992). The *Annual Review* article by Shao and Colavita (1992) is an excellent introduction into the principles and applications of ground-based and space-borne interferometry. Many of the concepts developed originally in the context of radio interferometry apply also at shorter wavelengths. The highly recommended textbook by Thompson, Moran and Swenson (1986) gives a comprehensive exposition of this field; it is a highly valuable resource for the practitioner of astronomical interferometry. The lectures from an NRAO summer school collected by Perley, Schwab and Bridle (1989) cover many aspects of synthesis imaging encountered in "real life".

References

Beckers, J.M. (1989), in *Diffraction-Limited Imaging with Very Large Telescopes*, Eds. Alloin, D.M., & Mariotti, J.-M., NATO ASI Series C, Vol. 274, p. 365

Buscher, D.F. (1988), MNRAS 235, 1203

Coudé du Foresto, V. (1994), Proc. IAU Symp. 158, p. 261

Di Benedetto, G.P., & Bonneau, D. (1990), ApJ 358, 617

ESO (1989), *VLT Report No. 59*

ESO (1992), *VLT Report No. 65*

Högbom, J.A. (1974), A&AS 15, 417

Hummel, C.A., Armstrong, J.T., Quirrenbach, A., Buscher, D.F., Mozurkewich, D., Elias, N.M., & Wilson, R.E. (1994), AJ 107, 1859

Jennison, R.C. (1958), MNRAS 118, 276

Lucy, L.B. (1974), AJ 79, 745

Mourard, D., Tallon-Bosc, I., Rigal, F., Vakili, F., Bonneau, D., Morand, F., & Stee, Ph. (1994), A&A 288, 675

Mozurkewich, D. (1994), Proc. IAU Symp. 158, p. 177

Mozurkewich, D., & Armstrong, J.T. (1992), in *High-Resolution Imaging by Interferometry II, ESO Conf. and Workshop Proc. Vol. 39*, ed. Beckers, J.M., & Merkle, F., p. 801

Mozurkewich, D., et al. (1991), AJ 101, 2207

Narayan, R., & Nityananda, R. (1986), ARAA 24, 127

Perley, R.A., Schwab, F.R., & Bridle, A.H., Eds. (1989), *Synthesis Imaging in Radio Astronomy*, ASP Conf. Series, Vol. 6

Peacock, A., et al. (1996), Nature 381, 135

Quirrenbach, A., Bjorkman, K.S., Bjorkman, J.E., Hummel, C.A., Buscher, D.F., Armstrong, J.T., Mozurkewich, D., Elias, N.M., & Babler, B.L. 1997, ApJ, in press

Quirrenbach, A., Elias, N.M., Mozurkewich, D., Armstrong, J.T., Buscher, D.F., & Hummel, C.A. (1993), AJ 106, 1118

Quirrenbach, A., Mozurkewich, D., Buscher, D.F., Hummel, C.A., & Armstrong, J.T. (1994), A&A 286, 1019

Reinheimer, T., Hofmann, K.-H., & Weigelt, G. (1993), A&A 279, 322

Richardson, W.H. (1972), JOSA 62, 55

Roddier, F. (1989), in *Diffraction-Limited Imaging with Very Large Telescopes*, Eds. Alloin, D.M., & Mariotti, J.-M., NATO ASI Series C, Vol. 274, p. 33

Shaklan, S.B., Colavita, M.M., & Shao, M. (1992), in *High-Resolution Imaging by Interferometry II, ESO Conf. and Workshop Proc. Vol. 39*, ed. Beckers, J.M., & Merkle, F., p. 1271

Shao, M., & Colavita, M.M. (1992), ARAA 30, 457

Shao, M., Colavita, M.M., Hines, B.E., Staelin, D.H., Hutter, D.J., Johnston, K.J., Mozurkewich, D., Simon, R.S., Hershey, J.I., Hughes, J.A., & Kaplan, G.H. (1988), A&A 193, 357

Tallon, M., & Tallon-Bosc, I. (1992), A&A 253, 641

Thom, C., Granes, P., & Vakili, F. 1986, A&A, 165, L13

Thompson, A.R., Moran, J.M., & Swenson, G.W. (1986), *Interferometry and Synthesis in Radio Astronomy*, Wiley Interscience, New York

von der Lühe, O., & Quirrenbach, A. (1995), in *ESO / ST-ECF workshop on calibrating and understanding HST and ESO instruments,* ed. Benvenuti, P., p. 173

Astrometry with Optical/IR Interferometers

M. Shao

Jet Propulsion Laboratory

Abstract. This paper is a tutorial on astrometry with long-baseline ground-based interferometers. There are two types of astrometric measurements with ground-based interferometers: global astrometry and narrow angle astrometry. The ultimate limits for both types of measurements are set by the atmosphere. However, the instrumental considerations for the two types of measurements are quite different. This paper gives an overview of the instrumental aspects of these two different types of astrometric instruments.

1 Introduction

Astrometry with an optical interferometer bears a much closer resemblance to astrometry with a radio interferometer than astrometry with a single telescope. Because of the significant differences, global astrometry and narrow angle astrometry will be treated separately.

An interferometer combines the light from two telescopes coherently by directing the light towards a central beam combiner. The light in one or both arms is usually delayed by an optical delay so that the light from the star via the two arms are equal to roughly a wavelength. Astrometry is performed by measuring the position of the delay line where stellar fringes are formed.

The delay line position is related to the position of the star and the baseline vector, by the well known relation from radio astronomy:

$$X = B * S + c \tag{1}$$

where B is the baseline vector (x, y, z), S is a unit vector to the star, and c is the delay offset. The delay offset represents the zero point of the delay line. If the zero point of the delay line is delay line position when fringes are observed for a star directly overhead, $c = 0$.

2 Global Astrometry

Global astrometry at optical and IR wavelengths is almost identical to VLBI astrometry at radio wavenlengths. Earth rotation can be used to change the baseline orientation of the interferometer. The baseline vector B in equation 1 is fixed to the Earth and rotates with a 24 hr period. The delay X also varies as a 24 hr sin wave. That sine wave has three measureable parameters, amplitude, phase and DC offset.

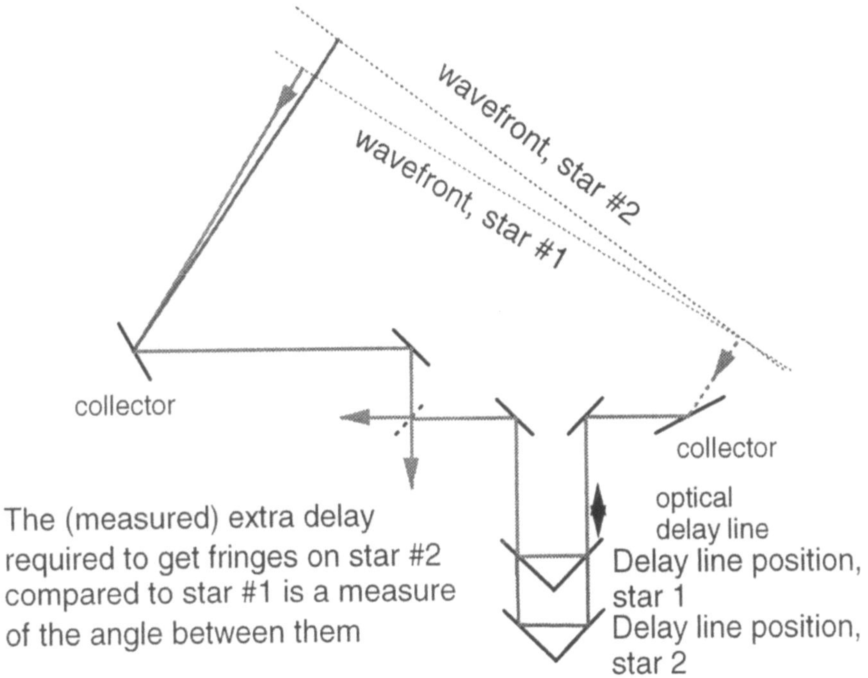

Fig. 1. Astrometry with an interferometer.

$$X = \text{amp} * \sin\,(2 * \pi * t\ /\ 24\ \text{hrs} + \phi)$$

If four stars are observed over ~ 24 hrs, that data set has 8 unknown variables, the baseline vector and delay offset represents 4 more. The result is 12 measurements (4 stars * 3/star) with 12 unknowns. Absolute R.A. can't be measured by this technique.

Data reduction for optical astrometry is virtually identical to its radio counterpart. A least squares software technique is used to solve for the positions of the stars and the baseline using as raw input the delay vs time data of each star. The least squares fit is actually to the delta positions (deviation from the catalog position) rather than the positions themselves.

More precisely, the least squares solution is the solution to the matrix equation

$$M(i) = A(i,j) * P(j)$$

where M is a vector of the difference in the delay where white light fringes are detected and the predicted delay for the nominal catalog position of the stars and the nominal baseline vector, P are the correction to the catalog positions of the stars and corrections to the nominal baseline, and A is a matrix of partial derivatives relating the stellar positions to delays.

2.1 Instrumental Errors

A typical interferometer used for global astrometry on the ground would have a $\sim$ 40m baseline. One milliarcsec (mas) accuracy would imply that the optical delay has to be measured with 0.2 μm accuracy. That level of accuracy is challenging from both an instrumental and an atmospheric point of view.

From an instrumental point of view, the most challenging aspect is baseline stability. Equation 1 and conventional radio interferometry analysis techniques make the assumption that the baseline of the interferometer is fixed to the solid Earth that rotates with a known angular velocity. For an interferometer with a 4000 km baseline, 1 mas accuracy implies the end points of the baseline are stable to 2 cm. So as long as the radio telescope's pivot point is stable to 2 cm, astrometry at the 1 mas level is not a problem.

In the optical with a 40m baseline, a number of effects can cause the baseline to shift by 0.2 μm. Thermal expansion of the telescope mount 1m above the surface of the Earth could shift the baseline by ten microns per °C. The air temperature over the course of a night can easily change by 10 °C. While the thermal problem is quite severe, other mechanical errors in the telescope mount can produce errors well in excess of 0.2 μm. Some examples are bearing imperfections. Ball bearings run in circular races. These races are not truly circular at the 0.2 μm level. Gravity causes the telescope structure to sag slightly. Telescopes are designed to point accurately and the gravity sag of a telescope is usually designed not to interfere with pointing. But of the 6 degrees of freedom of the telescope optics, pointing only represents 2. Interferometry is sensitive to motion of 5 of the 6 degrees of freedom.

Because of the myriad sources of mechanical error in the telescopes, making a telescope mechanically perfect for $\sim$ 1 mas global astrometry is a daunting task. The approach taken by groups doing global astrometry is not to make a perfect telescope but to build a metrology system to measure the imperfections of the telescope/mount at the 0.2 μm level. Then apply those corrections to the data analysis.

The basic approach is to place an optical fiducial, such as a retroreflector at the center of the telescope. Laser interferometers measure the position of that fiducial with respect to corner cubes buried 5–10 meters below the surface of the Earth.

Ultimately the accuracy of this is limited by the accuracy of the optical fiducial. If the fiducial is an open face corner cube with $\lambda/4$ wavefront accuracy, the potential accuracy is $\sim$ 0.15 μm.

2.2 Atmospheric Errors

The 2nd major class of error in global astrometry is the atmosphere. Even for long baseline interferometers, the turbulent atmosphere will cause the position of a star to wander at $\sim$ 0.3 arcsec rms values for typical $\sim$ 1 arcsec seeing conditions.

While the instantaneous position can fluctuate a large amount, the positional uncertainty does decrease with longer integration time. Again for nominal seeing conditions at a good observatory, the atmospheric noise is typically $\sim$ 10 mas for a 1 hr integration.

Again there is a potential for improvement. The technique is two color observations. The principle of operation is that the index of air is slightly different at different wavelengths. The index of refraction of air is higher at shorter wavelengths. The position of the star is then:

$$\text{Apparent}-X = \text{true}-X + \text{Wedge} * (N - 1) \tag{2}$$

Apparent position is the position corrupted by the atmosphere and Wedge is the atmospheric wedge over the interferometer at the time of the measurement. N is the index of refraction of air. If we assume that we know the index of refraction of air at two wavelengths where the position of the star is measured, then we have two equations in two unknowns.

$$\begin{aligned} \text{A}-X - 1 &= \text{T}-X + \text{Wedge} * (N(l1) - 1) \\ \text{A}-X - 2 &= \text{T}-X + \text{Wedge} * (N(l2) - 1) \end{aligned} \tag{3}$$

The two color technique is not a complete solution. The index of air is not very different at different visible wavelengths. Because of the difficulty of doing interferometry in the blue, two color astrometry usually means using 500nm and 800nm as the two colors. The index of refraction at those two colors are 1.000294 and 1.000290. In equation 3, we solve for the true position using the measurements of the apparent positions at two colors. The noise in the measurement of the apparent position due to photon noise for example, is multiplied by a large factor (294–290)/(290) because the indices at the two wavelengths are not very different.

A second problem is that the index of air is not known exactly due to the presence of water vapor. In practice the gain in two color measurement accuracy has been limited to a factor of 3–5 better than one color measurements. Potentially accuracies of a few (2–5) mas is possible for a one hour integration time.

3 Narrow Angle Astrometry

Narrow angle astrometry, over fields of view of a few 10's of arcsec to a few 10's of arcminutes, is potentially much more accurate both from an instrumental and atmospheric point of view. The light from two stars near each other will traverse the same turbulent atmosphere before entering the interferometer. A differential measurement of the relative positions of the two stars should then not suffer most of the errors due to atmospheric turbulence. This is indeed the case when we measured the separation of a double star system in 1991. The rms error for a 1 sec integration for two stars 3 arcsec apart was 1.2 mas (20 μas in 1 hr).

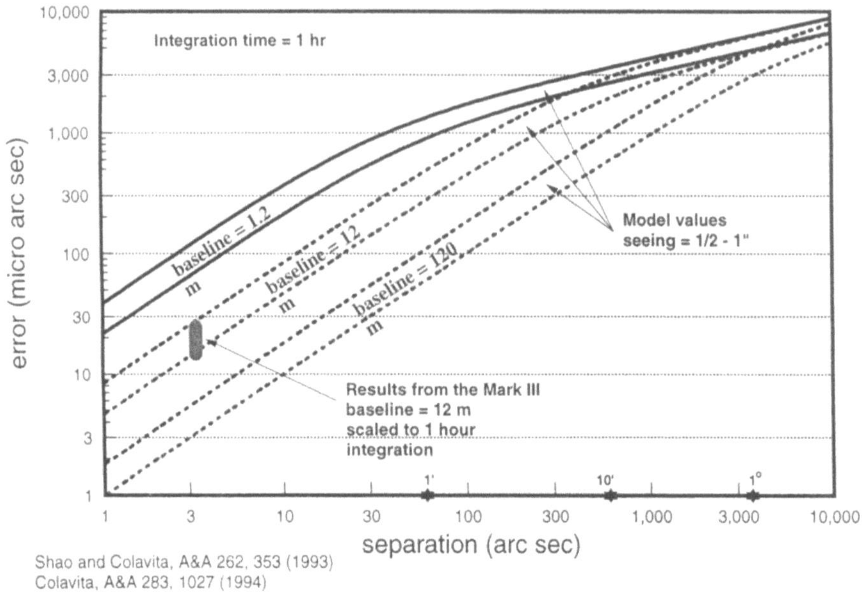

Fig. 2. Accuracy of narrow-angle astrometry.

3.1 Radio vs Optical Narrow Angle Astrometry

Narrow angle or relative astrometry with interferometers is fundamentally the same at radio and optical wavelengths. In the radio, the antenna is switched between two objects rapidly, faster than the coherence time of the atmosphere. The differential delay is then a measure of the angle between the two astrophysical objects, projected onto the baseline vector.

$$dX = X1 - X2 = B * (S1 - S2)$$

dX is the differential delay, B is the baseline vector and S1, S2 are unit vectors to the two stars. Note that the delay offset in equation 1 is no longer present when two delay measurements are differenced. In relative astrometry, we are only interested in measuring the vector $S1 - S2$. If we make two measurements of dX with two different baseline vectors (preferably two roughly perpendicular baselines) that is sufficient to solve for $S1 - S2$, if we know what the two baseline vectors are.

Phase reference interferometry is a powerful technique first developed for VLBI to increase the sensitivity of an interferometer. Observationally it is identical to relative astrometry, antennas are alternatively pointed at a bright object, then at a faint object. By using the bright object as a phase reference, one can extend the coherent integration time for the faint object significantly. The difference between the radio and optical comes from their different wavelengths.

If we take 1 cm and 1 μm as the nominal wavelengths of radio and optical

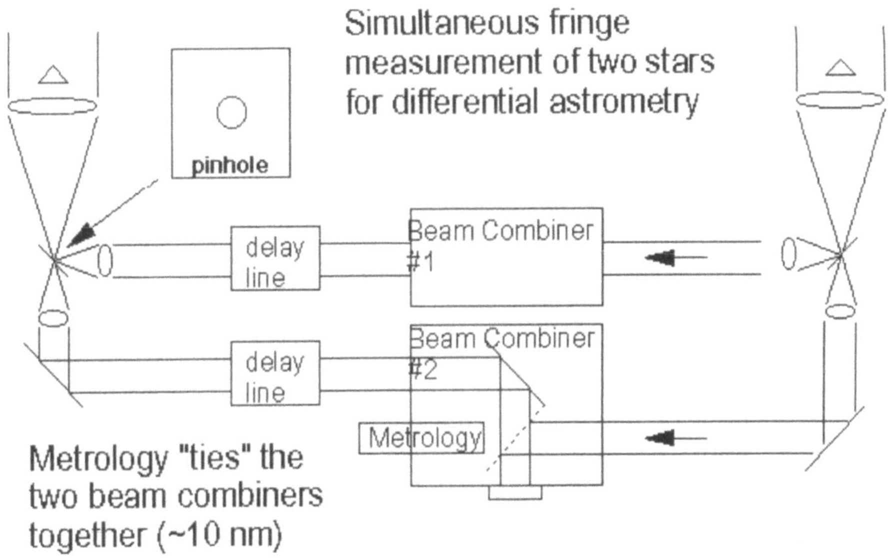

Fig. 3. Dual object interferometry.

techniques, we find that roughly speaking the isoplanatic angle, the angle in the sky where the atmospheric phase fluctuations can be considered identical is roughly 10,000 times larger for 1 cm than 1 μm wavelength. At 1 μm, the isoplanatic angle is roughly $\sim$ 10 arcsec. At 1 cm, the angle is roughly 30°. The other difference is the switching time interval. The atmospheric coherence time at 1 μm is roughly 20 msec, the coherence time at 1 cm is roughly 200–400 sec. The radio telescopes must switch between the faint star and the bright phase reference on a time scale faster than $\sim$ 400 sec. At 1 μm, the switching must be faster than 20 msec. Because it is not practical to re-point even a small 50 cm telescope on a 10 msec time scale, optical phase reference interferometry has to use two separate optical systems to make simultaneous observations of the fringes of both stars.

The basic concept is the same. Both radio and optical interferometry make differential astrometric measurements using the phase reference technique to increase sensitivity to find "nearby" references. The difference is in two areas, narrow angle in the radio is a few degrees, narrow angle in O/IR is a few 10's of arcsec and fast switching in the radio is 100's of seconds and fast switching in O/IR is 10's msec.

3.2 Baseline Stability

In global astrometry, we normally solve for the baseline vector along with the star positions because in general there is no way to determine the baseline vector with mas or sub-mas accuracy, other than by looking at the stars. But the requirements for differential astrometry is much less severe especially at optical wavelengths. Equation 2 is the reason why.

If we look at equation 2 more closely we can calculate the astrometric error that would result in having the wrong baseline. To first order, $S1 - S2 = dX/B$. A 1% error in B results in a 1% error in $S1 - S2$. But if $S1 - S2$ is 20 arcsec (for an IR astrometric interferometer), then 1% of 20 arcsec is 0.2 arcsec. If we are looking for 20 μas astrometry over a 20 arcsec field of view, the baseline vector must be known to one part in a million. The length of B (nominally 100m) has to be known to 1 part in a million (100 μm) and the angle to 1 μrad, or 0.2 arcsec.

Contrast this to 1 mas global astrometry where the baseline vector must be derived with 5 nano-radians, and 0.5 μm accuracy for a 100m interferometer.

In differential astrometry, the baseline vector is not derived from the differential measurements, rather the interferometer does global astrometry over 30–40 degrees of known stars (eg from the Hipparcos catalog) to measure the baseline vector to at least 0.2 arcsec (and 100 μm). That baseline is then used in the data reduction for the differential measurement, at an accuracy to 20 μas for the differential measurement.

The same relaxed baseline stability applies to radio differential astrometry, however, because the angle between the target and phase reference in the radio is typically a few degrees, the tolerance is relaxed by a factor of ~ 10 instead of $\sim 10,000$ in the O/IR.

4 Summary

Ground-based global astrometry will be limited to accuracies to a few milliarcsec by the atmosphere. However to achieve this level of accuracy, elaborate laser metrology is needed to monitor the baseline vector to ensure that it rotates with the Earth at the milliarcsec level.

Differential astrometry potentially can be as accurate as 10–20 μas using long baseline 100–200m IR interferometers. Phase reference techniques are needed to extend sensitivity to detect stars within ~ 20 arcsec of a bright target for relative astrometry. This small field of view is dictated by atmospheric turbulence. But the small field results in much relaxed tolerance for baseline stability, several orders of magnitude less precise than what is needed for global astrometry.

Part 2

SOLAR SYSTEM, EXTRASOLAR PLANETS, LOW MASS STARS AND BROWN DWARFS

Solar-System Observations with the VLTI

Thérèse Encrenaz

DESPA, Observatoire de Paris, F-92195 Meudon

Abstract. Solar-system objects show a wide range of physical dimensions (from a few kilometers to more than 100000 km) and a wide range of heliocentric distances (from 0.3 AU to more than 50 AU). VLTI observations of Solar-system objects will concentrate on targets with angular dimensions in the range of an arcsec or smaller. Major objectives are expected to be the search for asteroid companions, the morphological study of asteroids, satellites of outer planets and the Pluto-Charon system, and the study of distant objects. A low-resolution spectroscopic facility will be needed to provide information on the mineralogy of asteroids and satellites.

1 Introduction

The Solar system is characterized by a large variety of objects: planets, satellites, asteroids and comets (Table 1). The study of giant planets' satellites, asteroids and comets can greatly benefit from high-angular resolution interferometric observations (Encrenaz, 1996).

Most of the satellites have no atmosphere. There are a few exceptions: Titan, Saturn's largest satellite, with a N_2 atmosphere and a surface pressure of 1.5 bar; Triton' Neptune's satellite with a tenuous N_2 atmosphere (surface pressure of a few μbars) and Io, Jupiter closest Galilean satellite, with a very tenuous SO_2 (surface pressure of a few nanobars). Pluto and its satellite Charon show many analogies with the outer satellites; in particular Pluto is presently surrounded by a N_2 tenuous atmosphere comparable to Triton's one.

Trans-neptunian objects (TNOs) constitute a newly discovered family of Solar-system objects (Hainaut, 1994). About 20 of them have been discovered since 1992, with diameters of about 100 km, at heliocentric distances ranging from 30 to 45 AU, and many more are likely to be discovered in the future. The existence of this new population, in the so-called "Kuiper belt", was suggested as a possible reservoir for short-period comets, which all show very low inclinations. Pluto might constitute the biggest object of the TNO family.

Asteroids are mostly found in the asteroidal belt, at a heliocentric distance of about 3 AU. Their sizes range between several tens and several hundred km (about a thousand for the biggest asteroid, Ceres). They are too small to retain a permanent atmosphere and show a bare surface. Comets are very small bodies, usually less than 10 km in size, with very elliptical orbits; they spend most of their lifes at very large heliocentric distances where they remain unaltered. As they approach the Sun, the ices of their surfaces sublimate, producing an outgassing of parent molecules and an ejection of dust, forming the coma and the dust tail. The parent molecules are then dissociated into radicals, atoms and

TABLE 1
ANGULAR SIZES OF SOLAR-SYSTEM OBJECTS
(from Encrenaz, 1996)

OBJECT	ANGULAR DIAM.(")	DIAMETER (km)	GEOCENTRIC DISTANCE (AU)	RES.(km) in 0.5"
MOON	1900	3500	0.0025	1
MERCURY (quadrature)	6.5	4900	1	380
VENUS (quadrature)	23	12100	0.7	263
MARS (opposition)	18	6800	0.5	190
JUPITER	40	143000	5	1875
IO	1.0	3600	5	1875
SATURN	16	120000	10	3750
TITAN	0.8	5100	10	3750
URANUS	3.5	50800	19	7125
NEPTUNE	2.16	48600	30	11250
TRITON	0.12	2700	30	11250
PLUTO	0.08	2300	40	15000
CERES	0.7	1000	1.8	675
TYPICAL ASTEROID	0.06	100	2	750
TYPICAL COMET D = 1 AU	0.013	10	1	375
TYPICAL COMET D = 5 AU	0.003	10	5	1875
TYPICAL TRANS-NEPT.OBJ	0.004	100	35	13125

TABLE 2

SOLAR-SYSTEM OBSERVATIONS WITH THE VLTI: EXPECTED ANGULAR RESOLUTIONS IN KM

(Expected angular resolution: 10 milliarcsec for $\lambda = 2\ \mu m$)
(from Encrenaz, 1996)

λ (μm)	GEOCENTRIC DISTANCE (AU)			
	1	5	10	50
0.5	2	10	20	100
2	7.5	37	75	375
5	19	95	190	950
10	38	190	380	1900
20	75	375	750	3750

TABLE 3
THE LARGEST ASTEROIDS

NAME		TYPE	Mv	D (km)	Semi-maj.Axis (AU)	Ang. Diam.(") (at opposition)
1	CERES	C	7.5	1032	2.77	0.8
2	PALLAS	U	8	588	2.77	0.4
4	VESTA	V	6.5	576	2.36	0.6
10	HYGIEA	C	10	430	3.13	0.3
704	INTERAMNIA	U	11	338	3.06	0.2
511	DAVIDA	C	11	324	3.17	0.2
65	CYBELE	C	12	308	3.43	0.2
52	EUROPA	C	11	293	3.10	0.2
451	PATIENTIA	C	11.5	280	3.06	0.2
15	EUNOMIA	S	9.5	260	2.64	0.2
3	JUNO	S	9.5	248	2.67	0.2

ions which form the plasma tail. These phenomena become usually observable at heliocentric distances smaller than about 2 AU, where the sublimation of water (the dominant cometary constituent) starts to take place.

Solar-system studies provide information on a wide variety of domains, including the dynamical and chemical history of the Solar system, the formation and evolution processes of Solar-system bodies and the physico-chemical conditions of the primordial solar nebula. Small distant Solar-system bodies, in particular, are more likely to have escaped differenciation effects due to internal or external heating, and can provide valuable information upon the early stages of the Solar-system history.

Assuming a typical spatial resolution of 10 milliarcsec at 2 μm, VLTI observations should be able to resolve a cometary nucleus at 1 AU (if the activity is low enough for the nucleus to be visible), all main-belt asteroids larger than 50 km, most of the outer satellites and Pluto (Table 2).

2 Asteroids

The study of asteroids includes the determination of shapes, rotation period and polar axis (in order to retrieve the early history of their orbits), the analysis of their morphology and their albedo variations (for a study of the history of their surfaces), and a thermal mapping (to derive their physical properties and thermal inertia). Broad-band photometry allows a taxonomic classification, based on the chemical nature of their surfaces. With low-resolution spectroscopy, it is possible to derive more accurately their mineralogy and their chemical composition, and thus to infer their chemical history and evolution. Table 3 lists the main characteristics of the brightest asteroids. All have rotation periods ranging from about 6 to 20 hours.

High-angular resolution observations have been achieved on a few bright asteroids, by means of the HST and ground-based instruments using adaptive optics. Images of Vesta's rotation have been obtained with the HST in the near-infrared with an angular resolution of about 0.1 arcsec (Zellner et al, 1996). Ceres (Saint-Pé et al, 1993a,b) and Vesta (Dumas et al, 1996) have been mapped at 2 μm using the adaptive optic system COME-ON and COME-ON+ at the 3.6m ESO telescope, with a resolution of 0.13 arcsec, corresponding to the diffraction limit. These images have been used for a determination of the shape and the pole orientation, and a study of the surface properties (mineralogy, thermal inertia).

VLTI observations of the brightest asteroids (about half a dozen objects brighter than Mv = 10) should be feasible with VISA, using the target itself as a reference for wavefront correction and fringe tracking. The expected number of pixel along a diameter would be in the range of a few tens. However the rotation period has to be taken into account for estimating the optimum integration time of a single observation. For example, in the case of Ceres, individual observations should be limited to about 1 minute to take full advantage of the spatial resolution.

All asteroids of the main asteroid belt with a diameter larger than 100 km could be observed with VISA, using an external reference for wavefront correction and fringe tracking; however, observations are likely to be difficult because of the need for an external reference and because of the rotation problem.

A question of specific interest which could greatly benefit from the VLTI is the search for asteroid companions. Knowing how many asteroids have multiple systems would provide a key information in our understanding of the dynamical history of these objects. In spite of a few attempts using stellar occultations, previous ground-based observations have failed to detect an asteroid companion in an unambiguous way. However, the Galileo probe which encountered two asteroids (Gaspra and Ida) discovered a satellite around one of them. Ida is a satellite located at about 3 AU, with a diameter of about 50 km. Its satellite, Dactyl, is located at about 100 km of its parent body, and has a diameter of 1.5 km. Its mineralogy, different from Ida's one, suggests that the system might result from an earlier collision.

A program for searching asteroids' companions could be considered with the VLTI. At 3 AU, a 100 km separation corresponds to 0.04 arcsec, which could be resolved with the VLTI. Assuming a magnitude difference of 8 to 10 between the asteroid and its companion, the visual magnitude to be reached is in the range 18-20, in order to have access to a sufficient number of candidates. Apart from the case of the few asteroids brighter than Mv = 10 which could be used as references for wavefront correction and fringe tracking, an external reference would be needed. Observations would have to be repeated to cover a rotation period (typically several hours). This program is ambitious and needs long- term observations, but addresses an important scientific question which requires a statistical study. It should be noted that observations using adaptive optics could provide an important first step to this research, which could be followed by VLTI observations.

3 Titan

Titan was very poorly known before its encounter by the Voyager spacecrafts in 1980. The Voyager experiments revealed the nature of its nitrogen- rich atmosphere, the presence of several complex hydrocarbons and nitriles, and determined the atmospheric T(P) profile. However, the nature of Titan's surface, hidden below a thick cloud, is still an open question. Recent observations have shown that the surface of Titan can be probed in a few near-infrared windows (0.9 μm and 2.0 μm). High-resolution images of Titan have been achieved at 0.9 μm with the HST (Smith et al, 1996) and in the 2 μm region using adaptive optics at ESO (Combes etal, 1996). There is evidence for brightness variations on Titan's surface, which might be the signature of altitude and possible compositional variations (Coustenis et al, 1996).

Titan's angular size is 0.8 arcsec and its visual magnitude is 8. Its orbital period is 16 days. VLTI observations should be able to provide images with about 80 pixels over a diameter. The satellite could be used as a reference for wavefront sensing and fringe tracking. There is no problem connected to the rotation; observations would have to be made at maximum eastern and western elongations for a full mapping of both leading and trailing hemispheres. The observing time would thus be sufficient for CVF observations to be performed. Observations at higher spectral resolution (R above 1000) would even be more appropriate, as they would allow to separate the flux coming from Titan's surface (outside the CH_4 bands) from the contribution coming from the atmosphere (within the CH_4 bands).

4 Galilean Satellites

Galilean satellites have been observed at high angular resolution from the HST and in the near-infrared using adaptive optic systems. All four objects (Io, Europa, Ganymede and Callisto) show a wide variety of surface features, ranging from active volcanism (Io), tectonics (Europa, Ganymede) and heavy cratering (Callisto). Recent images of Ganymede and Callisto using adative optics have provided a H_2O mapping of both satellites, with a patchy distribution on Ganymede and a more uniform distribution on Callisto (Forni et al, 1996).

The galilean satellites are going to be extensively studied by the Galileo spacecraft, so their study with the VLTI is likely to be of minor scientific interest. However, it should be mentioned that a monitoring of Io with the VLTI could allow to study the evolution of its volcanic activity after the end of the Galileo mission. Io's diameter is 1.1 arcsec at opposition and its visual magnitude is 5. A VLTI observation at 2 μm could provide about 100 pixels along the equatorial diameter, which is the region of maximum volcanic activity. Like Titan, Io would have to be observed at maximum eastern and western elongation (rotation period: 1.7 days), and could be used as a reference for wavefront sensing and fringe tracking. There is no special need for high spectral resolution; an appropriate wavelength range would be the 4-5 μm region, in order to measure the thermal emission coming from the volcanoes.

5 Pluto, Charon and Triton

Pluto and Charon have angular diameters of 0.1 and 0.03 arcsec rsepectively, and are separated by 0.6 arcsec. Their visual magnitudes are 13.7 and 16.8 respectively. HST images of Pluto have been recently achieved (Stern and Buie, 1996), showing evidence for bright spots over Pluto's surface. Observations with the VLTI should be feasible in principle, at fixed times to account for the rotation period (6.4 days), with an external reference.

Triton, Neptune's satellite, has been encountered by Voyager 2 in 1989. Its surface shows signs of volcanic activity. Triton's visual magnitude is 13.7. VLTI observations at 2 μm would provide about 12 pixels over a diameter. Observations would have to be performed at maximum elongations (P= 6 days). An external reference (possibly Neptune if the planet is not too big) will be needed for wavefront correction.

6 Distant Asteroids and Cometary Nuclei

At a heliocentric distance of 3 AU, cometary nuclei have a visual magnitude of about 19. They should be observable with the VLTI with an external reference. The objective would be the study of the onset of the cometary activity, prior to H_2O outgassing.

Centaur-family asteroids are distant objects on very excentric orbits. Typical examples of this family are 5145 Pholus (D=140 km, a= 20 AU, Mv = 22) and 2060 Chiron (D=160-300 km, a= 8-19 AU, Mv= 20-22). These objects could be observable with VISA with an external reference. The objective would be the search for activity and the study of the comet- asteroid transition. As the cometary observations, these measurements should be performed along the Sun-object axis, where most of the activity, if exists, is likely to take place.

Finally, the trans-neptunian objects, with a typical visual magnitude of 23 (Hainaut, 1994), are presently at the limit of the VLTI capabilities, using the VIMA/hybrid configuration and external references for wavefront sensing and fringe tracking.

References

Combes, M. et al (1996): 2-micron images of Titan by means of adaptive optics. The Messenger, **83**, 40–42

Coustenis, A. et al (1996): Titan's surface in the near-infrared. Submitted to Icarus.

Dumas, C. et al (1996): Private communication.

Encrenaz, Th. (1996): Solar-system observations at high angular resolution Proceedings of the Les Houches workshop "High Angular Resolution in Astronomy", A.-M. Lagrange ed', April 1996.

Forni, O. et al (1996): Private communication.

Hainaut, O. (1994): Selected Observational Studies of Minor Bodies in the Solar System. Ph D dissertation, Institut d'Astrophysique de Liège and European Southern Observatory.

Saint-Pé, O. et al. (1993a): Demonstration of adaptive optics for resolving imagery of solar-system objects: preliminary results on Pallas and Titan. Icarus, **105**, 263–270

Saint-Pé, O. et al. (1993b): Ceres surface properties by high-resolution imaging from Earth. Icarus, **105**, 271–281

Smith, P. et al. (1996) Titan's surface revealed by HST imaging. Icarus, **119**, 336–349

Stern, A. and Buie, M. (1996): Private communication.

Zellner, B. (1996) Private communication.

Giant Planet Formation and the Masses of Extrasolar Planets

Günther Wuchterl[1,2]

[1] Institut für Astronomie der Universität Wien,
Türkenschanzstraße 17,
A-1180 Wien, Austria
wuchterl@amok.ast.univie.ac.at
[2] Institut für Kernphysik der Technischen Universität Wien
Wiedner Hauptstrße 8-10/142
A-1040 Wien, Austria

Abstract. Differential astrometry with the VLTI promises the measurement of the reflex motion of nearby stars around the barycenter of the star-planetary systems for possible planets downto Uranus' and Neptune's masses. I review the theory of planet formation with and emphasis on the resulting masses of the giant planets. It is argued, that planets with Uranus/Neptune-masses are the standard outcome of the giant planet formation process, while Jupiter-type planets require special protoplanetary nebula conditions. Finally, a hydrodynamical calculation of the formation of a giant planet at an orbital distance of 0.05 AU is presented.

1 The Planetesimal Hypothesis

According to the 'planetesimal hypothesis' (see e.g. Lissauer 1993) planets grow within circumstellar disks via pairwise accretion of small solid bodies, the so-called planetesimals. Sufficiently massive planetary bodies embedded in a gas-rich disk can gravitationally capture large amounts of gas, thereby producing Jovian type planets (Mizuno 1980, Hayashi et al. 1985, Bodenheimer and Pollack 1986, Wuchterl 1991a,b, 1995).

The accumulation of gas by condensible element planetary embryos in circumstellar discs ultimately results in giant planets with masses that cause measurable reflex-motions of the parent-star around the barycenter of the extrasolar system. A Neptune mass planet at a distance of 5.2 AU from a solar-mass star, that is 10 parsecs away causes a reflex motion with an amplitude of 24 microarcseconds (see e.g. Black 1995 for a review of the search for other planetary systems). Reflex motions of this amplitude may be measured by differential astrometry with optical interferometers (see chapters by Shao and Kulkarni, resp., in this volume).

I review the process of nebula gas-accumulation by condensible element planets (terrestrial and icy) with respect to the following questions:

1. How much gas from a circumstellar nebula can be bound to a growing 'core' (i.e. a condensible element planetary embryo) to form a proto giant planet?

2. How do the masses of proto giant planets depend on protoplanetary nebula conditions, i.e. densities and temperatures?
3. Are any masses favored?

I will conclude, that most nebula conditions lead to Uranus/Neptune- masses for the proto giant planets. High (jovian) mass protogiant planets need special, high nebula densities for their formation.

2 The Masses of Proto Giant Planets

2.1 How Much Nebula Gas is Bound to a Planetary Embryo?

The question (1) of how much gas is bound in an envelope of a condensible element planetary embryo has been investigated with an increasingly detailed description of the envelope structure. In the following I will call such an embryo, that grows by impacting planetesimals, a 'core'. Below a few earth masses the gaseous envelopes are negligible, consistent with the formation of terrestrial planets. At higher core-masses the resulting planetary masses depend on the assumptions made about the gaseous envelopes surrounding the cores. Two equilibria play a key role in the discussion of the gaseous envelopes of proto giant planets and their masses: a) the assumption of hydrostatic equilibrium, a force-balance, where the combined gravity of the core and its gaseous envelope is balanced by pressure gradients; b) the so-called 'thermal equilibrium', an energy-balance, where the energy liberated by the impacted planetesimals is balanced by heat losses to the ambient protoplanetary nebula.

When both equilibria are assumed protoplanets can only grow to a maximum mass. Mizuno 1980 found this 'critical mass' to be about 20 earth masses ($M_\oplus$), independent of the location and conditions in the solar nebula. While the critical *core mass* of about 13 $M_\oplus$ is consistent with the condensible element mass inferred for the present giant planets, the total mass falls short of Jupiter's mass by more than one order of magnitude but it is significantly larger than Uranus' and Neptune's. Beyond the critical mass one of the equilibria cannot hold.

Quasi-hydrostatic studies (Bodenheimer and Pollack (1986), Lissauer et al. (1995), Pollack et al. (1996)) allowed the protoplanetary envelopes to depart from thermal equilibrium but assumed the hydrostatic equilibrium to hold. Quasi-hydrostatic envelopes may contract or expand slowly, i.e. they are not allowed to move at velocities higher than a small fraction of the sound speed. Under this assumption forces almost balance and the net acceleration on the gas remains close to zero. The general result of those studies is that proto giant planets may grow to masses larger than the critical mass found for static proto giant planets. The accretion process is assumed to be terminated by limited supply from the solar nebula. Limiting processes may be nebula dissipation and/or tidal truncation of the gas flow from the nebula into the envelope, caused by the growing proto giant.

The most general, hydrodynamic studies concerning envelope accretion of proto giant planets (Wuchterl 1991a,b, 1995) do not assume energy- or force-balance. Here, the *motion* of the gas is calculated together with energy transfer, heating and cooling. Wuchterl (1995) found, that the evolution of a proto giant planet beyond the critical mass depend on the details of the envelope structure. 'Radiative' protoplanets eject parts of their envelopes at the critical mass an evolve to masses that are comparable to Uranus' and Neptune's. In contrary, 'convective' protoplanets continue to grow beyond the critical mass and accretion proceeds up to Jupiter's mass. In summary: Below a few earth masses planetary embryos gravitationally attract only small amounts of nebula gas. As a consequence the gaseous envelopes have negligible masses. Overall they resemble terrestrial type planets. From a few earth masses to the critical an increase in the core mass significantly increases the envelope mass. Core and envelope masses become comparable. Beyond the critical mass of about 15 earth masses the further growth of proto giant planets depends on the interior structure of the envelopes.

2.2 How do Planetary Masses Depend on Nebula Conditions?

We now turn to the second question, concerning the dependence of planetary mass on protoplanetary nebular conditions. A key result of the study of giant planet formation was the insensitivity of protoplanetary structure to the conditions in the protoplanetary disk. Given the mass of a core, the mass of its envelope in hydrostatic and thermal equilibrium was found to be independent of the density and temperature in the solar nebula and the distance from the star (Mizuno 1980). Apart from differences in the dust opacity (Mizuno 1980), the mean molecular weight of the envelope and the core's planetesimal accretion rate (Stevenson 1982), proto giant planets should have the same mass irrespective of their location in the nebula. The reason for this non-dependence of protoplanetary masses on the conditions in the ambient nebula is the radiative structure of the outer parts of the gaseous envelopes (Stevenson 1982). Wuchterl (1993) examined the convective stability of these radiative outer envelopes. According to his results, the outer envelopes become convective for nebula densities slightly higher than in standard minimum reconstitutive mass solar nebulae. Once the outer envelopes are convective, the masses of the proto giant planets start to *depend* on the nebula conditions. Wuchterl (1993) found envelopes from < 6 – 48 $M_\oplus$ for the same core mass but different nebula conditions. The critical mass for protoplanets with convective outer envelopes increased up to a total mass of 60 $M_\oplus$.

Even the largest critical 'convective' proto giant planets with 60 $M_\oplus$ still fall short of Jupiter and Saturn, an indication, that gas-accretion has to continue beyond the critical mass and will involve quasi-hydrostatic and/or dynamical phases. Wuchterl (1995) hydrodynamically calculated the envelope accretion beyond the critical mass for various solar nebula conditions. For conditions resulting in a protoplanet with a (standard) radiative outer envelope, all evolutions are very similar: a fraction of the envelope is ejected at the critical mass. In all cases, after the end of the dynamical phase, core and envelope masses approach values similar to Uranus' and Neptune's mass. For nebula conditions resulting in convective outer envelopes, the accretion histories are qualitatively and quantitatively different: for midplane nebula densities that are enhanced by a factor of eight relative to minimum mass values, the envelopes continue to grow beyond a Jupiter mass.

In summary: the envelope masses of proto giant planets with a given core are independent of the conditions in the nebula and the position at which they occur as long as the outer parts of their envelopes are radiative. When convection sets in, the masses start to depend on the nebula density and the dynamical accretion behaviour changes in a way so that Jupiter-mass planets can form. 'Radiative' protoplanets ultimately result in Uranus/Neptune-mass planets, 'convective' protoplanets accrete to larger (Jupiter) masses.

2.3 Are any Planetary Masses Favored?

In static and quasi-hydrostatic studies of giant planet formation the diversity of giant planet masses is usually attributed to processes that limit the supply of gas from the solar nebula. There are essentially to ways to limit the supply: (1) to remove the solar nebula and/or (2) to cut off the flow from the nebula onto the planetary envelope, e.g. by tidal truncation. Both of this mechanisms determining the masses of giant planets rely on the implicit assumption, that giant planets accrete gas essentially without intrinsic limits. The hydrodynamical studies outlined in the last two section show, that this assumption is justified only for protoplanets with convective outer envelopes. By examining the conditions that possibly produce such envelopes we may discuss the question (3) of how frequently accretion of massive, i.e. Jupiter-like planets occurs.

Wuchterl (1993) has rewritten the Schwarzschild criterion to determine the nebula properties that result in convective proto giant planets at the critical mass. The resulting condition essentially determines a minimum nebula density for every temperature (for given constitutive relations and planetesimal accretion rates). Above that nebula density the outer-most parts of a protoplanetary envelope become convective. Concerning the *envelope dynamics*, this is only an approximate condition (lower limit for the critical density), since the accretional behaviour changes only after a 'significant' fraction of the outer envelope has actually become convective (see Wuchterl 1993, 1995). However, even with the lower limit a considerable collection of solar nebula models (cf. Wuchterl 1993, Fig. 2) turns out to produce 'radiative' proto giant planets at the critical mass and, therefore, an accretional behaviour, that results in Uranus/Neptune-type planets. To find a convective proto giant planet, that accretes a Jupiter mass,the nebula midplane density had to be enhanced by a factor of 8 relative to minimum mass value.

In summary: Most solar nebula models in the literature produce 'radiative' proto giant planets. In all those nebulae Uranus/Neptune-mass objects should form since accretion to larger masses is hydrodynamically inhibited. Only in sufficiently dense nebulae producing 'convective' proto giant planets, the formation of Saturn and Jupiter is possible. Therefore, relying on the existing nebula models, hydrodynamic giant planet formation theories apparently favor

Uranus/Neptune-mass planets. As a consequence, Jupiter-mass objects should be the exception.

3 Formation of Giant Planets at Small Orbital Distances

Recently, a planetary mass companion of 51 Peg has been discovered at an orbital distance of 0.05 AU (Mayor and Queloz 1995, see chapter by Mayor). Guillot et al. (1996) showed, that a giant planet can be stable over 51 Peg's age. In this section I want to address the question whether a giant planet can form so close to a star.

Let me apply the arguments from the theory of planet formation given in the last sections to the case of 51 Peg's companion. To allow the onset of gas accretion at the critical mass, the outer parts of the protoplanetary envelope have to be convective. Working out the criterium for a minimum mass solar nebula (e.g. Hayashi et al. 1985) shows, that critical protoplanets should be convective inside of Mercury's orbit (cf. Wuchterl 1993, Fig. 2, upper left corner). It leads to the expectation, that a detailed hydrodynamical calculation of giant planet formation will show the onset of rapid gas accretion at the critical mass. This has not been considered previously since the terrestrial planets are well below the critical mass.

In order to calculate the growth of a proto giant planet we will assume, that the core is a rigid body of constant density which grows in mass according to the 'two body, particle-in-a-box' planetesimal accretion rate:

$$\frac{dM_{\rm core}}{dt} = 4\pi R^2(1+2\theta)\frac{\sigma}{P} \tag{1}$$

$4\pi R^2$ is the geometrical cross section of the core, used here for simplicity, $(1+2\theta)$ is the gravitational focusing factor, θ the Safronov Number, σ the planetesimal surface density and P the orbital period. The temporal evolution of the planetesimal surface density is calculated by assuming it to be constant across a spatially fixed 'feeding zone' and reducing the total planetesimal mass according to the core growth rate.

The Hill sphere of the planet is embedded in the Hayashi et al. (1985) solar nebula at 0.052 AU, i.e. a reservoir of gas with a temperature of 1252 [K] and a density of $5.296\,10^{-6}$ [g/cm^3].

The structure of the envelope is calculated between the Hill-sphere and the core surface by solving the equations of radiation hydrodynamics in spherical symmetry (cf. Wuchterl 1991a,b) together with a time dependent model for turbulent convection (Kuhfuß (1986),(1987)). The convection model reduces to the mixing length theory in the static local limit. The equations are solved with a self-adaptive grid technique (Dorfi and Feuchtinger 1995) with the convection model discretized according to Wuchterl (1995b). The detailed solar composition constitutive relations are as in Wuchterl (1991a,b) with the exception of the molecular opacities that are according to Alexander et al. (1989) and the high temperature opacities due to Weiss et al. (1990).

Model parameters of the calculation are given in Table 1. The temporal evolution of the protoplanetary masses is shown in Figure 1 for two different assumptions about the nebula temperature.

Model parameters for giant planet formation calculations at 0.05 AU

$a = 0.05,\ \varrho_{\rm neb} = 5.296\,10^{-6}[{\rm g/cm^3}]$				
	$T_{\rm Neb}/[{\rm K}]$	$\sigma_0/[{\rm g/cm^3}]$	$M_{\rm solids}/[{\rm M_\oplus}]$	θ
Case 1, Standard	1252	30	15	2
Case 2, Reduced T	600	30	15	2

Table 1. Model Parameters for the giant planet formation calculations at an orbital distance of 0.05 AU that are shown in Fig. 1. $\varrho_{\rm neb}$, $T_{\rm Neb}$ are the midplane densities and temperatures of the solar nebula in which the proto giant planet is embedded. The 'standard' values are taken from the Hayashi et al. (1985) minimum reconstitutive mass nebula. σ_0 is the initial planetesimal surface density, $M_{\rm solids}$ the total mass of planetesimals in the 'feeding' zone and θ the Safronov number.

In both cases significant gas accretion sets in. At core masses of 7.5 $M_\oplus$ and 13.5 $M_\oplus$, respectively, envelope accretion takes over as the dominating mass accumulation process. The proto giant planets grow beyond a Jupiter mass despite the high nebula temperatures and the small Hill-sphere caused by the strong solar tidal forces. This is apparently overcompensated by the enhanced self-gravity of the envelope gas due to the fact, that the nebula gas-density at 0.052 AU is orders of magnitude higher than at Jupiter's position. When considering the high mass end of the protoplanetary evolutionary curves, above 100 $M_\oplus$ say, it has to be kept in mind, that gas-supply-limiting-processes, as tidal truncation or nebula dispersal have not been taken into account.

4 Conclusion

Hydrodynamical studies of planetary envelope accretion show, that Uranus / Neptune-type planets are the standard outcome of the giant planet formation process. Jupiters require nebula densities that are enhanced relative to present nebula models to hydrodynamically enable the onset of efficient envelope gas accretion. The formation of 51 Peg type giant planets at small orbital distances occurs if condensible element cores grow to a few earth masses in the presence of a standard solar nebula. The search for extrasolar giant planets by differential astrometry with interferometers promises the detectability of the theoretically tentatively most abundant Uranus/Neptune-type giant planets around nearby stars. Giant planet formation theory will then be quantitatively confronted to the masses of extrasolar planets, probably as the first piece of our understanding of the formation of planets.

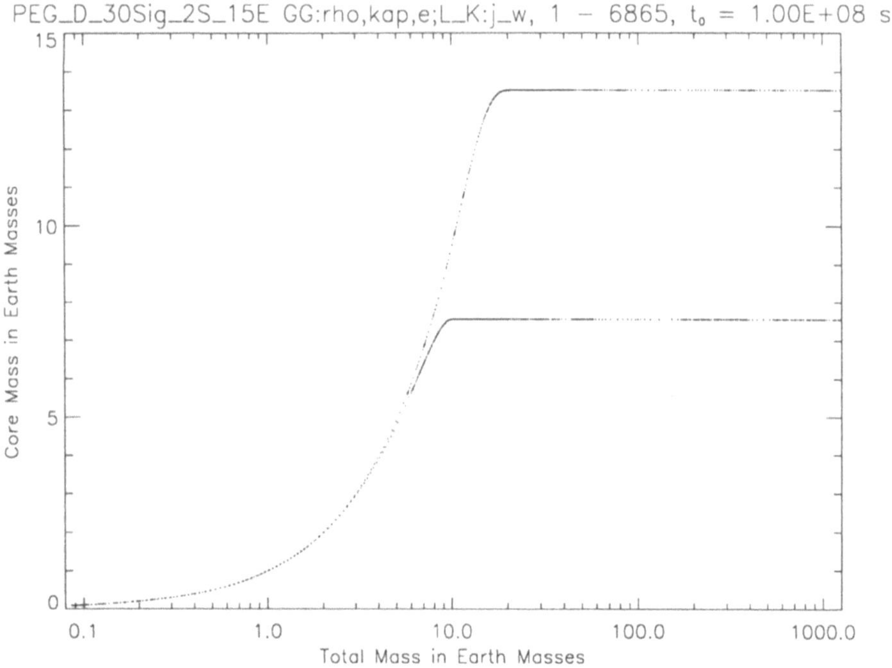

Fig. 1. Formation of giant planets at 0.052 AU distance from a solar mass star. The results of two radiation hydrodynamical calculations with time dependent convection are shown in the total-mass-core-mass diagram. Model parameters are given in Table 1. At low total masses the envelopes are negligible and the core mass is equal to the total mass; both curves overlap. At a few earth masses the curves separate and turn to the right as the envelopes begin to contribute a significant fraction to the total mass. At core masses of 13.5 $M_{\oplus}$ and 7.5 $M_{\oplus}$ for cases 1 (Standard) and 2 (reduced nebula temperature) respectively the curves turn sharply to the right as efficient gas accretion sets in and rapidly increases the envelope mass. The further growth in total mass proceeds at essentially constant core mass.

5 Acknowledgements

This work has been supported by the Österreichischer Fonds zur Förderung der wissenschaftlichen Forschung (FWF) under project number S-7307-AST.

References

Alexander, D.R., Augason, G.C., Johnson, H.R. (1989): Astrophys. Journ. **345**: 1014

Black, D.C. (1995): Completing the Copernican Revolution: The Search for Other Planetary Systems. Ann. Rev. Astron. Astrophys. **33**: 359–380

Bodenheimer, P. and Pollack, J. B. (1986): Calculations of the Accretion and Evolution of Giant Planets: The Effects of Solid Cores. Icarus **67**, 391-408

Dorfi, E.A. and Feuchtinger, M.U. (1995): Adaptive Radiation Hydrodynamics of Pulsating Stars Comp. Phys. Comm. **89**: 69–90

Guillot, T., Burrows, A., Hubbard, W.B., Lunine, J.I., Saumon,D. (1996): Giant planets at small orbital distances. Astrophys. Journ. **459**: L35–L38.

Hayashi, C., K. Nakazawa and Y. Nakagawa (1985): Formation of the Solar System. In *Protostars and Planets II* (D.C.Black and M.S.Matthews, Eds.) Univ. of Arizona Press, Tucson.

Kuhfuß, R. (1986): A model for time-dependent convection Astron. Astrophys. **160**, 116—120

Kuhfuß, R. (1987): *Ein Modell für zeitabhängige, nichtlokale Konvektion in Sternen.* Ph. D. thesis, Tech. Univ. Munich.

Lissauer, J.J. (1993): Planet Formation. Ann. Rev. Astron. Astrophys. **31**: 129–174

Lissauer, J.J., Pollack, J.B., Wetherill, G.W. and Stevenson, D.J. (1995): Formation of the Neptune System. In *Neptune* (Univ. Arizona Press, Tucson).

Mayor, M., Queloz, D. (1995): Nature **378**, 355.

Mizuno, H. (1980): Formation of the Giant Planets. Prog. Theor. Phys. **64**, 544-557

Pollack, J.B., O. Hubickyj, P. Bodenheimer, J.J. Lissauer, M. Podolak and Y. Greenzweig (1996): Formation of the Giant Planets by Concurrent Accretion of Solids and Gas. Icarus **submitted**.

Stevenson, D.J. (1982): Formation of the Giant Planets. Planet. Space Sci. **30**, 755-764

Weiss, A., Keady, J.J., Magee, N.H.,Jr. (1990): A Collection of Recent Los Alamos Opacity Tables for All Temperatures. MPA-Print 1990.

Wuchterl, G. (1991): Hydrodynamics of Giant Planet Formation II: Model Equations and Critical Mass. Icarus, **91**, 39-52.

Wuchterl, G. (1991): Hydrodynamics of Giant Planet Formation III: Jupiter's Nucleated Instability. Icarus, **91**, 53-64.

Wuchterl, G. (1993): The Critical Mass for Protoplanets Revisited: Massive Envelopes Through Convection. Icarus **106**, 323–334.

Wuchterl, G. (1995a): Giant Planet Formation. A comparative View of Gas-Accretion. In *Comparative Planetology* (M.T. Chahine, M.F. A'Hearn, J.H. Rahe, Eds.). (Kluwer Acad. Publ., Dordrecht) and
Earth, Moon, and Planets **67**, 51–65.

Wuchterl, G. (1995b): Time dependent convection on self adaptive grids. Comp. Phys. Comm. **89**, 119–126.

Theory of Very Low Mass Stars, Brown Dwarfs and Extra-solar Giant Planets

Isabelle Baraffe and Gilles Chabrier

Centre de Recherche Astrophysique de Lyon (UMR 5574 CNRS),
Ecole Normale Supérieure, 69364 Lyon Cedex 07, France

Abstract. Theory of very low mass stars, brown dwarfs and extra solar planets has considerably improved within the past years. The main efforts have been devoted to the equation of state (EOS), screening factors of thermonuclear reactions, opacities and non-grey atmosphere models, and yield a better understanding of such objects. Satisfactory agreement between observations and models now assesses the maturity of the theory.

1 Introduction

Over the past decade considerable effort, both observational and theoretical, has been directed towards deriving a more accurate determination of the stellar lower main sequence and of the sub-stellar domain covered by Brown Dwarfs and Planets. Astronomers have been looking for brown dwarfs for more than a decade, either with standard astronomical technics or with microlensing experiments. A breakthrough in the search for brown dwarfs was very recently achieved with the discovery of the first cool brown dwarf GL 229B with Adaptive Optics observations and IR direct imaging (Nakajima et al. 1995). At the same epoch, the search for planets blossomed with the discovery of a Jupiter - mass companion of the star 51 Pegasi by radial velocity survey (Mayor and Queloz 1995; Mayor, this conference). Now, the number of faint, cool stars and substellar objects is rising rapidly, promising in the near future a jump in the understanding of a wide variety of astrophysical problems, from planet and star formation, stellar structure to galactic formation and evolution.

2 Theoretical Improvements

Accurate modeling of the mechanical and thermal properties of very-low-mass stars (VLMS), Brown Dwarfs (BD) and Extra-solar Giant Planets (EGP) is of prior importance for a wide range of physical and astrophysical reasons, from the fundamental physics point of view to the astrophysical and cosmological implications. They provide natural laboratories to test the different theories, equations of state, nuclear reaction rates, model atmospheres aimed at describing the physics of dense and cool objects.

VLMS, or M-dwarfs, defined hereafter as objects with masses below $0.6M_\odot$, are compact objects, with characteristic radii in the range $\sim 0.1 - 0.6\, R/R_\odot$.

Their central densities and temperatures are respectively of the order of $\rho_c \approx 10$–10^3 gcm^{-3} and $T_c \approx 10^6 - 10^7$ K, so that the stellar interior represents a strongly correlated plasma. BD are defined as objetcs not massive enough for their central temperature to sustain hydrogen burning. This characteristic stems from the onset of degeneracy in the contracting protostar, which prevents an increase of the temperature as contractions proceeds. A BD therefore never reaches a state of equilibrium and cools down its whole life. The upper mass given by theoretical models is $\sim 0.07 - 0.09 M_\odot$, depending on the initial composition. The minimum mass depends on stellar formation theory and is still uncertain (cf. Hubbard, 1994). Brown dwarfs would bridge the gap in the observed mass-distribution of astrophysical objects between the lowest-mass star and the largest-mass planet, Jupiter ($m = 0.001\, M_\odot$). The distinction between BD and Giant Planets is based on different formation scenarios, but their inner structure and spectral signatures are both governed by essentially the same physics.

The effective temperatures of those objects are below $T_{eff} \approx 5000$ K, and surface gravities $g = GM_\star/R_\star^2$ are in the range $log\, g \approx 3.5 - 5.5$ for VLM, down to some 100 K for BD and EGP. The low effective temperatures allows the presence of stable molecules (H_2, H_2O, TiO, VO, CH_4, NH_3,...), whose bands constitute the main source of absorption along the characteristic frequency domain. The presence of these bands complicates tremendously the treatment of radiative transfer, not only because of the numerous transitions to include in the calculations, but also because the molecular absorption coefficients strongly depend on the frequency, so that the grey-approximation is no longer valid under the present atmosphere conditions.

These conditions show convincingly that the modeling of such objects requires a correct description of non-ideal effects in dense plasmas, as well for the equation of state (EOS) as for the screening factors of the nuclear rates, and a derivation of accurate models for dense and cool atmospheres, where molecular opacity becomes eventually the main source of absorption.

Only recently has a new EOS been derived, devoted to the description of low-mass stars and giant planets, which presents a consistent treatment of pressure ionization (Saumon, Chabrier & VanHorn 1995). Improvement in the field of cool non-grey atmosphere has blossomed in recent years with the work of Allard and Hauschildt (1995; 1996), Brett (1995), and Saumon et al. (1994), due to recent interest in the extreme lower main sequence ($T_{eff} \sim 2000K$), and the necessity to derive accurate atmophere models to identify the luminosity and the colors of M-dwarfs. Non-grey atmosphere models now extend down to $T_{eff} = 900$K, since the discovery of the first cool Brown Dwarf GL 229B offered a priceless test for atmosphere models (Allard et al. 1996, Tsuji et al. 1996a).

The modelisation of VLM stars, BD and EGP and the confrontation with observations will now be examined.

2.1 VLM Stars

In spite of considerable progress in stellar theory - internal structure, model atmospheres and evolution - all the VLMS models so far failed to reproduce accurately the observed color-magnitude diagrams (CMD) of disk or halo stars below ~ 4000 K, i.e. $\sim 0.4-0.6\,M_\odot$, depending on the metallicity. All the models predicted too hot an effective temperature for a given luminosity, i.e. were too blue compared to the observations by at least one magnitude. Such a desagreement stemed essentially from shortcomings both in the physics of the interior, i.e. equation of state (EOS) and thus mass-radius relationship, and in the atmosphere, since all models were based on grey atmospheres and approximate outer boundary conditions. Important progress has been made recently in this field with the derivation of the afore-mentioned EOS, non-grey model atmospheres for M-dwarfs (Allard and Hauschildt 1995; Brett 1995) and evolutionary models based on a consistent treatment between the interior and the atmosphere profile (Baraffe, Chabrier, Allard & Hauschildt 1995; Chabrier, Baraffe & Plez 1996). The effect of the outer boundary condition on the mass - T_{eff} relationship is shown Figure 1. In this figure, comparison is made between models with outer boundary conditions based on non grey atmosphere models (Chabrier & Baraffe 1997) and models using grey atmospheres (Burrows et al. 1989; D'Antona & Mazzitelli 1994, 1996).

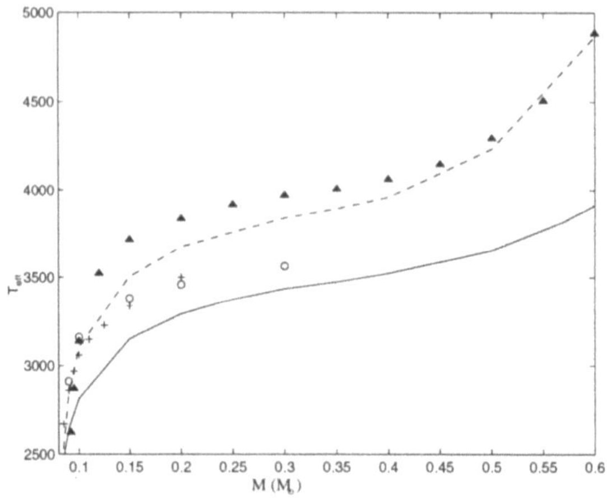

Fig. 1. Mass - T_{eff} relationship for different metallicity. Models based on non-grey atmosphere (Chabrier & Baraffe 1997) are represented by solid ([M/H]=0) and dashed ([M/H]=-1) curves. Comparison is made with grey models at solar metallicity from Burrows et al. (1989, +) and D'Antona and Mazzitelli (1994, o) and for [M/H]=-1 (D'Antona and Mazzitelli 1996, triangles).

Models based on the improved physics mentioned above now reach quantitative agreement with the observations for disk and halo populations, as shown by the agreement of the models with the dwarf and halo subdwarf sequences of Monet et al. (1992) and Dahn et al. (1995) (cf. Fig. 2). Same agreement is now reached with HST observations of Globular Cluster Main Sequences of low metallicity (Baraffe et al. 1997) (cf. Fig. 2). This allows now the derivation of reliable mass luminosity and mass-functions for these objects down to the brown dwarf limit (Méra and Chabrier 1997).

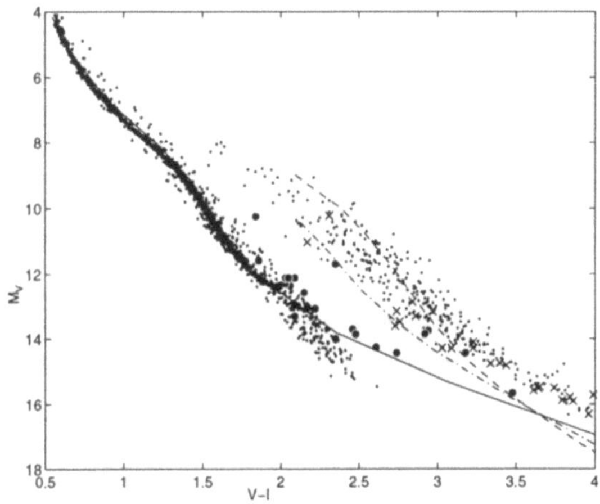

Fig. 2. Color - Magnitude diagramms for different metallicity: [M/H]=-1.5 (Baraffe et al. 1997, solid line), [M/H]=-0.5 (dash-dot) and [M/H]=0 (dash) from Baraffe et al. (1995). The Main Sequence of the Globular Cluster NGC6397 from Cool et al. (1996) is shown on the left part. Subdwarf halo field stars from Monet et al. (1992) are indicated by full circles, as well as disk stars of Monet et al. (1992) (crosses) and Dahn et al. (1995) (dots on the right part)

2.2 Brown Dwarfs

The recent discovery of substellar objects allows now to test the reliability of the theory in the Brown Dwarf regime. The first young brown dwarfs were identified in the Pleiades star cluster (Rebolo et al. 1996), namely Teide 1 and Calar 3, which both have retained their original Lithium. Their extremely low luminosity ($logL/L_\odot \sim -3.1$) combined with the presence of lithium yields unambiguously to masses ($m \sim 0.05 M_\odot$) lower than the lowest *stellar* mass given by theoretical models. Determination of the luminosity function in the Pleiades or any young cluster offers a fantastic possiblity in the near future of deriving a mass function for BD. As the two BD above mentioned remain quite "hot" ($T_{eff} \sim 2500K$), because of their relative youth ($\sim 10^8$ yrs), the most stringent test for extremely cool models is provided by Gl 229B (Nakajima et al. 1996). The presence of methane in its infrared spectra clearly reveals the substellar nature of Gl229B, since CH_4 absorption appears below 1600 K (Tsuji, 1996a) and a star at the hydrogen burning limit has an effective temperature of about 2000 K (Baraffe et al. 1995). Comparison of synthetic and observed spectra yields an effective temperature of $\sim 1000K$ (Allard et al. 1996, Tsuji et al. 1996a, Marley et al. 1996). Uncertainties in the age of the system and in the temperature of Gl229B yield some indetermination for the mass of Gl229B. The most likely solution $M \approx 0.04 - 0.055\ M_{\odot}$ is supported by the quality of the most reliable fit to the water bands, and would be consistent with an age similar to that of our solar system (cf. Allard et al. 1996).

In order to guide a search for substellar objects in the near infrared, we show in figure 3 different isochrones as a function of metallicity (Allard et al. 1996; Baraffe et al. 1997). The limit magnitudes in K required to reach the brown dwarf domain are indicated by full circles on the figure. The blue loop displayed by the metal-poor isochrone stems from collision-induced absorption of H_2 in the K-band (cf. Saumon et al., 1994; Allard & Hauschildt, 1995) which shifts back the flux to shorter wavelengths. For the solar metallicity isochrones, the blue shift is due to strong CH_4 absorption in the K bandpass. Such important blueshift is a photometric signature in the infrared of the transition from the stellar to the sub-stellar domain, whereas optical colors keep reddening almost linearly.

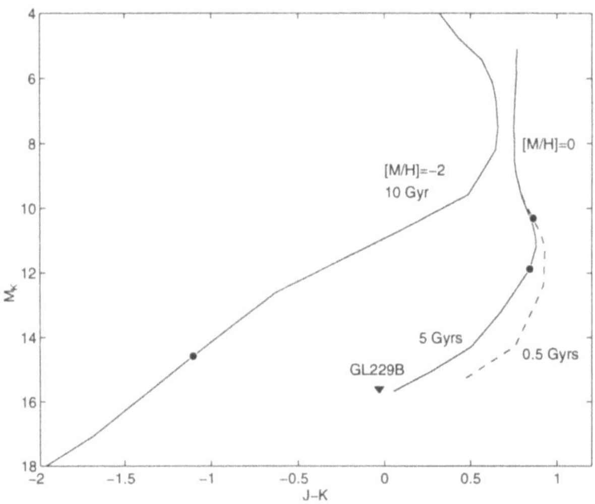

Fig. 3. Color - Magnitude diagramm in the near IR for the metallicity and ages indicated on the figure. The full circles on the curve indicate the stellar/sub-stellar transition.

2.3 Extrasolar Giant Planets

An extended analysis of EGP, ranging from 0.3 to 15 Jupiter masses ($1\ M_J = 0.001 M_{\odot}$) has been recently conducted by Saumon et al. (1996). The thermodynamics of the interior is described with the same afore-mentioned EOS, and effects of rotation and presence of an icy core have been explored. The still remaining uncertainties of the models reside in the estimation of the spectral emission of EGP, which is approximated by a black body distribution. Although weakened by such an approximation, the predictions made in Saumon et al. (1996) represent the first benchmark in search strategies for the detection of EGP. Improvements of solar and extrasolar giant planet evolutionary models have been developed by Guillot et al. (1995; 1996), exploring the influence of radiation in the outermost layers on the evolution. This modifies (reduces) the internal temperature, and the luminosity, and thus the evolution of the planet, compared to conventional models which assume convection throughout the entire interior. This bears important consequences for our understanding of planetary evolution and the stability of giant planets orbiting nearby stars, as 51Peg (Guillot et al., 1996).

The predicted absolute fluxes of BD or EGP (Allard et al. 1997) are illustrated in Fig. 4 (kindly provided by F. Allard) and compared to the sensitivity of ground and space-based observing platforms. The predictions are shown for different effective temperature and compared to a black body emission at the same T_{eff}. Strong departure from the black body emission illustrates the necessity of synthetic spectra for this type of objects.

3 Conclusion

The recent improvement in the description of the mechanical and thermal properties of cool compact objects, and of their photometric signature now provides solid grounds to analyse the observations and make reliable predictions. This allows the derivation of reliable mass-functions for VLM stars down to the brown dwarf limit. Forthcoming development of observations in the IR should increase our knowledge of the substellar domain and bring new insight on the still unsolved problem of stellar and planet formation.

Elaboration of cool atmosphere models and synthetic spectra, including grain formation which already affects the evolution of objetcs below $T_{eff} \sim 2600K$ (Tsuji et al. 1996b), represents now the next step of theoretical improvement. Work is already under progress in this field and promise most exciting results in the next future.

References

Allard, F., and Hauschildt, P. H. (1995): ApJ, 445, 433
Allard, F., Hauschildt, P.H., Baraffe, I., Chabrier, G. (1996): ApJl, 465, L123
Allard, F., Hauschildt, P.H., Alexander D. R., Starrfield, S. (1997): Ann. Rev. A&A, submitted
Baraffe, I., Chabrier, G., Allard, F., Hauschildt P. (1995): ApJl, 446, L35
Baraffe, I., Chabrier, G., Allard, F., Hauschildt P. (1997): A&A, submitted
Brett, J.M. (1995): A&A, 295, 736 ApJ, 406, 158 (BHSL93)

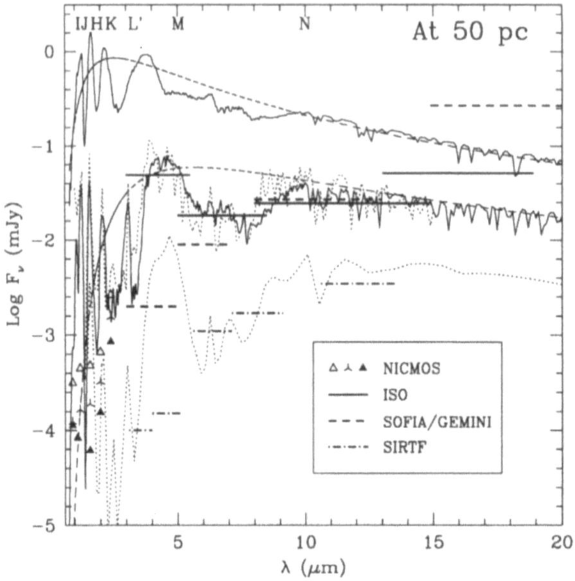

Fig. 4. Predicted absolute fluxes of BD or EGP at 50 pc compared to the sensitivity of ground and space-based platforms. Models of both Allard et al. (1996) (full) and Marley et al. (1996) (dotted) are shown which simulate (i) a brown dwarf near the hydrogen burning limit (topmost spectrum $T_{eff} = 2000K$), (ii) an evolved brown dwarf similar to GL 229B (central spectra: $T_{eff} = 900K$ and 960K), and (iii) an extremely cool object (lowermost spectrum: $T_{eff} = 500K$). Dashed curves give the corresponding black-body.

Chabrier, G., Baraffe, I., Plez, B. (1996): ApJ, 459, L91
Chabrier, G., Baraffe, I. (1997): A&A, submitted
Cool, A. M., Piotto, G., King, I.R. (1996): ApJ, 468, 655
Dahn, C.C., Liebert, J., Harris, H.C., Guetter, H.H. (1995): *The bottom of the main-sequence and below*, Ed. C. Tinney
D'Antona, F. and Mazzitelli, I, (1994): ApJ. Suppl., 90, 467
D'Antona, F. and Mazzitelli, I, (1996): ApJ., 456, 329
Guillot, T., Chabrier, G., Gautier, D., Morel P. (1994): ApJ, 450, 463
Guillot, T., Burrows, A., Lunine, J.I., Saumon, D. 1996, ApJ, 459, L35
Hubbard. W. B. (1994): *the Equation of State in Astrophysics*, Ed. G. Chabrier and E. Schatzman
Marley et al. (1996): Science, 272, 1919
Mayor, M., Queloz, D. (1996): Nature, 378, 355
Mera, D., Chabrier, G., 1997, A&A, submitted
Monet D.G. et al. (1992): Astron. J., 103, 638
Nakajima et al. (1995): Nature, 378, 463
Rebolo, R., Zapatero Osorio, M.R., Martin, E.L. (1995): Nature, 377, 129
Saumon, D., Bergeron, P., Lunine, L.I., Hubbard, W.B., Burrows, A. (1994): ApJ, 424, 333

Saumon, D. et al., ApJ, 1996, 460, 993
Saumon, D., Chabrier, G., VanHorn, H.M. (1995): ApJs, 99, 713
Tsuji, T., Ohnaka, K., Aoki, W., Nakajima. T. (1996a): A&A, 308, L29
Tsuji, T., Ohnaka, K., Aoki, W. (1996b): A&A, 305, L1

On the Detection of Exo-zodiacal Light by Nulling Interferometry with the Magellan Telescopes

Wesley A. Traub[1], Nathaniel P. Carleton[1], and J. R. P. Angel[2]

[1] Harvard-Smithsonian Center for Astrophysics, 60 Garden St., Cambridge, MA 02138, U.S.A.
[2] Steward Observatory, University of Arizona, Tucson, AZ 85721, U.S.A.

Abstract. We discuss a potential experiment to search for zodiacal dust around nearby stars, by the technique of nulling interferometry at mid-infrared wavelengths, using the Magellan twin 6.5 m telescopes at Las Campanas. The experiment requires an interferometer that can operate in nulling mode with very high on-axis rejection ability, for the central star, and very good detection sensitivity at 10 μm wavelength, for the zodiacal thermal emission. In its non-nulling mode, working in the range 1 to 10 μm, the interferometer could address a wide range of astrophysical issues.

1 Introduction

The process of planet formation around a young star is poorly understood, notwithstanding the astonishing fact that we currently know of more exo-solar planets than we do solar planets (Boss, 1996). Most of these exo-solar planets have been discovered with ground-based radial-velocity measurements, but two have been found using classical astrometry, and one was discovered with coronagraphic imaging.

Ground-based interferometry and adaptive optics promise to provide even more sensitive tools. Ground-based narrow-angle differential astrometry with the Palomar testbed interferometer (Colavita et al., 1994) will permit deep searches for reactive star displacements caused by Jupiter and Saturn mass objects, using reference stars with separations up to 20 arcsec. Very narrow angle differential astrometry has also been proposed (Traub et al, 1996) for the Infrared Optical Telescope Array (IOTA) using binary stars, where one star acts as a reference for the other, with separations up to 2 arcsec. Ground-based direct imaging of planetary systems, using adaptive optics, has been proposed for the converted 6.5 m MMT telescope (Angel, 1994; Stahl and Sandler, 1995).

In space, the proposed Space Interferometry Mission (SIM) (see the web site at http://www.ess.sunysb.edu/~simswg/index.html) will be able to search for stellar wobble due to planets using wide angle astrometry, and may provide an orbital test of star nulling at visible wavelengths. A focused "planet finder" mission has been proposed (Angel and Woolf, 1997), using four telescopes configured as a deep-nulling interferometer, giving the ability to produce a one-pixel image of an Earth-like planet at 10 pc and 10 μm wavelength, and also with broadband spectroscopic ability. If the planet finder is stationed 5 AU from the sun (to reduce background noise from solar zodiacal emission), then the largest noise source will be photon noise in the zodiacal emission around the target star itself. Early knowledge of this flux is clearly important for mission planning. In this paper we explore how one might detect extra-solar zodiacal dust around nearby stars using a ground-based interferometer.

2 Extra-solar Zodiacal Signal and Detection

We use the IRAS model of the solar zodiacal intensity (Good, 1994) as a benchmark. This model specifies the volumetric emission cross section as a function of cylindrical coordinates (r, z), where r is radial distance from the sun, and z is distance from the plane of symmetry of the dust. The temperature $T(R)$ is a function of the spherical radial coordinate R. We show in Fig. 1 the face-on optical depth $\tau(r)$ and in-plane temperature $T(r)$ of the IRAS model from the solar surface to 10 AU. If the sun were removed to 10 pc distance, then the inset scale in arcsec and milli-arcsec (mas) would apply. For comparison we show the face-on optical depth and temperature of the β Pic disk (Backman et al., 1992), with its inferred opacity discontinuity at 80 AU. The solar and β Pic opacities provide us with a rough guide to the range of expected conditions.

A nominal detection wavelength of 10 μm will reduce the contrast ratio between star light and dust emission, and also provide sufficient angular resolution. For the sun at 10 pc, the 10 μm signal is about 2 Jy. From the IRAS model, the dust flux out to 1 AU (which represents about 90 % of the total) is 0.2 mJy, giving a flux ratio of 10^{-4}. For comparison, planet fluxes at 10 AU are much weaker than an integrated IRAS-type flux, e.g., Earth $\sim 0.3\,\mu$Jy, and Jupiter $\sim 0.1\,\mu$Jy. The measured 10μm signal in an 8 arcsec beam from β Pic is inferred to break down as stellar $\sim$ 1.6 Jy and disk $\sim$ 1.2 Jy.

What is the potential of ground-based telescopes, and specifically the Magellan telescopes, for observing zodiacal emission around nearby stars? The two planned Magellan telescopes are each 6.5 m in diameter, with Magellan II located 60 m northwest of Magellan I (see later section). At 10 μm wavelength, the single-telescope diffraction limit is $1.22\lambda/D = 0.39$ arcsec, and the peak to peak fringe spacing using both telescopes is $\lambda/B = 0.034$ arcsec. Clearly the diffraction wing of a single telescope would swamp the IRAS model in Fig. 1. However, as an interferometer Magellan could operate in nulling mode, with the delay paths adjusted such that the output of one side of the beamsplitter reinforces constructively (0 rad phase difference) at the position of the star, and the outputs to the other side of the beamsplitter add destructively (π rad phase difference). The sky image from each telescope would be passed through a spatial filter to isolate a single diffraction-limited mode (see, e.g., Coudé du Foresto, et al., 1992). In the ideal case a one-dimensional fringe pattern is projected onto the sky, within the envelope of a single-telescope diffraction pattern. The finite extent of the star means that even if the cancellation is perfect at the center of the star, there will still be some leakage from the stellar limbs. The zodiacal source is distributed across the fringe pattern, with about one-half of the total signal appearing inside 0.15 AU, or 15 mas in Fig. 1. The first Magellan fringe transmission peak is at 17 mas, closely matching the expected distribution.

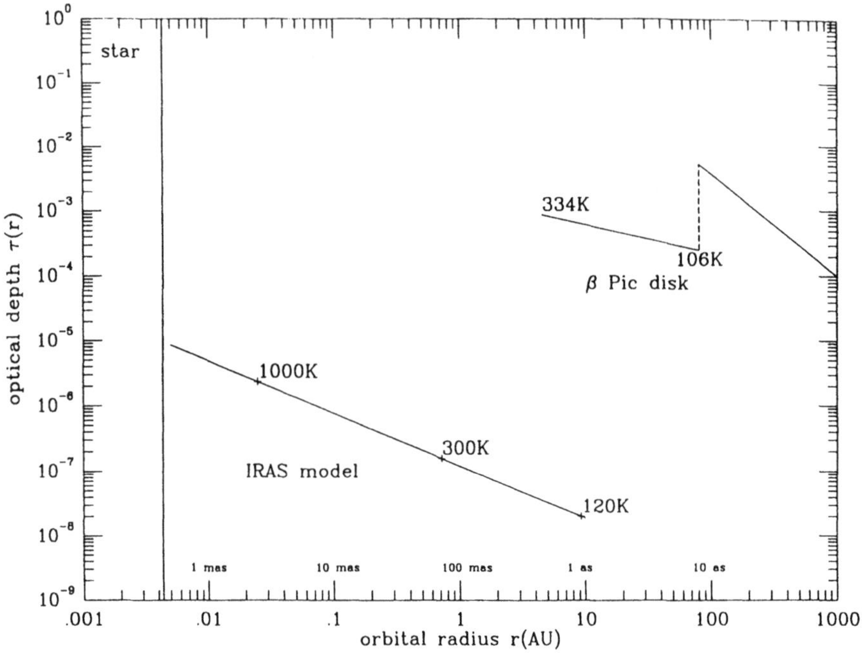

Fig. 1. The optical depth of the IRAS model of the solar zodiacal light, seen face-on, is shown as a function of radius in the solar system, and also as a function of angular radius at a distance of 10 pc. The IRAS dust model emissivity is independent of wavelength up to 38 μm. The solar limb is indicated. For comparison, the β Pic emission optical depth at 10 μm is shown, also face-on, from model no. 11 of Backman et al. (1992) which gives the best fit to a combination of ground-based and IRAS data. This β Pic model has three radial zones: a relatively clear zone from the star out to about 4.5 AU (uncertain by a factor of up to 5); an "inner" zone from 4.5 to 80 AU, with emissivity varying as λ^{-1}, probably composed of refractory particles; and an "outer" zone from 80 to 1000 AU or more, again with λ^{-1} emissivity, and probably composed of icy particles. Since β Pic is at a distance of 16.6 pc, the inset angular scale values should be reduced correspondingly.

A simple calculation with a circular star of diameter d centered on a sinusoidal fringe pattern with period p, both in angular units, shows that the relative leakage (i.e., ratio of $star \times fringe$ to $star$ alone) is $0.7(d/p)^2$. For $d = 0.86$ mas (e.g., the sun at 10 pc) and $p = 34$ mas the leakage is 4×10^{-4}, which reduces the star signal to 0.8 mJy. Assuming a signal reduction factor of 0.5 due to the fringe pattern, the dust signal becomes 0.1 mJy. The ratio of transmitted dust/star signals is then about 0.1. This means that with careful comparison to dust-free stars, the dust signal could be detected as a roughly 10% enhancement of the total star-system signal with an ideal nulling interferometer.

Two obvious non-ideal factors are wavefront aberration and photon noise. We assume that the wavefront will be corrected using an active secondary mirror, such as is now being built for the converted Smithsonian Astrophysical Observatory and University of Arizona 6.5 m MMT telescope on Mt. Hopkins. Following Woolf and Angel (1995) we assume that this correction reduces the stellar wavefront error to $\sigma_x \simeq 80$ nm rms with an effective subaperture scale of 0.5 m. In this case, the aberrated star light spreads over a projected cone angle of 4 arcsec. The fraction of starlight falling within the diffraction angle is given by the Strehl ratio $S = \exp(-\sigma^2)$, where $\sigma = 2\pi\sigma_x/\lambda \simeq 0.05 rad$ is the phase noise. Thus the scattered fraction in the halo will be $1 - S = 25 \times 10^{-4}$, and the amount entering the diffraction angle will be further diluted by the geometric factor $(0.39/4.)^2 = 0.010$, so the unmodulated fractional intensity will be 0.25×10^{-4} or 0.05 mJy. Thus the scattering leakage is less than the nulled disk leakage, and can be removed in the data analysis.

Fluctuations in the telescope and sky background fundamentally limit ground-based observations at 10 μm wavelength. From a figure in Traub and Stier (1976) showing sky emissivity *vs* wavelength from a mountain top (Mauna Kea) site, we find that the theoretical sky emissivity in the 8–13 μm band is 0.11, dominated by the 9–10 μm ozone band; at an effective sky temperature of 222 K, the flux in a diffraction-limited solid angle of a 6.5 m telescope is 20 Jy. If we arbitrarily assume a telescope emissivity of 0.10, and a temperature of 273 K, then the telescope flux in the same solid angle is 62 Jy. (Note that the telescope emissivity would have to drop to 0.03 in order to match that of the sky, and it would have to go even lower if the ozone band were excluded.) The net signal is 82 Jy, which yields a photon fluctuation noise of 1.8 mJy rms in 1 sec, assuming an effective detection efficiency of 0.25 electron/photon and a narrowed relative bandpass of 0.20. This compares to the star leakage 0.8 mJy and the model dust signal 0.1 mJy. The time to reach a signal to noise ratio of unity on the IRAS model dust emission is then about $(1.8/0.1)^2 \simeq 320$ sec. If we further ask for a 3σ detection, and assume a 50% duty cycle from chopping to the sky or dithering on the fringe, we find a total detection time of under 2 hours.

Current experience approaches this performance. For example, the MIRAC detector is a 128×128 Si:As hybrid BIB array, operated at liquid helium temperature, with a measured sensitivity of 50 mJy in 1 arcsec square, 1 sec integration, 10% bandwidth, at the IRTF telescope (3.0 m), with an assumed emissivity of 0.20 (G. Fazio, private communication, 1996). Under these conditions the expected background flux is 960 Jy, and the calculated fluctuation level is 19 mJy in 1 sec, again assuming a detection efficiency of 0.25. The MIRAC detector is therefore operating close to the theoretical photon noise limit, within a factor of 2 or 3. There are several possible reasons for this factor of a few difference in noise. The larger value could result from a lower detection efficiency (~ 0.04), or from temperature fluctuations of 0.1 K in the relatively weak sky background, or from fluctuations of 0.002 K in the stronger telescope background.

In summary, it appears feasible to search for zodiacal light emission around nearby stars using the Magellan telescopes in interferometric mode at a wavelength of 10 μm. To be able to detect a solar-type zodiacal light signal will require careful wavefront control and low background emission from the telescopes.

3 Magellan Telescopes and Interferometer

The Magellan Project will put two 6.5 m diameter telescopes at the Las Campanas Observatory in Chile (de Jonge, 1994). Construction of Magellan I started in 1993, and it is expected to open in 1998. The follow-on telescope, Magellan II, already has its foundation in place, 60 m northwest, connected to Magellan I by a below-grade tunnel. Funding is in place to complete both telescopes and their initial instruments; an interferometry capability, as discussed in this paper, is still in the pre-planning phase. The partners in the Magellan consortium are the Carnegie Institution, the University of Arizona, Harvard University, the University of Michigan, and the Massachusetts Institute of Technology.

The potential for interferometry with Magellan is excellent. The Cerro El Manqui site has nominal 0.5 arcsec seeing (FWHM) in the visible, which projects to $r_0 \simeq 9\,m$ at 10 μm wavelength. The secondary mirror will be a deformable active element, so there will be only three surfaces leading to a corrected image at the Nasmyth focus, giving a compact, low background image. A low-background beam relay should be achievable by placing all of the relay mirrors (after the primary, secondary, and Nasmyth diagonal) in a vacuum environment, where they may be cooled, and remain clean.

The hardware necessary to bring the beams from the two main telescopes together starts at the f/15 Nasmyth focus, which will be near an elevation trunnion. Each beam must be collimated (probably to about 25 cm diameter), run through a conventional coude relay system down to the tunnel level, and then be directed back and forth along the tunnel by an optical delay scheme, so that the beams finally come out with equalised paths. From there, we have to enter a beam combination system. Cooling the mirrors in this path (8 reflections after the Nasmyth focus) can be done by surrounding each mirror with a cooled shroud with appropriate baffling for the beam entry and exit. This should be possible in the vacuum environment, which will be necessary anyway, to eliminate internal seeing and reduce longitudinal dispersion at shorter wavelengths.

A large beam-combination area will be required, for dichroic separations of the beams and multiple detections, for image stabilization and fringe tracking, and for the primary interference measurement. We will draw upon our current experience (Carleton et al., 1994) with the IOTA interferometer in designing the beam combination system, including our use of single-mode fibers and fiber couplers at 2.2 μm wavelength (Coudé du Foresto, 1994). In addition, we will consider bringing in beams from auxiliary smaller telescopes, which would be used for alignment and testing of most of the interferometric system, and could do substantial scientific work in interferometry, independently of the main telescopes. One or more of these could also be used in conjunction with the main ones, to provide phase information.

Other science projects which need the resolution and sensitivity of the Magellan interferometer at 10μm include characterizing the diameter of the hot dust clouds in active galactic nuclei, searching for possible infrared emission from superluminal sources, and examining the nature of infrared sources toward the Galactic Center; in the 1–10μm range the interferometer would also be able to look for multiplicity and dust distribution around BN-type objects, search for multiplicity at small separations in T Tauri stars, determine the distribution of dust around late-type stars, and other projects.

In summary, we believe that there is an excellent potential for doing interferometry at the Magellan telescopes. With its large collecting area, high angular resolution, and low background, a Magellan interferometer would be able to make unique science measurements on the zodiacal thermal emission around nearby stars. This work should extend our understanding of the processes of star and planetary system formation. These observations will be critical to the design of a future "planet finder" spacecraft.

We gratefully acknowledge the support of the Smithsonian Astrophysical Observatory in this work, and thank the Magellan Project and ESO for making possible the presentation of this paper.

References

Angel, J.R.P. (1994): Ground-based Imaging of Extrasolar Planets Using Adaptive Optics. Nature **368**, 203–207.

Angel, J.R.P., Woolf, N.J. (1997): An Imaging Nulling Interferometer to Study Extra-Solar Planets. ApJ, in press for Jan. 1997.

Backman, D.E., Gillett, F.C., Witteborn, F.C. (1992): Infrared Observations and Thermal Models of the β Pictoris Disk. ApJ **385**, 670–679.

Boss, A.P., (1996): Extrasolar Planets. Physics Today, Sept. 1996, 32–38.

Carleton, N.P., et al. (1994): Current Status of the IOTA Interferometer. SPIE **2200**, 152–165.

Colavita, M.M. et al., (1994): ASEPS-0 Testbed Interferometer. SPIE **2200**, 89–97.

Coudé du Foresto, V., Ridgway, S.T., Mariotti, J.-M. (1992): in: Volonté, S. (ed.), Targets for Space-Based Interferometry. ESA, Beaulieu, 731.

Coudé du Foresto, V. (1994): Interferometrie Astronomique Infrarouge par Optique Guidee Monomode. Ph.D. Thesis, Université de Paris VII.

de Jonge, M.J. (1994): Status of the Magellan Project. SPIE **2199**, 22-30.

Good, J. (1994): Appendix G, Zodiacal Dust Cloud Modeling Using IRAS Data. in: Wheelock, S. et al., IRAS Sky Survey Explanatory Supplement, JPL Publication 94-11, Pasadena, CA.

Stahl, S.M., Sandler, D.G. (1995): Optimization and Performance of Adaptive Optics for Imaging Extrasolar Planets. ApJ **454**, L153-L156.

Traub, W.A., Carleton, N.P., Porro, I.L. (1996): A Search Technique for Planets in Nearby Binary Stars Using a Ground-based Interferometer. JGR, **101**, 9291–9295.

Woolf, N.J., Angel, J.R.P. (1995): Use of the LBT to Study Zodiacal Emission Around Other Stars. OSA Conf. on *Adaptive Optics*, Technical Digest **23**, 44–46.

Direct Observation of Extrasolar Planets with an Infrared Interferometer

Vincent Coudé du Foresto[1], Jean-Marie Mariotti[2], and Guy Perrin[2]

[1] Max-Planck-Institut für Astronomie, Königstuhl 17, D-69117 Heidelberg, Germany
[2] Observatoire de Paris, DESPA, 5 place Janssen, F-92195 Meudon Cedex, France

Abstract. Direct observation of extrasolar planets with an infrared interferometer is a difficult task because of the very high dynamic range required (typically 10^4 – 10^5 in the best cases), but not impossible if a very careful spatial filtering and calibration is imposed upon the recombined beams.

We propose an observing strategy to monitor hot extrasolar planets of the 51 Peg class, for which the VLTI provides sufficient sensitivity and adequate baselines.

The unexpected discovery of several Jupiter-sized planets in the close vicinity of their parent stars (two examples of which are 51 Peg and τ Boo) creates a challenge for theories of planetary formation and structure. Only a direct measurement of their mass and luminosity (not accessible via the current radial velocity techniques) can help discriminate between the different planetary models that have been proposed for this new class of objects.

These new planets revolve around their stars in circular orbits at a typical distance of 0.05 AU, and they are expected to have a temperature in the range 1000–1500 K. For an observer at 15 pc, this translates into a binary system that can be well modeled by a bright circular disk of known diameter (determined by the magnitude and the spectral type of the parent star), and a much fainter point source within a few milliarcseconds.

It should be noted that the period of the orbit, as well as its semimajor axis a and its ephemeris are precisely known from radial velocity measurements. At 1000–1500 K, the planet's luminosity peaks in the infrared K band, at a fraction $\alpha = 3 \times 10^{-5}$ to 3×10^{-4} of the stellar brightness. The brightness ratio, the orbit orientation Ω, and its inclination i (which leads to mass determination since $M \sin i$ is known from spectroscopy) are the only three parameters to retrieve from the interferometric observations.

The visibility function of the binary system shows, on top of the traditional stellar Airy function, a fringe pattern whose peak to peak amplitude is 2α, and whose periodicity and orientation in the (u, v) plane depend on the instantaneous position of the planet, i.e. the known phase ϕ of the radial velocity oscillations. Thus any visibility point can be located into a (u, v, ϕ) data cube (or, rather, a data tore), which can be filled in progressively as the source is observed during several (not necessarily consecutive) nights and with different baselines. Once enough data have been accumulated, they can be solved for α, i and Ω.

To obtain a high dynamic range, the visibilities need to be measured with a high degree of accuracy, though if a lot of data points are collected, the error on each individual point can be substantially larger than the brightness ratio α. We define here the accuracy as the absolute statistical error on fringe contrast measurements obtained in 5 min of observing time. On bright sources interferometers are limited by seeing perturbations. Using a spatial filter (e.g., a single-mode fiber) to clean the corrugated beams already improves the accuracy of 0.3 % or better, as was demonstrated with the FLUOR unit on IOTA (see the paper in this conference). Yet this instrument is still limited by one remaining turbulence mode, the differential piston between the pupils, which is insensitive to spatial filtering. It is not known how much more would be gained by removing the piston with a fringe tracker.

Expected performances are shown in Figures 1 and 2, derived from simulations using ($\alpha = 10^{-4}$, $i = 75$ deg, $a = 3.6$ mas) for 51 Peg, and ($\alpha = 3 \times 10^{-4}$, $i = 75$ deg, $a = 2.4$ mas) for τ Boo. Each source is observed, alternating with a reference in sequences of 5 min, for as long each night (6 h) as its elevation permits. At the current 0.3 % accuracy level, both planets could be detected, though i would not be determined for all possible orbit orientations.

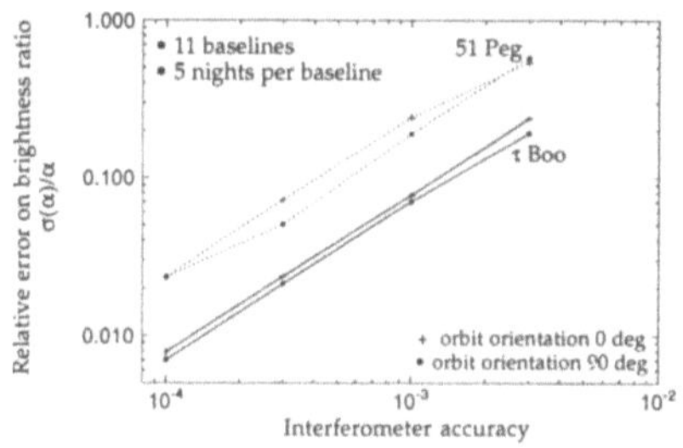

Fig. 1. Performance obtained on the planet photometry.

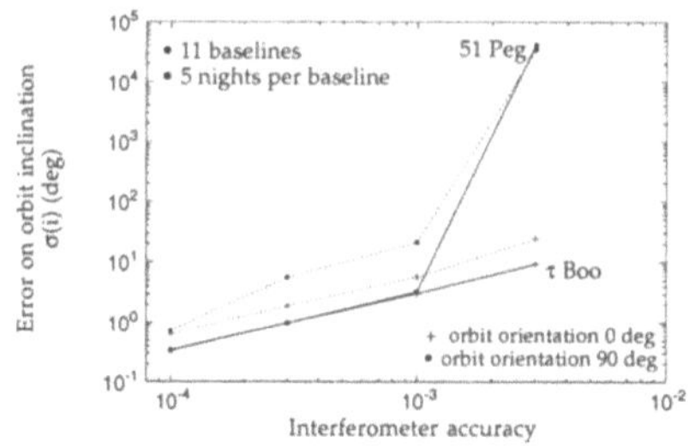

Fig. 2. Performance obtained on the orbit inclination measurement.

Conclusion Direct determination of the mass and luminosity of 51 Peg-like exoplanets obviously represents a very difficult problem, due to the high dynamic range required to observe the binary systems. It is, however, one for which interferometry is the most efficient, because the scientific information can be extracted by modeling observations with a very small number of parameters. Indeed the simulations show that, provided a large number of nights can be dedicated to this program, a VLTI equipped with a single-mode 2 μm instrument and a fringe tracker has the sensitivity, the precision and the baselines required to directly observe selected exoplanets.

Part 3

AGN AND STARBURST GALAXIES

VLTI Observations of Radio-Loud Active Galaxies

George K. Miley

Leiden Observatory, Postbus 9513, 2300 RA Leiden, The Netherlands

Abstract. I shall consider the relevance of VLTI for studying the nuclear regions of radio-loud galaxies and quasars. Various scientific drivers for observing these objects will be discussed and the feasibility of such observations will be appraised, given the expected angular resolutions and sensitivities.

1 Introduction

On behalf of the European astronomical community let me first thank the ESO Council for last week restoring VLTI into the official future programme of ESO. We should be grateful to Oskar von der Lühe for his enthusiastic support of VLTI during the last three dark years and to Francesco Paresce and Oskar for organising this workshop.

I have been asked to talk about the prospects of studying radio-loud AGN with VLTI. Inevitably my discussion will duplicate to some extent the material covered by the excellent discussion of the previous speaker, Martin Ward, on the general topic of AGN, irrespective of their radio properties.

I shall begin by mentioning some fundamental questions about nuclear activity that are relevant to VLTI. Then I shall consider the various emission components of radio AGN that have characteristic angular scales appropriate to VLTI and give my assessment of what we might expect to learn by studying each of these, given the presently expected sensitivities. After describing some preparatory work already underway I shall conclude by presenting some exciting new results that we might hope to obtain from VLTI.

2 Some Fundamental Questions

From studies of active galaxies and radio sources that have been made over the past 40 years, it has become increasingly evident that nuclear activity plays an important role in galaxies and their evolution. However, there is still a great deal about this phenomenon that we do not fully understand. Three important questions regarding the nature of AGN are relevant to this present discussion.

1. How do the various types of observed AGN differ from one another?

In his talk Martin Ward laid considerable emphasis on orientation-based unified models as the only accepted framework in which AGN should be currently interpreted. As many of you know I am less certain of this than he is. Although it is clear that the orientation of the fundamental AGN axis (radio jet) to the observer sometimes plays an important role in determining what we observe, two other factors must also be important. First, there is evolution. Active nuclei are known to vary on a time scale of years. Nobody seriously believes that these variations are due to wobbling of the orientation axis of the AGN jet/beam. During their lifetime of $> 10^8$ years, it is would be surprising if the objects did not undergo profound changes which would affect their observed properties. Secondly, the environment in which the active nucleus occurs is known to have substantial effects on what we see. For example, powerful radio sources are never associated with active nuclei in spiral galaxies. My preferred framework is therefore one in which orientation, evolution and environment all affect the observed properties of active nuclei, but where it is not yet clear which of these three factors is the dominant one in producing the observed population. Mapping the nuclear regions with VLTI should allow interaction between the various components to be studied and it may thereby constrain the different factors involved.

2. By what process is the nuclear energy produced and collimated?

Our current picture is that an accretion disk feeds a rotating black hole which produces a collimated jet of relativistic plasma. Although there is evidence that such ideas are generally correct, the details of the processes involved are only murkily understood despite the many theoretical schemes that have been proposed. By providing information on the kinematics of the nuclear material on the sub-parsec scale, VLTI studies of the broad-line regions will offer a new tool for studying the processes involved.

3. What is the relation of nuclear activity to galaxy formation and evolution?

The space density of luminous (radio-loud and radio-quiet) AGN underwent a dramatic "delta-function" rise and fall when the Universe was about 1 to 2 Gyr old, i.e. during the first 20% of the time since the Big Bang. This quasar/AGN era (redshift 1.5 to 4) coincides with the epoch during which most galaxies were formed, and agrees with estimates for the ages of globular clusters. Nuclear activity at some level exists in almost all nearby galaxies and it is possible (in my view likely), that all galaxies went through an active nuclear stage early in their existence. Radio-loud AGN can be identified and observed out to the highest redshifts. By studying how the structure of the various nuclear components changed systematically with redshift, VLTI may provide clues to the role of nuclear activity in galaxy evolution.

3 Linear Scales, Sensitivities and "Building Blocks"

The characteristic linear scales of several components of active galaxies are matched to the resolution of VLTI and are of interest in addressing some of the above fundamental questions. The VLTI baselines (to $\sim$ 130 m) will give structural information on scales down to $\sim$ 15 mas at N-band, 3.5 mas at K-band and 0.9 mas at V-band. The corresponding maximum linear resolution is of course a function of redshift and varies from 100 pc for high-redshift objects at N-band to better than 0.1 pc for a nearby AGN at $z \sim 0.001$. Taking their characteristic linear sizes into account, three of the nuclear "building blocks" of

AGN are particularly relevant for study with VLTI. These are (i) the mid-IR warm dust component, (ii) synchrotron jets and (iii) the broad-line region.

Throughout this talk I shall rely on the sensitivity estimates from previous VLTI studies, that were sent to us by Oscar and Francesco before the workshop. I shall consider three separate sensitivity modes: (I) the "bright-source mode" using VISA, (ii) the "blind operation mode" using the hybrid array, where the interferometer parameters are sufficiently well known and controllable to permit fringe tracking and (iii) the "offset star" mode with the hybrid array, where a nearby bright star is available for adaptive optics corrections and fringe calibration. I have not gone into the assumptions on which the expected sensitivities were based, and I expect that these will be discussed by other speakers. However, I thought it useful to translate the sensitivities into units which are relevant for the AGN discussion. Table 1 shows the predicted limiting sensitivities at N-band, K-band and V-band for these three modes, expressed in terms of magnitude, millijanskys and erg cm^{-2} s^{-1}. Let us now examine the relevance of these sensitivities for relevant components of AGN.

Table 1. Predicted Sensitivities of VLTI

		Bright Source Mode (VISA)	Blind Operation Mode (VIMA)	Offset Star Mode (VIMA)
N-Band	Mag	5	10	12
	mJansky	430	4	0.7
	erg cm^{-2} s^{-1}	4×10^{-13}	4×10^{-15}	7×10^{-16}
K-Band	Mag	9	15	23
	mJansky	170	0.5	0.0004
	erg cm^{-2} s^{-1}	8×10^{-13}	2×10^{-15}	2×10^{-18}
V-Band	Mag	13	20	25
	mJansky	23	0.04	0.0004
	erg cm^{-2} s^{-1}	4×10^{-13}	7×10^{-16}	7×10^{-18}

4 The Nuclear Dust Shroud

IRAS revealed that active galaxies typically posseses an emission component that peaks in the mid-infrared close to 25 micron and also radiates in the near-IR (e.g. Miley and De Grijp 1985). This component dominates the IR emission of Seyfert 2 and narrow-line radio galaxies and is produced by dust at a temperature of 100 - 300 K. Simple arguments, using the observed luminosities and temperatures together with Stefan's Law, give a characteristic scale few tens of parsecs, between the presumed characteristic scales for the broad and narrow-line regions and for NGC 1068 a component with such a size has been measured using speckle techniques. With the VLTI baselines, such components should be well resolved in the case of low-redshift AGN

The IRAS spectrum is a composite of several different components (Golombek et al. 1989). It is a reasonable assumption that the 25 micron fluxes are dominated by mid-IR components with spectra similar to that of the "naked" peaked IR emission from Tololo 0109-38 (Miley and De Grijp 1985). Although a few bright objects such as 3C 390.3 (Miley et al. 1984) should be observable at N-band in the bright source observing mode, higher sensitivity will be needed to study a significant sample. An isolated component with a 1 Jy peak at 25 micron having a spectrum similar to that of Tololo 0109-38 would have a signal to noise of 1000 in 10 minutes at both N-band and K-band for our hybrid VLTI operated in "blind" tracking mode. IRAS observations have shown that several tens of radio-loud AGN have 25 micron flux densities in excess of 500 mJy. Hence if the blind tracking mode is made to work, VLTI should be a powerful tool for studying the nuclear dust component in AGN.

The characteristic scales of the mid-IR dust components and the relationship of these components to the broad and narrow line regions are of interest. In the case of radio-AGN, the dust may well have an influence on the propagation of the synchrotron jets. A comparison of the results of VLTI with those of VLBI will allow interaction between the jets and the nuclear medium to be probed and allow two unique and complementary sets of diagnostics be derived and compared. Studying the morphology of the nuclear dust and its relation to the other components should provide information about possible anisotropic propagation of the nuclear photons and the effects of blocking and orientation.

5 Superluminal Synchrotron Jets

One of the triumphs of very long baseline radio interferometry has been to show that radio jets are widely present in galactic nuclei on the parsec scale and that these often contain components that move outwards, at apparent speeds which are faster than the velocity of light. The apparent superluminal motion is generally interpreted as a relativistic effect, where subluminal ballistic motion is viewed close to the line of sight. The baselines of VLTI at K-band are equivalent to a transatlantic VLBI baseline at 5 GHz. In some cases (e.g. the classic radio sources Centaurus A and Cygnus A), moving jets have angular scales which are resolvable with K-band VLTI and optical VLTI would have the resolution capable of measuring motion in superluminal synchrotron jets out to a redshift of a few tenths.

Will VLTI have sufficient sensitivity to detect the IR and optical counterparts of parsec-scale radio jets? We can estimate this from a knowledge of the distribution of radio flux densities of compact components, with some assumptions about how the radio spectrum extrapolates to shorter wavelengths. From surveys made with radio VLBI at 5 GHz at VLTI resolutions, there are at least 100 accessible objects whose fringe visibilities correspond to effective flux densities in excess of 1 Jansky (Taylor et al. 1996). Taking a radio-to-optical spectral index of $\sim$ -0.35, which is midway between that of typical radio compact components ($\sim$ 0) and extended optical jets such as M87 ($\sim$ -0.7), 1 Jansky at 5

GHz extrapolates to several tens of millijanskys between 10 microns and the optical. I conclude that although perhaps a few synchrotron jets might be detectable in the bright source, there will be considerable opportunity for studying synchrotron jets if the "blind" operating mode can be made to work. The most sensitive wavelength for studying these jets will be K-band, but observations in the optical are desirable to resolve superluminal motion.

The scientific drivers for studying synchrotron jets with VLTI are many. The mechanisms which convert kinematic energy of the jets into observed synchroron emission are not well understood. The radiative lifetime of synchrotron-emitting relativistic electrons, t_{synch} depends on the frequency emitted, according to $t_{synch} \sim$ -0.5. Hence the electrons responsible for the optical/near IR emssion are typically a factor of $\sim$ 100 younger than the radio-emitting electrons. Comparison of the detailed propagation of jets in the optical and the radio can provide information on the importance of jet orientation and interaction of the jet with the ambient medium in determining the observed properties.

6 Broad-Line Regions

Because of the potential information about the kinematics of the nuclear regions, mapping broad-line emission regions (BLRs) is perhaps the most important driver for optical and infrared interferometry of AGN. The broad-line spectra of radio-loud and radio-quiet AGN are very similar. Slight differences include weaker FeII emission and broader squarer and lumpier line profiles that are exhibited by radio-loud AGN.

Although it is now generally accepted that the characteristic scales of the BLRs lie in the region of a few tenths of a parsec, I regard these canonical sizes with some scepticism. The sizes have been estimated by two indirect methods. In the first of these, classical photoionisation theory is applied. It is assumed that the BLRs are photoionized by a point "nuclear source of continuum photons. The volume needed to produce the observed broad recombination lines is derived and provides a scale size. The second more direct method also assumes nuclear photoionisation and applies the technique of reverberation mapping, using the observed variability of the line and continuum emission. Attempts are made to identify corresponding features in variability of the line and the continuum fluxes. Once a delay, τ between corresponding features has been measured, τc gives the size of the BLR. Many assumptions are made in using such methods. Of course there is spherical symmetry (invariably assumed by theorists and almost as invariably found to be wrong when morphologies can be measured). Another questionable assumption is that the nuclear continuum photons are emitted by a single source which is much smaller than the BLR (see Appendix). So although there is certainly some evidence that BLRs typically have sizes of a few tenths of parsec, these estimates could well be wrong by an order of magnitude.

Although planned VLTI baselines will probably be too short to resolve the BLRs in the infrared, the mas-scale resolution of the VLTI in the optical, (e.g. V-band) should be sufficient to resolve the BLRs of AGN out to a redshift of about

0.03, even if the scale estimates of our present pundits are correct. Although the VLTI may have difficulty in resolving the BLR proper, it is well suited for probing the transition zone between the broad and narrow line regions whose is existence is hinted strongly in many cases.

The prospects of VLTI for studying BLRs and BLR/NLR transitions are of course a considerable function of the sensitivity achievable. The bright-source mode will be useless for such observations, except perhaps for one or two extremely bright nearby AGN (e.g. NGC 4151). Several tens of radio AGN have $H\beta$ fluxes in excess of 10^{-13} erg $cm^{-2}s^{-1}$ and when observed in 1000 km/s wide channels should produce a signal to noise of about 100 in 10m at V-band, taking the sensitivity estimates for the "blind operation" hybrid mode. Advertised blind-operation sensitivities should also be sufficient to detect quasars over a large range of redshifts, including Ly α from z > 2 QSOs, for which the flux often exceeds 10^{-14} erg $cm^{-2}s^{-1}$.

In designing VLTI instruments, studies of BLRs need sufficient velocity resolution to provide limited kinematic information, but sufficient bandwidth to detect them. The broad lines have widths of several thousand km/s, so that a velocity resolution of about 500 - 1000 km/s would be optimum for their study.

Although such observations will clearly be difficult, there are numerous scientific considerations which should drive us to make great efforts to detect and study the emission line regions. It is said that "a map is worth a thousand models" and there is indeed some truth to this. Exploring whether or not the BLRs are resolved and studying their geometries will provide constraints for photoionisation models and scenarios for the propagation of synchrotron jets. Secondly, the morphologies would be carefully studied for the evidence of accretion disks which are believed to feed matter into the central black hole. Thirdly, mapping the compact line emission may help to answer a question which is at present not well understood, namely why there is the distribution of line-widths is two-peaked, corresponding to physically distinct NLRs and BLRs, each presumably with their preferred length scales.

In the case of radio AGNs, kinematic information and the relation between this and the radio morphologies would provide information about the details (energetics etc) of how superluminal syncrotron jets interact with the nuclear gas . The most fundamental driver for studying the kinematics of the BLR is the information it might provide about the enclosed mass on the scale of a few tenths of parsec, M and the resultant constraints on the presumed black hole at the centre. Taking the simplified formula, the mass, $M \sim$ R v^2/G, where v is the velocity measured at radius R. Scaling R by the resolution difference between VLTI and the HST, would imply that VLTI can measure masses a hundred times more effectively.

7 Conclusions

Table 2 summarises my conclusions about the usefulness of the various modes of VLTI for studying radio-loud AGN. Although the attainable spatial resolution is well suited for tackling several problems, it may be too small for some, notably broad-line regions, where observations in the optical will be essential.

Table 2. Number of Accessible Objects for VLTI Studies

VLTI Mode	Warm Dust (N and K)	Synchroton Jets (K , V and possibly N)	Broad-Line Region (V)
Bright Source (VISA)	Possibly few bright objects	Possibly few bright objects	Possibly few bright objects
"Blind Operation" (Hybrid)	> 50 objects	> 100 objects	> 100 objects
Offset Star (Hybrid)	>100 faint objects	> 100 faint objects	> 100 faint objects

The usefulness of VLTI for studying AGN will be determined by the sensitivity attained. The bright-source mode is just not sensitive enough. However, much can be done if the blind operating mode can be made to work at close to the advertised sensitivities. Because of the ability to combine phase referencing with adaptive optics, AGN near bright stars should be observable to the lowest flux levels and promise to be amongst the most fruitful targets. Malcolm Bremer (Leiden), together with Guy Monnet (ESO), Mark Wieringa (CSIRO) and myself have cross-correlated several catalogues of radio sources with catalogues of bright stars to define a sample of several hundred radio AGN suitable for adaptive optics and VLTI studies. Preparatory studies are underway on this sample, including identifications, spectroscopy and observations with ADONIS.

Since the nature of a workshop such as this invites speculation about the future, I shall finish with an invitation for you to imagine yourselves at a conference to be held in this same room, fourteen years from now.

8 Broad Emission Line Structure in 2428-416: Excerpts from the Third IAU Symposium on "Optical Interferometry of AGN", November 2010

"Last year, during the second symposium on OIAGN, we showed that the K-band continuum jet in 2428-416 expanded with an apparent velocity of 2.92c $\pm$ 0.07c during the period from 2005.3 until 2008.4. The positions of the near-IR components coincided approximately with those of the VLBI radio blobs. We interpreted the slight differences ($\sim 0.04''$) as due to localised particle acceleration as the jet propagates through shocks in the ISM of the nuclear regions (Beethoven et al. 2008). We concluded that our measured differences between the phase and group velocities supported the screen-type models of superluminal jets that supplanted the old purely ballistic models nearly a decade ago. Our latest VLTI and VLBI observations show that the superluminal motion of the synchrotron jet continued through September 2010, with an apparent velocity of 2.87c ± 0.12 c.

Today I will present the first information about the structure of a broad-line region, obtained from the new VLTI spectrographic mode. The observations were made between October 2nd and 5th of this year. Last night we managed to obtain the first narrow-band maps of Hα. The results I shall present are therefore very preliminary. The three separate channel maps correspond to channel separations of -2500 km/s, 0 and +2500 km/s relative to the systemic velocity of 2428-416 and the fourth map is the composite of the three channel maps. The maps were derived using the latest version of the VLTI pipeline (Mozart et al. 2006, 2007) running on a Dell P49 Intel supercomputer with Windows NT 2009. The reduction makes heavy use of model fitting techniques, so the complex details in the maps should be interpreted with caution. However, there is no doubt that the four main components are real. They are present on each of the channel maps which were processed independently by the pipeline. They are distributed within 10 deg of the well established radio axis and their total extension of 28 mas corresponds to a projected linear size of 3.2 pc at the redshift of 2428-416. The component on the left is extended perpendicular to the radio axis by 0.4″. The peak intensity changes regularly with position he three maps ∼ 2700 km/s. This component coincides in position with the peak component in the radio emission and we identify this component with an accretion disk surrounding the nucleus. On the usual simplified assumptions, we derive a mass of 2×10^8 solar masses for the mass of the black hole. In a more detailed analysis which is at present underway (Schubert et al. 2010), Franz Schubert, a graduate student of mine, has begun studying the gravitational potential in more detail using numerical-analytic quintaxial models of nuclear dynamics recently developed at Leiden (Rembrandt et al. 2009).

One of the biggest surprises for us was the location of the second and third components. These components are each relatively compact (0.2±0.1″) and are present in all three channels maps, implying that they both have velocity widths of several thousand km/s. Their positions coincide closely with those of the superluminally expanding radio/optical blobs, implying that they are clouds which are excited by the outward propagating synchrotron jet. In each case, a slight offset (0.05″) was measured and we believe this offset to be significant. A fourth broad-line component is also present within 30 deg of the radio axis, but not obviously associated with a radio component. In addition, there is evidence on all three channel maps of a highly aligned "diffuse" component extended over the 3.2 Mpc occupied by the four compact components.

These measurements finally rule out the simplified picture of the BLR that has been in vogue for the last 25 years (point nuclear source of photons which is surrounded by a spherical symmetric BLR). In the case of this radio AGN, comparison of the radio continuum (VLBI), optical continuum (VLTI) and optical line (VLTI) data leads naturally to a jet-screen model, in which a beam of gamma rays is emitted close to the nuclear black hole. The beam moves outwards at the speed of light interacting with clouds in the nuclear region. During the interaction relativistic particles are produced and the site of these particles moves outward at the speed of light with the gamma beam. The phase velocity differs from the group velocity and the observed propagation resembles that of a screen. The pattern speed is observed to exceed c. The BLR is a multi component entity, composed of the accretion disk, blobs of material ionized by the gamma rays and a possible diffuse component whose origin still needs explanation.

We have just submitted a proposal to the OPC for a Key Programme designed to monitor 2428-416 at monthly intervals in line and continuum during the next five years. Although this will be time consuming, we believe that monitoring the propagation of the jet and the development of the accretion disk to be of fundamental astrophysical importance, particularly in view of the kinematic information that it will provide about the transport of matter towards the outer regions of the black hole and the constraints on ionization models.

Finally, we note that the observations presented here provide additional arguments for the proposed VLT/VLTI upgrade that is at present being considered by the new joint ESO/ESA Council (ESGAA). As you know, this project, based on the latest developments in fibre-optics technology and mirror design. The proposed ten "new-technology" 8m telescopes and thirty 2m telescopes will form an array with a maximum baseline of 2 km, spaced around the Paranal summit, with the light transported to the correlator via phase-stabilised optical fibers. The additional resolution (factor of 15), sensitivity (up to factor of 4) and speed (factor of 50) will be sufficient to probe the kinematics of BLRs in several lines in unprecedented detail and will allow objects like 2428-416 to be studied out to a redshift of 6".

9 Acknowledgements

I thank Leonid Gurvits, Neil Jackson, Julian Krolik, Rudolf Le Poole and Huub Röttgering for help and/or discussions in preparing this talk. The observations discussed here were made using the VLTI, a facility run by the European Space and Ground-based Astronomical Agency (ESGAA) and funded by the European Union. I acknowledge support by a programme subsidy of the Netherlands Organisation for Research (NWO).

References

Golombek, D., Miley, G., Neugebauer, G. (1988): *IRAS Observations of Radio Galaxies* (Astron. J.), **95**, 16

Miley, G.K., Neugebauer, G., Soifer, B.T., Young, E., Beichman, C.A., Clegg, P.E., Habing, H.J., Harris, S., Low, F.J., and Rowan-Robinson, M. (1984): *A 25-Micron Component in 3C390.3*" (Ap. J. Letters. **278**, L83

Miley, G., De Grijp, R. (1985): *IRAS Observations of Active Galaxies - A Review* (Proc. First International IRAS Symposium, Ed. F.P. Israel, p. 41 (Reidel))

Taylor, G.B., Vermeulen, R.C., Readhead, A.C.S., Pearson, T.J., Henstock, D.R., Wilkinson, P.N. (1996 in press): *A Complete Flux Density Limited VLBI Survey of 293 Flat-Spectrum Radio Sources.* Ap.J. Suppl. In press

Observations of Extragalactic Sources Near Bright Stars

Bertrand Théodore[1] and Patrick Petitjean[2]

[1] Service d'Aéronomie du CNRS, BP 3, F-91371 Verrières-le-Buisson cedex, France
[2] Institut d'Astrophysique de Paris – CNRS, 98bis Boulevard Arago, F-75014 Paris, France

Abstract. We examine some possibilities to achieve extragalactic observations with the VLTI using close bright stars for fringe tracking and wavefront sensing.

1 Introduction

Observations using VLTI are constrained by the necessity of finding a bright compact reference object to be used for fringe tracking. In case of observation of faint objects a bright star can possibly be used instead, provided that it is close enough to lie in the so-called isoplanatic field. This however limits the number of potential targets. Indeed, the observable fraction of the sky has been previously estimated in the case of the adaptive optics system of the VLT (for which the same constraint exists) to be of the order, at most, of 5 % (Théodore et al., 1994).

2 Searching for pairs star/object

In order to identify potentially observable sources, we have cross-correlated catalogs of extragalactic objects (Veron, Burbidge,...) with the Guide Star Catalog, searching within 40 arcsec around each object for the presence of a star brighter than magnitude 13. We come to a small sample of only a few tens of couples star/object suitable for observation.

As an illustration, we present some observational results for one of our selected targets using the ADONIS adaptive optics system associated with the SHARPII camera mounted on the ESO 3.6 m telescope. We have observed the radio–source Molonglo 1411-192 that lies at 20 arcsec from a $m_v = 10$ star. This object was studied by Gopal–Krishna et al. (1992) who determined its redshift to be $z = 0.477$. They also have mapped the radio emission, revealing the presence of two radio lobes which geometrical center corresponds (within 2 arcsec) to the position of the galaxy observed in the visible. Figure 1 shows our image in the K' band of this object. We have detected a faint object ($m_{K'} = 18.3$), pointed by an arrow on Fig. 1, corresponding precisely to the maximum of the nothermost lobe (see Fig. 1 in Gopal–Krishna et al.).

A bright star is also required when the observed object is too extended. This problem arises when studying the very central part of the galaxies by observation

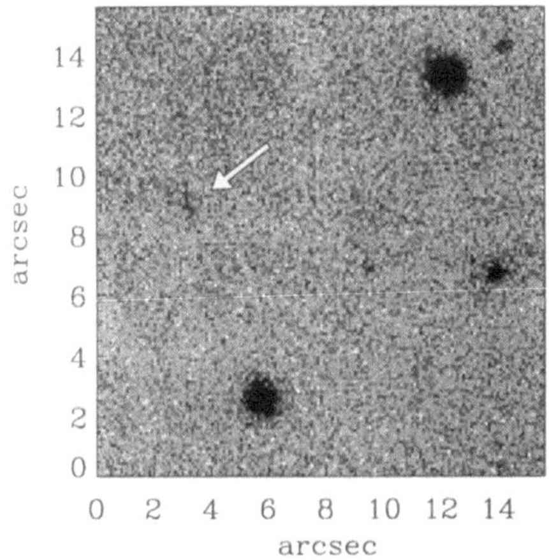

Fig. 1. K' band image of the radio-source 1411-192. North is up, East is right. The galaxy identified by Gopal-Krishna et al. (1992) to be the radio-source is at the bottom. The arrow locates the maximum of the radio emission, which coincides with a faint object.

of the brightness profile. Indeed, obtaining a clear picture of the central light distribution would lead to a better understanding of the central mass distribution. This has been done by Lauer et al. who used the HST to study the inner core of M32. But observations on a larger sample are required. We thus have selected galaxies from the ESO catalog on which is superposed a bright star that could be used for fringe tracking.

3 Conclusion

The need of a bright star for observing faint or extended object reduces the number of potential targets. We have compiled a list of possible observations by cross-correlating the catalogs. Its smallness stresses out the necessity for a systematic search of other objects by doing preliminary observations, particularly in the galactic plane where the probability to find a pair star/object is high.

References

Gopal-Krishna, Giraud E., Melnick J., Steppe H (1992): A Powerful, Ultra-steep Spectrum Radio Galaxy Having Low-excitation Emission Line Spectrum. A&A **254**, 42–48

Lauer et al. (1992): Planetary Camera Observations of the Central Parsec of M32. AJ **104**, 552–562

Théodore B., Petitjean P., Hubin N. (1994): Scientific Capabilities of the VLT Adaptive Optics System. The Messenger **77**, 20–27

AGN Models and Fringe Visibilities

Fabien Malbet[1], Patrick Petitjean[2,3], and Gilles Henri[1]

[1] Observatoire de Grenoble - CNRS & Univ. Joseph Fourier, UMR 5571
414 rue de la Piscine, BP 53, F-38041 Grenoble cedex 9, France
[2] Institut d'Astrophysique de Paris - CNRS UPR 341
98bis, Boulevard Arago, F-75014 Paris, France
[3] DAEC, CNRS URA 173, Observatoire de Paris–Meudon
F-92195 Meudon, France

Abstract. The ultimate goal of VLTI (VISA or VIMA) is to produce very high resolution images each of which will need good UV coverage, thus a large number of baselines and a large amount of observing time. Yet a small number of baselines is enough to obtain visibility curves that can be used to give decisive constrains on models (see e.g. Stee et al. 1995). We illustrate this point by computing fringe visibilities for very simple models of AGNs. The purpose here is not to discuss AGN models but to show how fringe visibilities can help to derive fundamental information on the surrounding of the central engine of AGNs at scales of a few mas.

1 Models of AGNs

We consider that *the* typical AGN is located at a distance of 18 Mpc (the distance to NGC 1068) and can be described as a central engine of radius $R = 10^{-3}$ pc and total luminosity $L = 10^{45}$ erg.s^{-1}, surrounded by a dusty disk. The dust is heated by the flux from the center and reradiates the energy at larger wavelength. The spectral energy distribution (SED) of the AGN emission is computed and the parameters are adjusted to reproduce the observed data. The latter are taken from NCC 1068 (Rieke & Low 1975; Roche et al. 1991).
We use two models:

Model 1: is an inclined optically thick flaring disk with inner and outer radius r_i, r_o. We compute the power P received by a unit surface of the disk as the energy radiated by the central engine in the solid angle subtended by the surface. The temperature is taken as the black-body temperature $T = (P/\sigma)^{0.25}$ where σ is the Stephan constant. The thickness of the disk is $H(r) = h(r/r_\mathrm{o})^z$. The fit of the SED gives $r_\mathrm{i} = 0.26$ pc, $r_\mathrm{o} = 125$ pc, $z = 0.6$ and $h = 1.22 r_\mathrm{o}$.

Model 2: is an inclined overall optically thin disk made up with small optically thick globules. The temperature at any radius is calculated as for Model 1. The optical depth perpendicular to the disk is $\tau(r) = \tau_\mathrm{o}(r/r_\mathrm{o})^z$. The fit of the SED gives $r_\mathrm{i} = 2$ pc, $r_\mathrm{o} = 300$ pc, $z = -0.05$ and $\tau_\mathrm{o} 0.026$.
These models are certainly not realistic but are rather extreme models *which both fit the SED* (see Figs. 1 and 2). It must be noticed also that the large outer radius is a consequence of our constraining the models to fit the SED up to 100 μm.

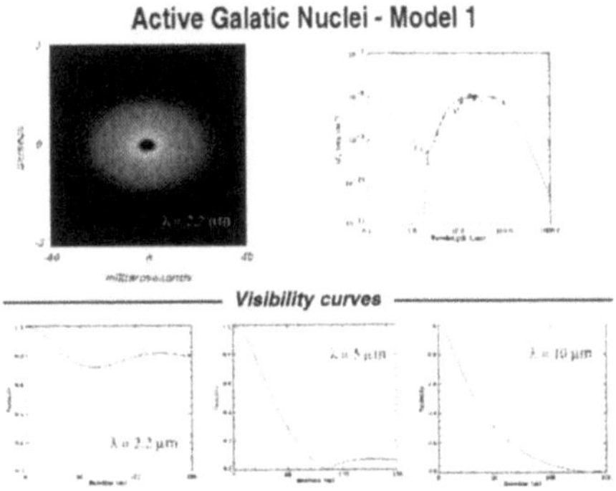

Fig. 1. Results from the AGN Model 1: inclined optically thick flaring disk of dust. Upper left panel: simulated image at 2.2 μm. Upper right panel: SED. Lower panels: visibility curves at 2.2, 5 and 10 μm.

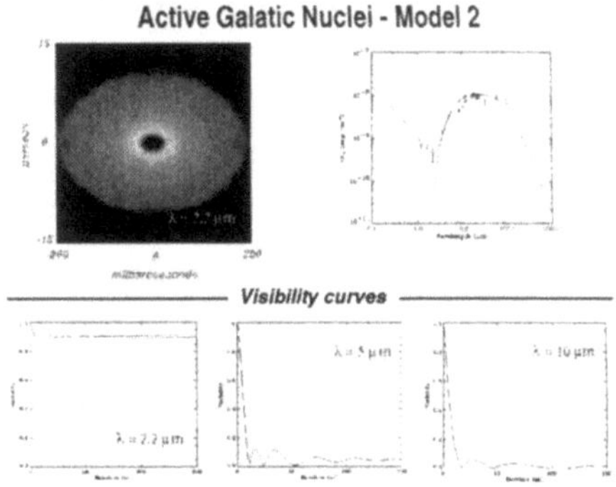

Fig. 2. Results from the AGN Model 2: inclined overall optically thin disk made up with small optically thick globules. Upper left panel: simulated image at 2.2 μm. Upper right panel: SED. Lower panels: visibility curves at 2.2, 5 and 10 μm.

2 Results

Figs. 1 and 2 show the simulated image at 2.2 μm, the SED and visibility curves at 2.2, 5 and 10 μm for models 1 and 2 respectively. Solid and dashed lines are for the major and minor axes respectively. It can be seen that the visibilities are definitively different. The typical resolution of these observations is

$$\left(\frac{R}{1\ \mathrm{mas}}\right) \sim 6\ \left(\frac{\lambda}{5\ \mu\mathrm{m}}\right)\ \left(\frac{B}{40\ \mathrm{m}}\right)^{-1} \tag{1}$$

(see Malbet & Bertout 1995) with λ the wavelength and B the baseline. At a distance of 18 Mpc, 1 pc corresponds to about 10 mas. It can be seen from Fig. 1 that it is possible to probe the structure of the innermost parsec of such an AGN providing a direct test for dust emission models.

References

Malbet F., Bertout C. 1995, A&AS 113, 369.
Rieke G.H., Low F.J. 1975, ApJ 199, L13.
Roche P.F., Aitken D.K., Smith C.H., Ward M.J. 1991, MNRAS 248, 606.
Stee P., de Araújo F.X., Vakili F. et al. 1995, A&A 300, 219.

AGNs at VLTI Scales

G. Mark Voit

Space Telescope Science Institute, Balitmore, MD 21218, USA

Abstract. The VLTI will be able to resolve 20 mas structure at 10 μm and 4 mas structure at 2 μm. Its resolution will be well matched to the critical regions in active galaxies where the violent environment surrounding the nucleus meets the interstellar medium of the host galaxy. With the VLTI we will be able to address how accreting matter flows towards the nucleus and whether irradiation of that matter produces the infrared continuum emission from quasars. In the more distant future, nulling interferometry might provide information about the nature of the broad line region and other structures just below the resolution limit.

1 Active Galaxies: Basics

Active galactic nuclei (AGNs) exhibit a wide variety of manifestations, including quasars, Seyfert 1 galaxies, Seyfert 2 galaxies, LINERS, and BL Lac objects, but their underlying architecture is probably quite similar. The standard model for AGNs posits a supermassive black hole girdled by an accretion disk that feeds it gas at rates ranging up to $10\,M_\odot\,\mathrm{yr}^{-1}$. These accretion disks, which radiate away about 10% of the incoming gravitational potential energy, can emit luminosities up to $10^{47}\,\mathrm{erg\,s}^{-1}$, mostly in the form of UV and soft X-ray photons. A minority of AGNs also propel radio jets that move at relativistic speeds, presumably along the rotation axis of the black hole. For a good introduction to AGNs, see Blandford *et al.* (1990).

The prevailing opinion among investigators of AGNs is that much of their seeming diversity can be traced to orientation effects. These so-called "unified models" of active galaxies suggest that a thick torus of molecular clouds orbits the nucleus between the narrow-line and broad-line regions (Krolik & Begelman 1986; 1988). When seen on-axis, the torus reveals the broad-line region and accretion disk at its center. When seen edge-on, the torus hides the innermost part of the AGN, and only the narrow-line region can be seen directly. In some models, cloud-cloud collisions within the torus facilitate accretion of the host galaxy's interstellar medium into the nucleus.

The newest line of evidence favoring the standard picture of AGNs has come from interferometry of H_2O masers in NGC 4258, a nearby spiral galaxy with a LINER nucleus (Greenhill *et al.* 1995; Miyoshi *et al.* 1995). A thin, warped disk of masing molecular clouds $0.13 - 0.25\,\mathrm{pc}$ (4 – 8 mas) in radius orbits a $3.6 \times 10^7\,M_\odot$ object at the center of NGC 4258 in perfect circular Keplerian rotation. The disk axis runs parallel to a large-scale Hα jet directed almost perpendicular to the galaxy's rotation axis. Polarized broad lines and UV continuum emission from NGC 4258 indicate that the active nucleus itself is shielded from us (Wilkes *et al.* 1995). Infrared interferometry of this galaxy, and numerous others, with the VLTI will be ideal for studying the mysterious interface between the accretion-dominated AGN environment and the galactic disk of the host galaxy.

2 Infrared Emission from AGNs

To within about an order of magnitude, the broad-band spectra of active galaxies are flat in νF_ν from 100 μm through the X-ray band. These spectra typically peak in the UV, drop to a local minimum at around 1 μm, and rise again to a local maximum somewhere in the mid- to far-IR (Sanders *et al.* 1989). At wavelengths longer than 100 μm, AGN fluxes usually decline sharply, suggesting that heated dust grains emit the IR continuum from most active galaxies. The virtually universal minimum at around 1 μm adds weight to this hypothesis because dust grains hot enough to emit efficiently shortward of 2 μm quickly sublimate (Sanders *et al.* 1989; Phinney 1989). Across these two orders of magnitude in wavelength, AGN fluxes frequently remain constant in νF_ν to within factors of a few, implying that the emitters, if thermal, must span a temperature range of $30 - 1500$ K.

The limited spatial resolution currently afforded by IR instruments is insufficient for testing the various hypotheses for the origin of the IR emission:

Warped Disk. Illumination of a galaxy by an embedded AGN heats its interstellar dust. To reproduce the broad-band spectral characteristics of AGNs, dust at each order of magnitude in distance must intercept roughly equal amounts of UV flux, totaling $\sim$ 30% of the UV output from the central engine (Sanders *et al.* 1989).

Torus + Galaxy. A well-defined obscuring torus, as depicted in unified models, sits at a particular distance from the nucleus, so the temperature range of the dust it contains is limited. In between the broad- and narrow-line regions, dust temperatures are $\sim 60 - 600$ K, leading to a peak flux in the mid-IR. If the torus shields the rest of the host galaxy from direct AGN illumination, the galaxy itself must supply the far-IR continuum (Pier & Krolik 1992; 1993).

Starburst. Certain heterodox models for active galaxies claim that there is no central engine (Terlevich & Melnick 1985). Instead, they argue that all the phenomena we associate with AGNs, except for jets, which are rare, can be explained as immense, centrally concentrated starbursts. In this case, the temperature of the IR continuum would not vary greatly with position, and the 2 μm continuum emission from quasars would spread over much more than a few parsecs.

Synchrotron. The IR continuua of some active galaxies, especially BL Lac objects, could be entirely non-thermal, with a far-IR/sub-mm decline owing to synchrotron self-absorption. Non-thermal far-IR synchrotron emission would not need to be extended at all, even on interferometric scales.

The VLTI, which will be able to resolve thermally emitting mid-IR regions in the brightest AGNs, will easily eliminate many of these hypotheses. If the mid-IR emission indeed comes from AGN-heated dust, VLTI will be able to tell us much more.

3 Resolution & Brightness

The VLTI baselines will extend to $\gtrsim$ 100 m. At 10 μm, the nominal resolving power will be $\sim$ 20 mas. Our ability to resolve a thermally emitting region using this beamsize depends primarily upon the optical/UV flux of the illuminating source. The equilibrium dust temperature, T_d, of optically thin dust at some angular distance δ from an optical/UV source of flux $F_{\rm obs}$ is determined by the flux the dust sees, $F_d = \delta^{-2} F_{\rm obs}$. Since dust grains generally absorb UV radiation much more efficiently than they absorb infrared radiation, a sort of "dust greenhouse effect" elevates T_d well above the effective temperature of the radiation field. Figure 1 illustrates how δ varies with $F_{\rm obs}$, given in terms of V magnitudes, for several interesting dust temperatures. Since emission from dust at $\sim$ 300 K peaks at 10 μm, instruments like those described above can resolve thermally emitting 10 μm regions around objects brighter than $V = 15$. Fainter 10 μm regions from somewhat cooler dust will be even more extended. Numerous nearby active galaxies satisfy this brightness criterion. The CfA survey identified tens of AGNs with $V < 14.5$ (Huchra & Burg 1992), including numerous Seyfert galaxies and LINERS, a BL Lac object, and one luminous quasar, 3C 273 at $z = 0.173$. The fluxes of these objects are typically greater than a few hundred mJy at 10 μm and a few tens of mJy at 2 μm, potentially bright enough for on-source fringe tracking.

Interferometric imaging of AGNs with $V \lesssim 15$ will enable the following kinds of studies:

2 μm Compactness. If AGNs are merely extreme starbursts, their 2 μm emission should be entirely resolved.

10 μm Compactness. If the IR continuum from AGNs is primarily non-thermal, their 10 μm emission will be unresolved.

Warped Disk Morphology. The signature of a thermally emitting warped disk will be an elongated IR image whose position angle changes systematically with wavelength.

Accretion-Disk / Galaxy-Disk Relationships. IR interferometers with 100 meter baselines will be tuned to spatial frequencies in between the VLBI/VLBA observations of masing disks in AGN and diffraction-limited IR imaging with single-element telescopes.

UV Anisotropy. Anisotropic emission of UV flux will systematically elevate T_d along certain position angles.

Torus Size/Morphology. IR interferometers could conceiveably image the putative obscuring tori around AGNs, whose sizes are currently not well constrained.

Starburst Knots. Regions of intense star formation near AGNs will show up as bright spots at both 2 μm and 10 μm.

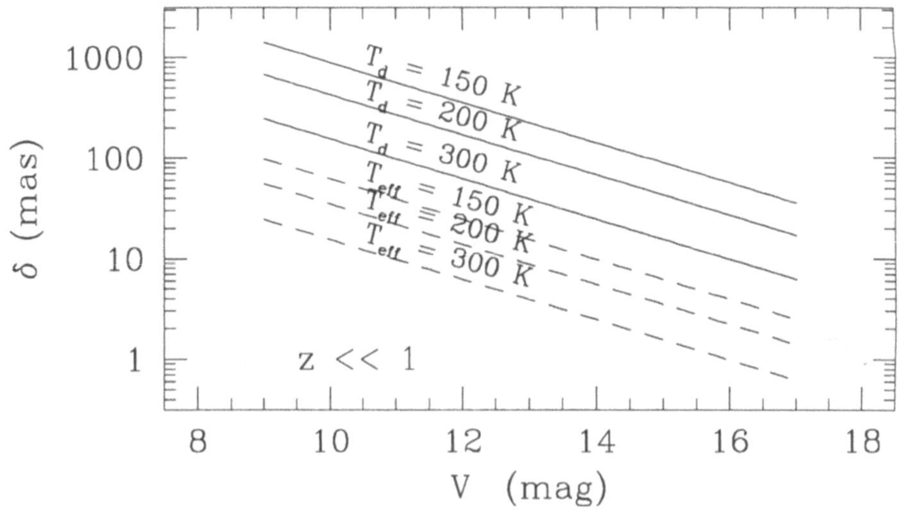

Fig. 1. Angular sizes of thermally emitting regions. The solid lines give the angular sizes, δ, of regions with typical dust temperature T_d around objects with varying V magnitudes. The dashed lines give the effective temperature of the radiation fields incident on these regions

Many higher-redshift quasars are slightly fainter than $V = 15$. The Palomar-Green (PG) survey, which searched $\sim$ 25% of the sky for UV-bright quasars, found over 60 quasars more distant than 3C 273 brighter than the survey's magnitude limit of $B = 16.16$, but only one with $B < 15$ (Schmidt & Green 1983). The IRAS 12 μm all-sky survey detected a number of these at the $\sim$ 100 mJy level (Sanders *et al.* 1989). At 10 μm, a 100 m interferometer can resolve the $T_d \approx 200$ K region around a $V \approx 15.5$ object when $z \ll 1$.

4 High-z Active Galaxies

At high redshifts, the relationship between resolvability and brightness is less straightforward than at low redshifts. The angular-size/distance relation in an expanding universe magnifies the circumnuclear regions, so that dust at a particular T_d subtends a much larger angle than in low-redshift quasars of comparable brightness. However, the thermal radiation from this dust redshifts to longer, less easily resolvable wavelengths. Figure 2 outlines the impact of redshift on resolvability for quasars of magnitude $R = 17 - 19$. The tradeoff between magnification and redshift turns out to be roughly even. Note, though, that at $z \approx 4$, the region corresponding to $T_d \approx 500$ K is several ten of mas across, potentially making it easier to resolve obscuring tori in high-z quasars than in all but the brightest low-redshift Seyfert galaxies.

Interferometry of bright high-z quasars at $\sim 2\,\mu$m images them in their rest-frame optical band. Because the fringes created by these objects will be too faint

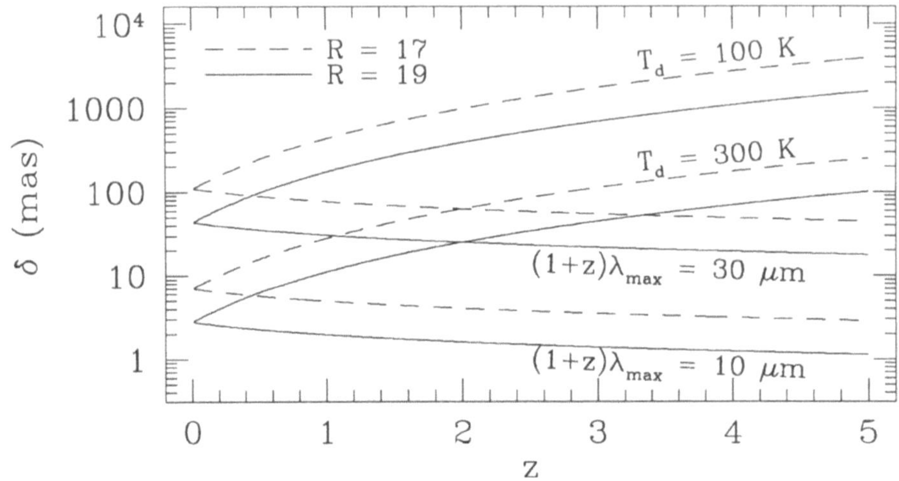

Fig. 2. Angular sizes of thermally emitting regions at high redshift. The lines indicate the sizes of thermally emitting regions around quasars of $R = 17$ and 19. At constant T_d, the expansion of the universe magnifies the apparent sizes but shifts the radiation to longer wavelengths. When the position of the observed peak in the thermal spectrum is held constant, the apparent size of the thermally emitting region drops gradually with increasing z.

to track with the VLTI, we will be able to observe only distant quasars that appear near bright stars. While we might hope to see what the host galaxies of high-redshift quasars look like in detail, reflection nebulosity will probably overwhelm the signal from starlight. Thermal emission models for AGN IR continuua require dust grains to intercept a sizeable fraction of the optical-UV continuum, and $\sim 10\%$ could easily be scattered. The reflected luminosity alone can exceed the host galaxy's total output and will be concentrated around the nucleus. Near-IR images of quasar reflection nebulae, in conjunction with Balmer-line mapping, might provide very useful information about the distribution of gas and ionizing radiation in high-z quasars as close as $10 - 20$ pc to the central engine.

5 Nulling & AGN Tomography

The broad-line regions of AGNs subtend angles of 1 mas or less, except perhaps in the very nearest Seyfert galaxies. Imaging of these regions will have to await interferometry on an even grander scale. Nevertheless, it might be possible to extract useful information about broad-line region structure by taking advantage of the special nulling designs invented for planet-finding interferometers (Bracewell 1978; Shao & Colavita 1992).

The most basic nulling interferometers consist of two identical telescopes and a beam combiner that inverts the phase of one telescope's signal before adding it to the other. If all the optical pathlengths are equalized through the intervention of active delay lines, the signal from an object in the symmetry plane between the two telescopes cancels at all wavelengths. As the object moves to one side or the other of the symmetry plane, the interferometer's throughput increases until it reaches a maximum at an angle similar to the beamsize.

Figure 3 shows how the transmission curves of the various nulls in a 4-element nulling interferometer vary with distance from the symmetry plane. Such instruments will detect significant fluxes when pointed at bright targets a few times smaller than the beamsize, such as quasar broad-line regions. Even though we will not be able to image these regions directly, well-planned observing strategies might still extract useful information about the velocity fields, sizes, and shapes of these regions.

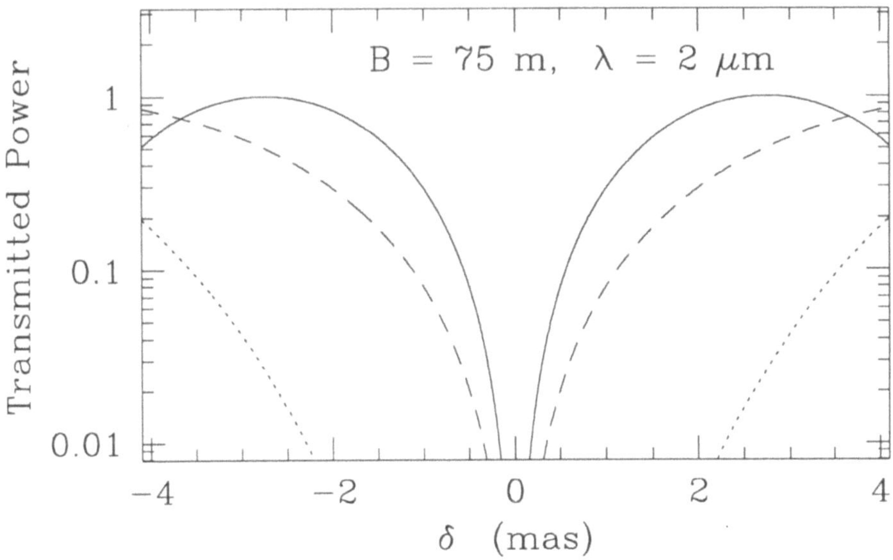

Fig. 3. Null shape in a 4-element planet-finding interferometer. The signals from four telescopes along a 75-meter baseline can be combined in various ways to produce nulls of different shapes. This figure shows how the 2 μm transmission through different combinations of telescope varies with the angular distance δ from the symmetry plane. Note the signficant amounts of transmission at angles smaller than the diffraction limit.

At $z \gtrsim 2$, the broad Balmer lines from quasars shift to $\sim 2\,\mu$m. The effective beamsize of a 100 meter nulling interferometer operating at $2\,\mu$m is ~ 4 mas, but the wings of the null transmit more line emission from the outer parts of the broad-line region than from the inner parts. A single emission-line observation with the nucleus centered on the null conveys information about the convolution of the null's shape with the projected velocity field. Scanning the null across the source or observing through nulls of different sizes and shapes, can therefore be used to deconvolve the circumnuclear velocity field, at least partially. These kinds of tomographic methods could potentially be used to weigh black holes in distant quasars, if the null characterization and pointing accuracy are sufficiently precise.

Figure 4 illustrates a simple example of how various telescope configurations respond to a hypothetical broad-line region. The underlying emission model assumes a line-emitting region with a velocity dispersion $\propto r^{-1/2}$ and a surface brightness $\propto r^{-2}$, truncated at 1/4 of the interferometer's beamsize. The assumed instrument is the kind of 4-element interferometer whose nulls are depicted in Figure 3. The transmitted flux drops rapidly with null width, indicating the size of the BLR, and lines measured through the nulls are markedly thinner than those measured by a single telescope, showing that $v(r)$ decreases with distance.

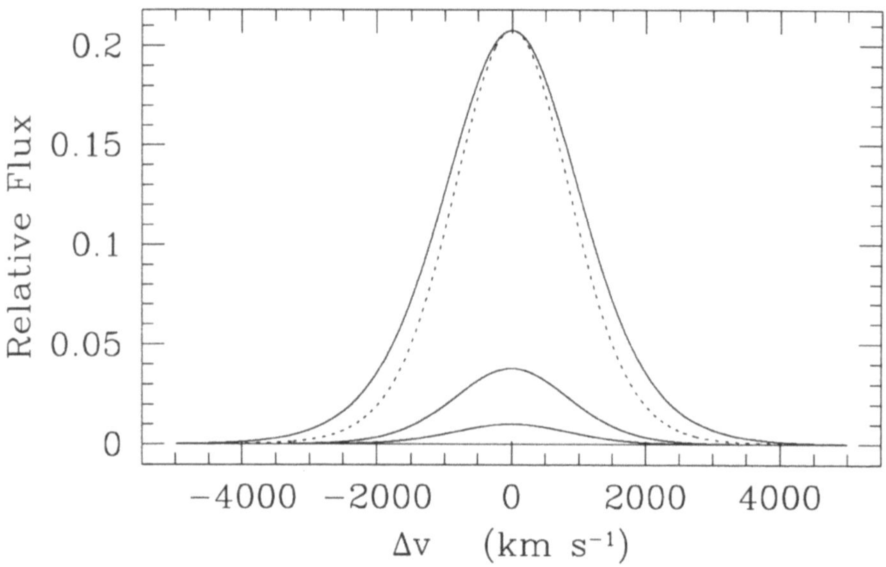

Fig. 4. Simulated broad emission line seen through different nulls. This figure shows the the transmission of light from a hypothetical broad-line region through the different nulls in Figure 3. The top line gives the flux a single-element telescope would measure. Each successive solid line below this one gives the line emission detected through a progressively wider null. The dotted line shows the line profile through the narrowest null, renormalized to have the same peak flux as the unmasked line, in order to make its narrower width more obvious.

References

Blandford, R. D., Netzer, H. and Woltjer, L. (1991): *Active Galactic Nuclei*, ed. T. Courvoisier and M. Mayor (Springer-Verlag, Berlin)

Greenhill, L., Jiang, D. R., Moran, J. M., Reid, M. J., Lo, K. Y. and Claussen, M. J. (1995): Detection of a Subparsec Diameter Disk in the Nucleus of NGC 4258. *Astrophys. J.* **440**, 619–627

Huchra, J. and Burg, R. (1992): The Spatial Distribution of Active Galactic Nuclei. I – The Density of Seyfert Galaxies and LINERS. *Astroph. J.* **393**, 90–97

Krolik, J. H. and Begelman, M. C.: (1986): An X-ray Heated Wind in NGC 1068. *Astroph. J.* **308**, L55–L58

Krolik, J. H. and Begelman, M. C. (1988): Molecular Tori in Seyfert Galaxies – Feeding the Monster and Hiding It. *Astroph. J.* **329**, 702–711

Miyoshi, M., Moran, J., Herrnstein, J., Greenhill, L., Nakai, N., Diamond, P. and Inoue, M. (1995): Evidence for a black hole from high rotation velocities in a sub-parsec region of NGC 4258. *Nature* **373**, 127–129

Phinney, E. S. (1989): Dusty Disks and the Infrared Emission from AGN. in *Theory of Accretion Disks*, ed. F. Meyer, W. Duschl, J. Frank, & E. Meyer-Hofmeister (Kluwer, Dordrecht), 457–470

Pier, E. A. and Krolik, J. H. (1992): Infrared Spectra of Obscuring Dust Tori around Active Galactic Nuclei. II. Calculational Method and Basic Trends. *Astroph. J.* **401**, 99–109

Pier, E. A. and Krolik, J. H. (1993): Infrared Spectra of Obscuring Dust Tori around Active Galactic Nuclei. II. Comparison with Observations. *Astroph. J.* **418**, 673–686

Sanders, D. B. *et al.* (1989): Continuum Energy Distributions of Quasars: Shapes and Origins. *Astroph. J.* **347**, 29–51

Schmidt, M. and Green, R. F. (1983): Quasar Evolution Derived from the Palomar Bright Quasar Survey. *Astroph. J.* **269**, 352–374

Shao, M. and Colavita, M. M. (1992): Long-Baseline Optical and Infrared Stellar Interferometry. *ARA&A* **30**, 457–498

Terlevich, R. and Melnick, J. (1985): Warmers: the missing link between starburst and Seyfert galaxies. *Mon. Not. Roy. Astron. Soc.* **213**, 841–856

Wilkes, B. J. *et al.* (1995): Optical Detection of the Hidden Nuclear Engine in NGC 4258. *Astroph. J.* **455**, L13–L16

The AGN Dance: Astrometric Variability as the Result of Supernova Explosions in Nuclear Starbursts

Jorge Melnick[1], Itziar Aretxaga[2], and Roberto Terlevich[2]

[1] European Southern Observatory, La Silla, Chile
[2] Royal Greenwich Observatory, Cambridge, UK

Abstract. The starburst model of AGN predicts the supernova rates in the nuclear star clusters of low luminosity AGN to be less than one per year. The compact remnants of these explosions evolve in regions of very high density and thus radiate most of the energy in less than one year. Statistically, therefore, no more than one compact supernova remnant is expected to be active at any given time. The broad line region in the starburst model is produced in the compact supernova remnants so, for low luminosity objects, the BLR will typically consist of only one compact remnant.

Thus, a strong prediction of the model is that the BLR of low luminosity AGN should move at random with a typical step which depends on the luminosity of the nucleus. We call this motion the AGN dance.

1 AGN Variability

One of the most powerful probes into the inner realms of AGN is their photometric variability which is observed to occur over a wide range of time scales and luminosities ranging from fast, low amplitude, flickering in X-rays on scales of hours, to flares of more than one magnitude lasting months to years in the optical. Variability poses one of the most stringent tests to any theory of the nuclear activity phenomenon and there is no single model that can account for all aspects of AGN variability.

Since the bulk of the energy radiated by AGN appears to be emitted in the blue/UV part of the spectrum [1], understanding the optical/UV variability appears to be the first and possibly the most important step towards understanding the energy generating mechanism in AGN.

1.1 Variability at Optical Wavelengths

Lyutyi and his group (Lyutyi et al., 1984 and references therein) were the first to undertake massive photometric monitoring programmes of several active galaxies aimed at characterizing their optical variability. Lyutyi found that AGN variability at optical wavelengths can be described as flares characterized by a rapid rise in luminosity followed by a gradual decay lasting months to years modulated by

[1] This may not be the case for BL-Lacs where the geometry and the relativistic boosting factors are still not well understood

smooth similar amplitude variation with time scales of years. This results have been substantially confirmed by international monitoring campaigns of a few selected objects with excellent time resolution, notably the studies of NGC4151 by Penston et al., (1981) and of NGC 5548 by Korista et al., (1995, and references therein). Based on these studies, the main aspects of the photometric variability of AGN can be characterized as follows:

- Most AGN are variable. For radio-quiet AGN, the amplitude of variation correlates with luminosity in such a way that the least luminous nuclei are the most variable (Pica & Smith 1993, Hook et al., 1994, Cristiani et al., 1996)
- Variability appears to occur in discrete events or *flares* characterized by a rapid rise in luminosity followed by a slow decay, with typical time scales of weeks to months. (Lyutyi, 1963; Aretxaga 1993 and references therein).
- The light curves of the flares appear to be similar from event to event and from galaxy to galaxy (Cid-Fernandez, Terlevich and Aretxaga, 1996).

In addition to photometric monitoring, some of these international campaigns were also done spectroscopically so there is a wealth of fascinating information on the spectral variability of a few AGN. The results of these observations can be summarized as follows:

- Generally (but not always) there is a delay between the variation of the continuum and the lines, the latter *lagging* the continuum by 1-2 weeks (see Peterson 1993 for a review).
- The lines change intensity and shape. In some cases broad components develop in the permitted lines which radically change the classification of an object from Narrow Line/Seyfert 2 during quiescence, to Broad Line/Seyfert 1 during outburst (eg. Gregory et al., 1991, see Aretxaga & Terlevich 1994b for a compilation).
- The emission-line spectrum of the flares (after subtraction of the low state component) show a striking resemblance to the spectrum of peculiar type 2 supernovae (Terlevich and Melnick, 1988; Filippenko, 1989; Stathakis and Sadler, 1991 and references therein).
- The total energy released in each flare is close to 10^{51} ergs (Alloin et al., 1986).

1.2 AGN Variability in the *Starburst* Model

The properties of AGN variability at optical wavelengths described above can be explained remarkably well by the so-called *Starburst model* that postulates nuclear activity to be the direct consequence of violent star formation activity in the nuclear regions of galaxies (Terlevich et al., 1987; 1992, 1995, Aretxaga & Terlevich 1994a, 1994b, Aretxaga et al., 1996). The model explains variability as due to supernova explosions in the high density environment of the nuclear regions of galaxies, and their subsequent evolution as (compact) supernova remnants (Terlevich et al., 1992).

1.3 The AGN Dance

The *Starburst* model tells us that the Broad Line Region (BLR) is formed by compact supernova remnants (cSNRs; Terlevich et al., 1992). The supernova rates (ν_{SN}) predicted by the model for low luminosity AGN (e.g. those with $M_B > -21$) are $\nu_{SN} \leq 1\ yr^{-1}$ (Aretxaga & Terlevich 1994a). Since cSNR radiate their energy in about one year, for these objects the BLR will typically consist of only one cSNR and, therefore, as the cluster evolves and new stars explode as supernovae, the nucleus should be observed to move around the center of the cluster. We will refer to this predicted *astrometric* variability of low luminosity AGN as *the AGN dance.*

In this contribution we examine the possibility of observing the *dance* in nearby low luminosity AGN with the VLT(i).

2 Observing the AGN Dance

In order to calculate the mean distance between consecutive supernovae in a starburst cluster we will make the following simplifying assumptions:

1. Starburst clusters are spherical with a mass distribution characterized by a de Vaucouleurs law.
2. The mass to (blue) light ratio does not change with radius
3. The IMF is a Salpeter power-law of slope $\alpha = -2.35$ in the mass range $0.3 < M/M_\odot < 50$.

The probability of distribution of SN explosions is equal to the surface brightness distribution normalized to the total light,

$$P(r)dr = \frac{1}{3.61 R_e^2} 10^{-3.33[(r/R_e)^{1/4}-1]} r dr$$

where R_e is the effective radius of the cluster, this is so because the supernova rate per unit blue luminosity is a constant regardless of age or IMF of the cluster (Aretxaga & Terlevich 1994a). Clearly half of the supernovae explode within $0.5R_e$. The mean separation between consecutive explosions can be computed using the Monte Carlo method with this probability distribution. The resulting distribution of separations from one million explosions is shown in Figure 1

The model predicts that the optical nuclei of low luminosities should dance with a typical step close to the effective radius of the mass distribution R_e. This radius can be calculated from the relation (Poveda, 1957; Young, 1976),

$$M_T = 2073 R_e \sigma^2$$

where σ is the projected velocity dispersion of the cluster stars and M_T is the total mass (obtained from the cluster luminosity via the IMF, Aretxaga & Terlevich 1994a).

The light curve simulations of Aretxaga (1993) show that the minimum superposition (in time) between consecutive flares, and the maximum luminosity

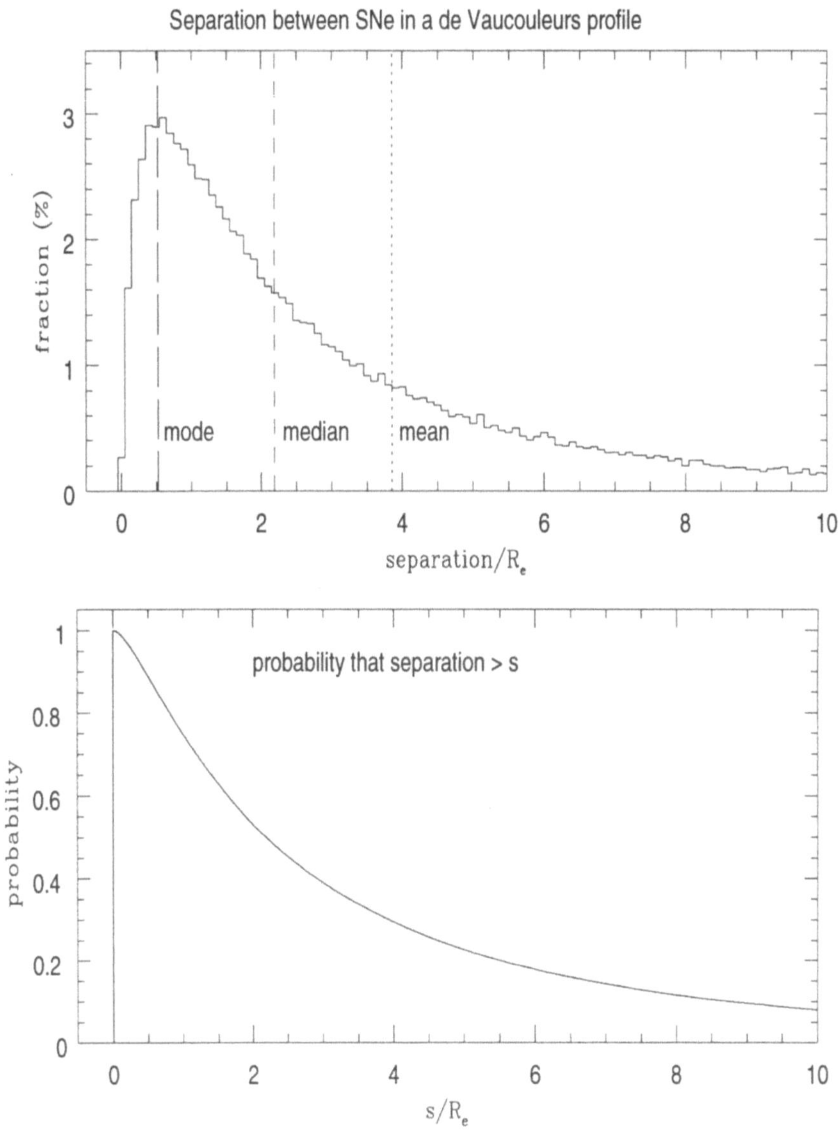

Fig. 1. *Top. Distribution of separations (in units of the effective radius R_e) of consecutive supernova explosions for a de Vaucouleurs mass distribution. Bottom. Cumulative distribution of separations for consecutive explosions.*

contrast of individual flares against the cluster occurs for supernova rates in the range 0.2 $<\nu_{SN}<$ 0.6 which correspond to blue luminosities between -19 and -21. These values are almost independent of the IMF parameters because the the blue luminosity is dominated by massive stars.

Table 1 gives the parameters for a few representative galaxies in the southern hemisphere selected from the list of Winkler et al. (1992) searching for southern objects of nuclear blue luminosities between -21 and -19 and redshifts lower than cz<20,000 km s^{-1}

Table 1. Target galaxies

Galaxy	Distance (Mpc for H_0=75)	M_B	ν_{SN} (yr^{-1})	M_T ($M_\odot$)	R_e (pc)	θ_e (mas)
H0307-730	111	-20.7	0.4	1.3×10^9	16	30
NGC3783	37.3	-20.1	0.2	6.5×10^8	10	55
NGC4593	34.8	-19	0.1	1.0×10^8	1.5	9
E141-G55	147	-20.7	0.4	1.3×10^9	16	22

The table gives the distance to each galaxy calculated from the observed redshifts assuming H_0=75 km s^{-1} Mpc^{-1}, the absolute blue magnitude of the nucleus (M_B), the implied supernova rates (ν_{SN}), the total mass (M_T) calculated using $L_B/M_T = 15 L_\odot/M_\odot$ (Aretxaga & Terlevich 1994a), the effective radius R_e of the mass profile (calculated assuming a central velocity dispersion of 200 km s^{-1}which appears to be typical of objects in this range of luminosities) and the corresponding angular radii in milli-arcseconds (θ_e).

The angular radii are well within the range of the VLTi and in fact could also be observed with a 8m unit telescope in the diffraction limited mode (active optics). However, in order to define an optimal observing strategy it is important to evaluate the uncertainties in these calculations.

2.1 Uncertainties ...

There are several important unknowns that may have led us to overestimate the effective radii by rather large factors.

The first one is the velocity dispersion. We have assumed $\sigma = 200$ km s^{-1} which corresponds to the observed velocity dispersion of NGC4151 in the central $2''$. This must be considered as an upper limit since the expected effective radius of the nuclear cluster in this galaxy is only about 30 mas.

A second fundamental unknown is the IMF of the central cluster. The mass-to-light ratios (used to compute the total mass) may vary by factors of 2–3 depending on the lower mass limit. The variation may be even larger for IMF slopes steeper than Salpeter, but a factor of 2 is a reasonable estimate of the maximum error for ages in the type II supernovae phase ($> 2 \times 10^7$yrs).

A third perhaps even more fundamental uncertainty is mass segregation. HST observations of 30 Doradus (Malamuth and Heap 1994) indicate that the IMF in the core of the prototype starburst 30 Doradus appears to be significantly flatter than the IMF outside the core. If confirmed, these observations (taken before the refurbishement of HST) indicate that in starburst clusters the effective radius of the supernova progenitors may be significantly smaller than that of the overall mass distribution.

Together, these three effects imply that the values of R_e given in Table 1 may be overestimated by as much as a factor of 10. If these effects are not relevant, on the other hand, it will be possible to observe the dance with a single 8m telescope equipped with adaptive optics. Otherwise, the VLT interferometer will be required.

2.2 Observational Strategy

The observational strategy will depend on whether the nuclear clusters can be directly resolved. If this is the case then individual supernovae (the optical nucleus) will stand out more clearly from the stellar background and their position will be easily measurable. Otherwise it will be necessary to measure the centroid of the light distribution, which will be the sum of an extended object (even in the worst case assumptions discussed above the characteristic size will be a few VLTi pixels) plus an unresolved spike. Thus, with a typical astrometric accuracy of $60\mu as$ the expected movement of the nucleus (a few mas) will be measurable. The questions of wavefront corrections and fringe tracking will require careful consideration. Most nuclei in the target list will be bright enough to allow fringe tracking and some even to do the wavefront corrections. But they will be resolved considerably reducing fringe contrast. Therefore, it will still be necessary to select AGN with bright nearby stars to use the interferometer. While there are galaxies in this configuration, it is clear that the availability of Laser Guide Stars is of considerable importance for this project.

The main problem will be the preparatory observations. VLT(i) observations of the AGN dance will require intensive monitoring with smaller telescopes. Spectra of all target objects must be obtained in order to classify the nuclear type (Seyfert 1 or 2). This requires a 2m class telescope. A systematic campaign of photometric monitoring must be conducted in order to determine ν_{SN} and thus characterize the variability of the nuclei with a sampling rate of one observation per AGN per month. A second epoch of spectroscopy together with a first epoch of astrometry with the VLT(i) should be conducted as soon as the photometric monitoring reveals galaxies that change luminosity above a certain threshold.

A 1m class telescope with adequate image quality is required for photometric monitoring. Such observations should begin as soon as possible in order to produce results by the time the VLT(i) is in full operation.

References

Alloin D. et al. 1986, *ApJ,* 308, 23.

Aretxaga, I., 1993, *Ph.D. thesis,* Universidad Autónoma de Madrid.

Aretxaga, I., Terlevich, R. 1994a, *MNRAS,* 269, 462.

Aretxaga, I., Terlevich 1994b, *""Violent Star Formation: from 30 Doradus to QSO"* Eds. G. Tenorio-Tagle et al. Cambridge University Press, page 343.

Aretxaga, I., Cid Fernandes R., Terlevich R.J., 1996, *MNRAS,* in press.

Cid Fernandes R., Terlevich R.J., Aretxaga I., 1996, *MNRAS,* submitted.

Cristiani S. et al. 1996, *AA,* 306, 395

Filippenko, A.V., 1989, *Ap.J.,* **97,** 736.

Gregory S.A., Tifft W.G., Cocke W.J. 1991, *ApJ,* 102, 1977

Hook I.M., Mc.Mahon R.G., Boyle B.J., Irwin M.J., 1994, *MNRAS,* 268, 305

Korista K. et al. 1995, *ApJS,* 97, 285.

Lyutyi, V.M., Oknyanskii, V.L., Chuyaev, K.K., 1984, *Sov. Astron. Lett.,*

Malamuth, E.M., Heap, S.R, 1994, *AJ,* **107,** 1054.

Penston M.V. et al., 1981, *MNRAS,* 196, 857.

Peterson B.M., 1993, *PASP,* 105, 247

Pica A.J. & Smith A.G. 1983, *ApJ,* 272, 11.

Poveda, A, 1957, *Bull. Ton. Tac.,* 17.

Stathakis, R.A., Sadler, E.M., 1991, *MNRAS,* **250,**786.

Terlevich, R., Melnick, J., Moles, M., 1987, in *IAU Symp. #121, Observational Evidence of Activity in Galaxies.* (eds. E.Ye. Kachikian, K.J. Fricke, J. Melnick), Dordrecht, Reidel.

Terlevich, R., and Melnick., J., 1988, *Nature, 333,* 239. **10,** *235.*

Terlevich, R. Tenorio-Tagle, G, Franco, J, Melnick, J., 1992, MNRAS 255, 713

Terlevich, R. Tenorio-Tagle, Roziscka, M. G, Franco, J, Melnick, J., 1995, *MNRAS, 272, 198* **255,** 713.

Winkler, et al., 1992, *MNRAS,* 257, 659

Young, P.J., 1976, *AJ,* 81, 807.

Adaptive Optics Observations of Starburst Galaxies

Olivier Lai[1], Daniel Rouan[1], Eric Gendron[1] and François Rigaut[2]

[1] DESPA, Observatoire de Paris-Meudon, 5 place Jules Janssen, F-92195 Cedex Meudon
[2] CFHT, P.O. Box 1597, Kamuela, HI 96743 USA

Abstract. High angular resolution can help to explore the link between starbursts and AGNs: by being able to identify the respective contribution of the starburst and the active nuclei, we can investigate how each affects the other. The proposition that starburst are the results of mergers and that starbursts are progenitors of AGNs can also be tested, by looking for tracers of mergers. Observations of individual galaxies start to answer some of these questions.

High Angular Resolution for Starburst Galaxies?

The goal of these observations is to investigate the link between starbursts and AGNs: the hypothesis of Sanders et al., (1988) that starburst could be progenitors of AGNs yelds a fairly coherent model, because interaction of galaxies which trigger the starbust is also what creates a non axi-symmetric potential required for the gas to collapse on the central engine and ignite the activity of the nucleus. Indirect evidence for the association of starbursts and AGNs comes from many common features, namely: a high concentration of molecular gas in both, mergers or bars often seem to associated with such activity and both types are often associated with each other.

The proper separation of the starburst and the AGN allows to probe the mechanisms that start each process and the way they interact. Mergers (or tracers thereof, such as tails, wisps, double nuclei) can be detected even at a very advanced stage, thus the question of there always being a merger can be investigated (Mirabel & Sanders, (1989)). Also, the angular momentum problem can be explored (i.e. bars within bars, Schlossman et al., (1989)).

The Advantages and Difficulties of Such Observations

Advantages. The nucleus of many active galaxies is pointlike and bright enough (compared to the bulge) to allow guiding on it. The number of sources available for a systematic program is very dependent on the limiting magnitude of the AO system. The number of sources with a nucleus brighter than V=14 (from Véron-Cetty & Véron, (1993)) is 205, whereas there are 765 with V<16.

Difficulties. These observations require an expertise in AO and a subtle methodology of observation to accomodate the extension and/or low contrast of the guiding source. Also, some heavy data processing is required to bring out some crucial features (deconvolution and 2D-photometry).

Three Galaxies with ADONIS

Markarian 565. This galaxy shows a double nucleus, so the question arises: is it a true merger or rather a giant HII region? The photometry shows that the western nucleus is stellar ($\Rightarrow$ probable misclassifiaction as a Seyfert) whereas the eatsern nucleus is a mixture of red super-giants and ionised gas, i.e. the site of active star formation. This is probably the result of an advanced merger where an AGN has not (yet?) ignited.

NGC 863. This nucleus shows a common association of Seyfert activity and circumnuclear starburst (e.g. NGC 1068 or NGC 7469). The starburst is in the shape of a ring, yet there seems to be a bridge of matter between the nucleus and the ring. Can it be a "bar within a bar" (Schlossman et al., (1989)) along which the gas of the starburst region can collapse on the nucleus?

NGC 1365. This well known barred spiral galaxy has a Seyfert nucleus and radio emission (Sandqvist et al., (1995)) reveals vigorous starburst activity. Observations in [OIII] show a fan shaped emission cone(Jörsater et al., (1989)), which our data reveals as two "holes", where the violent interaction of the energetic flow with the ISM could deplete it. There is some extended emission perpendicular to the [OIII] cone, and the nucleus is double. Its orientation is not aligned with the rest of the nucleus, making any interpretation very difficult.

Arp 299 Observation with PUEO

A mosaic of (6×6) 1 min. exposures of this galaxy was obtained, with a superb resolution: the raw images have a 0.2″FWHM and the Strehl Ratio is 17%. We identify the nucleus and two probable star forming region. We also see very small nodules that we interpret as very young OB clusters, similar to those seen by HST in M82 (O'Connel et al., (1995)), brighter than any young cluster in the Local Group.

These results illustrate how adapative optics is now mature enough to allow tackling AGN/starburst physics.

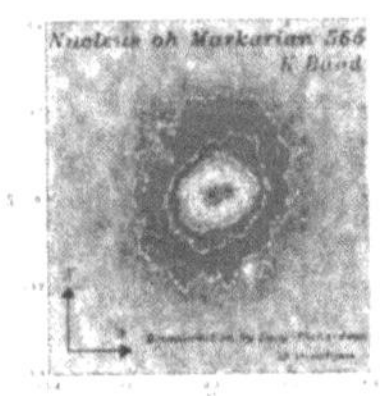
Markarian 565

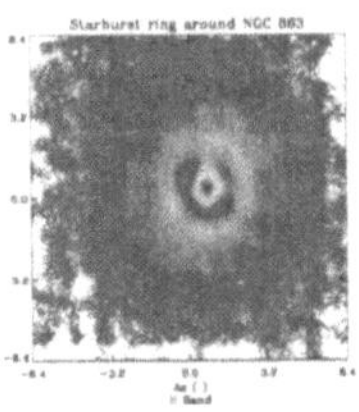
NGC 863

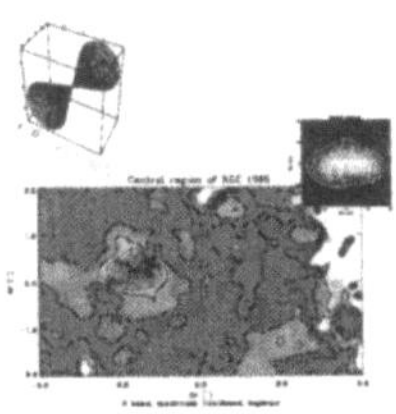
NGC 1365

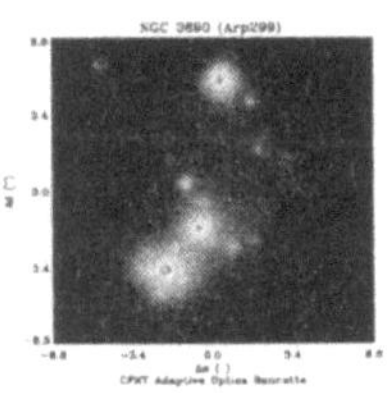
Arp 299

References

Jörsater S., Lindblad P.O., 1989, *The [OIII] plume in NGC 1365*, in Meurs E.J.A., Fosbury R.A.E.(eds), ESO conf and workshop *Extranuclear Activity in galaxies*, 39

Mirabel F. & Sanders D.B., 1989, ApJ, 340, L9

O'Connell R.W., Gallagher III J.S., Hunter D.A., Colley W.N., 1995, ApJ, 446, 1

Sanders D.B., Soifer B.T., Elias J.H., Madore B.F., Matthews K., Neugeubauer G., Scoville N.Z., 1988, ApJ, 325, 74

Sanqvist A., Jörsater S., Lindblad P.O., 1995 A&A, 295, 585

Schlossman I., Frank J., Begelman M.C., 1989, Nature, 338, 45

Véron-Cetty M.P. & Véron P., 1993, *A catalogue of quasars and active Nuclei*, 6th Edition, ESO Press.

Cosmology with the Very Large Telescope Interferometer: Using a Space Based Astrometric Reference Frame

David Tytler[1]

University of California San Diego, CASS 0424, La Jolla CA 92093-0424, USA

Abstract. Cosmology with large interferometric telescopes is a rich and largely unexplored subject, involving three types of measurement: astrometric measurement of absolute distances and proper motions, dispersions of relative proper motions, and images. The ground based interferometers can have huge apertures, which are necessary for faint cosmological targets. But, alone, they are limited to astrometry within the isoplanatic patch, and hence to relative positions, which are of little use for parallaxes and proper motions because reference stars have unknown parallaxes and huge (500 μarcsec) unknown motions. We propose that space missions should measure global astrometric parallaxes and proper motions for (V> 16) reference stars within the isoplanatic patches of important cosmological and Galactic targets. Ground based interferometers can then measure absolute distances (parallaxes) and proper motions to 10 μarcsec, tied to these reference stars. In combination, space and ground based interferometers can make a wide variety of measurements, some of which were believed to be restricted to space missions, and others where not considered possible because space missions lack light gathering power: absolute distances accurate to <10% for most globular clusters and about ten near by galaxies; proper motions of stars in near by dwarf galaxies and stars near to giant black holes; the masses and distances to individual MACHOS which cause microlensing events in the halo and bulge of our galaxy; proper motions of 1000 galaxies out to Virgo; and images of giant black holes, AGN and distant galaxies.

But cosmological observations stretch the VLTI technically. To observe the few best targets, we need to be able to measure positions to < 10μarcsec over a large portion of the sky. Since natural guide stars are too far apart, or too faint, laser guide stars are needed to correct the wavefronts of the individual 8-m unit telescopes, and the fringe tracking system must have an extremely high throughput to work on the brightest stars (V > 16) near to important targets. Most of the science is at 1 –2 microns, where excellent adaptive optics will be needed on the 8-m telescopes.

1 VLTI Characteristics

I preview cosmological observations which could be made with ground based interferometers which use large aperture telescopes, with specific reference to the VLTI. I have borrowed heavily from Peterson et al (1996) who reviewed science with a space interferometer, because I show that the ground based telescopes can undertake some projects which were considered for space alone. Science with interferometric images with 1 – 10 μ-arcsecond (μas) resolution was discussed by Begelman & Krolik (1991). I take an optimistic view, skip over the many technical and observational difficulties, such as integration times and confusion in complex fields, and omit many relevant references. Let us begin with an introduction to the technical issues, which require close attention if the VLTI is to do cosmology.

Following the Palomar Testbed Interferometer (Shao & Colavita 1992), the VLTI will have a dual feed, which allows high precision narrow angle (or differential) astrometry (Kovalevsky 1995). One feed looks at a relatively bright reference star, while the other looks at the target, which can be very faint. The two can be separated by up to 60 arcsec, but astrometric accuracy improves linearly with the inverse separation (Shao & Colavita 1992), as the paths through the atmosphere to the target and reference become nearly identical. Each feed covers about 2 arcsec diameter, the potential imaging area.

To correct the wavefront of the light gathered by the individual 8-m unit telescopes, we need a bright star V,K $<$ 13 within 10 arcsec of the target to sense the wavefront. This restriction rules out all but a tiny fraction of the sky, and must be overcome using laser guide stars, one for each unit telescope. Lasers allow us to correct the wavefront anywhere in the sky.

To provide stable interference fringes (fringe tracking), we can use slightly fainter reference stars (V,K $<$ 16 assuming the wavefront is already corrected to near the diffraction limit of the unit telescopes). Sky coverage as a function of angle to the reference star is reasonable at the Galactic plane: 15% at 10 arcsec, 45% at 20 arcsec and 75% at 30 arcsec, but unacceptably small at the pole: 0.8% at 10 arcsec, 3% at 20 arcsec and 7% at 30 arcsec. The ideal fringe tracking system would work on fainter stars, with higher throughput, wider bandwidth or perhaps better open loop stability.

The key advantage of the VLTI over most other interferometers is its sensitivity: $V = 25$ and $K = 23$ in 600 sec with signal to noise ratio 10 in 0.7 arcsec seeing. This is why it can play an important role in cosmology.

Tip/tilt correction is sufficient to make the auxiliary 1.8 m telescopes diffraction limited at $>$ 1 micron, and the unit 8-m telescopes at $>$ 5 microns. But adaptive optics is essential for cosmology, which will use the 8-m unit telescopes (UTs) at 1 –2 microns.

Baselines are $B < 200$ m for the 1.8 m telescopes (0.001 arcsecond = 1 mas point spread function at 1 micron) and $B < 130$ m for the UTs ($<$1.6 mas).

2 Ground Based Astrometry Using Stars on a Global Reference Frame from Space

In an extremely important paper which opened up a whole new astronomical technique, Shao & Colavita (1992) pointed out that ground based telescopes can make extremely high precision relative astrometric measurements whenever references stars are a few arcsec from targets: 10 μas with the VLTI for reference stars within $<$ 10 arcsec, when the signal to noise in the fringe visibility is about $>$55 at 2.2 microns. The smaller then angle to the reference, the more similar the light paths through the atmosphere, and the better the relative astrometric accuracy. There are many uses for relative astrometry: binary stars, orbits of unseen brown dwarfs and planets, masses and distances to microlenses, dispersions of the proper motions of stars in clusters and galaxies.

However distances require absolute parallaxes. Parallaxes relative to a reference star with an unknown parallax are of little use because we do not know how to proportion the motion between the two stars, and if both are at the same distance, no relative parallax motion is seen. Similarly, absolute proper motions are much more useful than those measured relative to stars with unkown motions. All faint reference stars move, by about 500 μas /yr (24 km/s at 10 kpc, which is much more than the 150 μas relative astrometric accuracy of a single 8-m aperture). There are far too few within an isoplanatic patch to give a reference frame stable to 10 μas , so ground based observations give relative not absolute parallaxes and proper motions.

We propose that ground based telescopes should use reference stars which are observed from space. Space observations can give even higher astrometric accuracy over both narrow and wide angles, and hence they can construct a global astrometric reference frame, with absolute parallaxes and motions. This synergy between the global reference frame from space, and the fainter targets from the ground based, allows large ground based telescopes to measure distances (absolute parallaxes) and absolute proper motions, tied to the space frame. This greatly increases the range of cosmological observations for the VLTI.

Of the various space astrometry missions, Hipparchos lacks accuracy: 1000 μas . The proposed ESA mission GAIA is close to what is needed: 10 μas to V =15, but it covers the whole sky, where as the VLTI would prefer fainter reference stars (V = 16 or 17, as faint as fringe tracking allows), selected because they are in the isoplanatic patches of important Galactic and cosmological targets. This is a task better suited to NASA's SIM mission, which should obtain 5 μas on preselected targets (V=16 in 2,000 sec, or V=18 in 10,000 seconds, with a baseline of B=7 m).

The use of a space based absolute reference frame is a natural way to use the VLTI, since there must be a relatively bright reference star within the isoplanatic patch for fringe tracking. We are proposing that this same star should also provide absolute position, motion and distance.

3 Comparison with Other Instruments

The VLTI has unique advantages, but only if it is fully equipped with laser guide stars, adaptive optics, and high throughput fringe tracking, and even then only for a few years, as many other instrument and missions are competitive.

HIPPARCHOS: 1000 μas (VLTI 100 times better), survey of 10^5 stars (VLTI gives 100 times better relative astrometry, but points at selected targets).

HST: point spread function of 100 mas at 1 micron (VLTI 50 times better), arcmin field of view (VLTI 1000 times smaller).

DIVA: A proposed small German interferometric satellite, which would survey 10^6 stars with V < 10 giving parallaxes to 800 μas (Bastian et al 1996).

SIM: NASA's Space Interferometer Mission (2003) (Peterson et al. 1996). V< 20 (VLTI readily goes 5 magnitudes fainter, but SIM may reach V = 26 in 20 hours), 5 yr life (for proper motion – not parallax – the VLTI can compensate for its lower annual accuracy with more years of observation) 0.3 arcsec field (VLTI is 40 times larger), 4 μas with a B=10 m (VLTI is 2 times worse, and needs a space reference grid for parallax), good UV plane coverage (VLTI is not as good for complex fields), dynamic range < 2000 and can null out bright sources to 10^4 at 70 μas or 10^6 at 7 μas (VLTI worse).

GAIA: ESO proposed sky survey to V <16 (VLTI 9 magnitudes better), 5×10^7 star sky survey (VLTI points), 10 μas at V=15 (VLTI similar, but needs space reference frame for parallax).

NGST: An 8-m version of NASA's Next Generation Space Telescope (www:// ngst.gsfc.nasa.gov/) gives 5 μas astrometry (VLTI 2 times worse), over 4 arcmin field (VLTI 10^4 times smaller area), with much higher sensitivity and dynamic range. The NGST is dramatic! It will compete strongly with all large ground based optical/IR telescopes and planned interferometers.

Keck Interferometer: Two 10-m telescopes and four 2-m outriggers, 2003. (VLTI much better UV coverage).

There are also many interferometers with smaller apertures including the Palomar Testbed Interferometer, I2I, ISI, OCAST, SUSI, IOTA, NPOI and CHARA, which will not compete on fainter cosmology targets.

4 Assumptions

For the remainder we will assume a cosmology friendly VLTI. 1) Laser guide stars on at least two 8-m UTs. 2) Adaptive optics on at least two UTs (movable to others), allowing fringe tracking on V> 16 to get good sky coverage, and science at 1 – 2 microns. 3) Parallax and proper motion from space and on a global reference frame for at least one reference star (V $\simeq$ 16) in the isoplanatic patch of all interesting targets. The VLTI then gives positions to about 14 μas when the reference star is 10 arcsec away, and 30 μas at 30 arcsec.

5 Extragalactic Distance Scale

The VLTI can be used to provide improved distances to standards. However space missions such as GAIA or SIM are needed to provide reference stars in the global reference frame, and those same missions can and will also observe the luminosity standards themselves, because they are bright. Indeed they are brighter than the reference stars. There are three main types of object:

Cepheids distances come from main sequence fits to open clusters. The VLTI could get >1% distances to < 20 stars with parallaxes of 200 – 1000 μas and V < 10. An excellent next step after the huge ground based studies from microlensing (Beaulieu & Sasselov 1997).

RR-Lyrae distances come from statistical parallax. The VLTI could get 1 – 10% distances to < 20 stars at < 5 kpc with $V = 8 - 11$.

Planetary Nebulae do not have good distances. The VLTI could obtain 1 – 10% distances for < 40 with parallax 100 – 1000 μas and V = 12 – 16.

5.1 Galaxy Distances from Rotation

Absolute distances to a few near by Galaxies can be obtained by comparing Galactic angular and radial velocity rotation rates (van Maanen, Reasenberg et al, 1988; Peterson et al. 1996). Measure both the angular and radial velocities at two places where the rotation velocity (relative to that galaxies center) is expected to be the same, then solve for the rotation velocity, inclination and distance. A transverse velocity of 100 km s^{-1} at 1 Mpc is 210 μas /10 years, which gives a distance to 5% per star, and 1% for 25 stars. The target A – F supergiant stars are bright, with $M_v \simeq -8.5$ and V =21.5 by 10 Mpc, but few stars in the galaxy will be near to V<16 reference stars.

6 Globular Cluster Distances and Ages

The next satellite measurements of the cosmic microwave background should determine the age of the universe to about 1%. VLTI could improve globular cluster ages from 10% today to a few percent. Improving distances improves the turnoff luminosity, which improves the mass and thence the age. A 12% distance error is a 22% age error (Renzini et al. 1996). The VLTI can give 1% distances at 1 kpc, and 3% distances with 10 stars at 10 kpc. Stars with $M_v \simeq -1$ are V $\simeq$ 14 at 10 kpc, providing excellent reference stars.

Meylan (this meeting) discusses relative proper motions inside clusters. If we assume globular clusters are isotropic, then a comparison of radial and angular velocities gives distances. No global reference frame is needed for this since we are measuring the dispersion in proper motions, not the absolute values. Consider a cluster like 47 Tuc with $\sigma_v = 11.5$ km s^{-1} , but at 10 times the distance: 46 kpc. The dispersion of proper motions will be $\sigma_\mu = 53$ μas /yr. Since stars are close together in the clusters, we can take advantage of the gains in relative astrometry with close pairs, perhaps obtaining 2 μas on pairs separated by 1.6 arcsec (V $\simeq$ 14). If we then measure 100 stars we get a 10% error on the σ_μ. But gains are slow, since $\sigma_\mu \gg$ measurement error, and crowding and changing blends may spoil the accuracy. Astrometric images might be competitive, because they capture many stars at once. A single 8-m should give errors of 150 μas /yr, or $\simeq 3.5\sigma$ detections of proper motion per star in 10 years, which would be difficult for distant clusters, because the physical dispersion is only 3.5 times the expected measurement error, which we need to measure accurately and subtract from the observed dispersion.

7 Dark Matter in the Galaxy

The VLTI is well suited to the measurement of the masses and distances to individual MACHOS in the galactic halo and bulge when they cause microlensing events. Microlensing events are seen where the density of stars is unusually high: the galactic bulge (detect 50/yr), the Magellanic clouds (6 – 8 events in 3 years), and in the near future, the inner regions of M31.

The bulge events have typical masses of 0.3 $M_\odot$, and may be caused by ordinary stars in the bar of our Galaxy. If we knew individual masses and distances we could identify the lens objects, and later measure masses for planets which we expect will be found orbiting the lenses (Rhie & Bennett 1996; Peale 1996).

Events towards the LMC are of unknown origin. Their masses are around 0.2 $M_\odot$, depending on the halo kinematics, and hence they might be old, cold white dwarfs.

Several possible measurements could be attempted (Miralda-Escudé 1996). The maximum image splitting seen in the sky (Miyamoto & Yoshii 1996) is

$$2\theta_E = \frac{4}{c}\sqrt{GM}\left(\frac{1}{D_L} - \frac{1}{D_S}\right)^{1/2} \leq 2\left(\frac{M}{M\odot}\frac{8\ kpc}{D_S}\right)^{1/2} \quad mas,$$

where M is the lens mass, D_L and D_S are the distances to the lens and source star. The event lasts about 40 days (typically 5-100 days, depending on the relative velocities and M), and will be hard to observe because the point spread function is about 1 mas at 1 micron with a 200 m baseline.

The centroid shift is 1 – 2% of the splitting, or $< 20 - 40$ μas for a 1 $M_\odot$ lens at $D_L = 4$ kpc towards a source at $D_S = 8$ kpc, which will also be hard to measure.

The proper motions of the lens and source are huge: $\simeq 4$ mas/yr, but measurements will be hard because of the blending of the lens, source and many other stars in these crowded fields.

Proper motion will separate the lens and source in the sky, to about 18 mas in 5 yr (transverse velocity 100 km s^{-1} , $D_L = 6$ kpc), so that they could be resolved. The lens might be faint, V=24 for $M_v = +10$, and we would like to measure colors to determine a spectroscopic distance. Hardware and techniques intended to observe planets near to stars may help with this observation, although it may be built to handle much larger separations (about 1 arcsec) and higher contrast ratios (10^9).

The lens and source parallax are large: 170 μas at 6 kpc, but again the field is crowded, and the lens is very faint.

8 Structure of Other Galaxies

The mass to light ratio in dwarf galaxies increases up to 100 as luminosity drops. The VLTI could measure the proper motions of many stars to check the isotropy of the velocities which might change M/L determinations by a factor of two (e.g. Eckart & Genzel 1996). The individual stars in a galaxy at 80 kpc (parallax 12 μas) with internal motions of 10 – 100 km s^{-1} would have proper motions of 25 – 250 μas . The distribution of the dispersion in proper motions across the galaxy can be used to map out the distribution of dark matter relative to light, while unusually large motions can indicate mass concentrations (black holes).

Alternatively, if we assume the velocities are isotropic, the comparison of proper motions with radial velocities gives a distance. Proper motions of stars in large galaxies are hard, but just possible out to Virgo: 41 μas in 10 yrs for 300 km s^{-1} velocities at 16 Mpc.

9 Proper Motions of Whole Galaxies

The proper motions of many whole galaxies are just within the range of the VLTI. A 500 km s^{-1} velocity at Virgo (16 Mpc) gives 6 μas /yr, or a 25% error in 7 years. The target stars are faint: V=22.5 for $M_v = -8.5$ A - F Population I supergiants in spirals and irregulars, and many should be measured per galaxy to map out and correct for the internal motions in the galaxies. A large project could be launched to map the motions of 1000 galaxies, to reconstruct orbits, identify groups, and map out the mass distribution. This project requires fringe tracking on very faint stars, since an 8 kpc galaxy at 16 Mpc covers 1.7 arcmin, and includes on 0.9 foreground stars brighter than V=16 near the Galactic poles, or 16 in the plane.

10 Search for Black Holes in Galaxies

The presence of a massive black hole in a galaxy can be deduced indirectly from the AGN activity (radio, IR, UV, X-ray and broad emission lines), and more directly by sensing the distribution and motions of stars and gas (e.g. Miyoshi et al 1995).

Images can be used to look for light cusps in the centers of distant galaxies. If an HST images can reveal a 10^9 M$\odot$ black hole at 16 Mpc in images with 100 mas resolution, then the VLTI might see the same at 400 Mpc with 4 mas images, far enough to sample 10^6 galaxies.

The VLTI has the astrometric sensitivity to measure the proper motions of individual stars near black holes out to Virgo, but this would be hard because of crowding. Low resolution spectra of the integrated light from within a few mas of a massive black hole would show huge velocities, allowing their detection to great distances (500 km s^{-1} at 4 mas for a 2×10^9 M$\odot$ hole at 1000 Mpc – where I scaled distance against angular resolution from the M87 observations of Ford et al. 1994), or the detection of lower mass holes in Virgo galaxies (500 km s^{-1} at 16 Mpc for a 3×10^7 M$\odot$ hole, where $v^2 \propto M/$[angular resolution]).

11 AGN and QSOs

Beyond Virgo, proper motions are too small for the VLTI, and images and spectra become the main cosmological measurements. At cosmological distances, 2 mas resolution at 1 micron corresponds to 24 h_{50}^{-1} pc at $z = 3$ for $q_0 =$ 0.5. Narrow emission line regions of massive black holes (QSOs) are 100 pc in size and resolved at all redshifts. Broad line regions are only 0.1 pc across

(2 mas at 10 Mpc) and hard to resolve, but important because their shape and homogeneity is unknown. Ward (this volume) discusses IR observations of molecular torii. Knots in the jets of superluminal sources are probably <pc in size, and move 0.1 – 1 mas/yr. Their dynamical range is unknown on mas scales. Comparison of optical with radio data relate to the physics of the radiation mechanism and the magnetic fields. If the light from some AGN is dominated by starbursts (Melnick, this volume), then the star cluster may be resolved. It would be especially interesting to null out the QSO light to image the inner structure of high z active galaxies. High angular resolution images of gravitational lenses provides accurate astrometry which is the key to the construction of accurate mass models.

12 Young Galaxies

Galaxies with cool gas and young stars have a lot of structure; compare HST images of the near by M100 spiral before and after the refurbishment. There types of galaxies look fuzzy from the ground because of the Earth's atmosphere, and we expect more substructure at high redshifts because there is more star formation, and lumps from which galaxies are made are not well mixed. Although the high galaxies are extremely faint, and they will have complex shapes, requiring many baselines, it would be exciting and presumably rewarding to attempt mas resolution images of even a few.

High redshift galaxies are common, so they can be chosen near to bright fringe tracking stars. Steidel et al (1996) find 0.4 galaxies at $z \simeq 3$, so that 3% of random positions on the sky (e.g near bright stars) have one such high redshift object.

It would be very interesting to make high resolution images of galaxies causing absorption lines seen towards bright high redshift QSOs. Most such QSOs have V$\geq$ 18, but a few at $z \simeq 2$ have V= 16 for fringe tracking. We could determine what types of galaxies cause various types of absorption.

13 Conclusions and Requirements

Cosmological observations with ground based interferometers are technically challenging, but the equipment also would benefit most other studies. The science is rich and largely unexplored.

There are two main types of measurement: differential astrometry, and images. The astrometry can give absolute parallaxes and proper motions when the reference star in the isoplanatic patch (of the target) has a known parallax and proper motion. Observations from space are needed to place these reference stars in a global (absolute) reference frame. Astrometry can also measure the dispersion in proper motion of stars in a globular cluster, dwarf galaxy, or near a massive black hole, without the need for a reference star in a global reference frame.

Adaptive optics is needed to work in the near-IR, while laser guide stars are needed on the unit 8-m telescopes to obtain useful sky coverage. We need two perpendicular baselines for astrometry and images, so three unit telescopes should be equipped, or only two if the hardware is moved between telescopes.

Every effort should be made to increase throughput and bandwidth so that stars fainter than V= 16 can be used for fringe tracking. Otherwise sky coverage at the Galactic pole is unacceptably small (3% with a reference star within 20 arcsec) and most individual cosmological targets will not be observable.

Ability to record spectra with a resolution of 100 km s^{-1} from light from areas of a few mas is important for the detection of massive black holes and the structure of AGN emission line regions.

It is a pleasure to thank Chas Beichman, Marc Colavita, Julian Krolik, Jordi Miralda-Escudé, Alvio Renzini especially Mike Shao for timely information.

References

Bastian, U., Roser, S., et al. Astron. Nachr., (1996) **317** 281
Beaulieu, J.P., & Sasselov, D.D (1997) (astro-ph/9612217)
Begelman, M.C. & Krolik, J.H., "High Resolution Imaging of Compact Phenomena", in *Workshop Proceedings: Science Objectives and Architectures for Optical Interferometry in Space*, M. Shao, S. Kulkarni, & D. Jones, eds. (JPL: Pasadena) (1991)
Bennett, D.P. & Rhie, S.H. (1997) ApJ **472** 660
Eckart, A. & Genzel R. (1996) Nature, **383**, 415
Ford et al. (1994) ApJ **435** L28
Kovalevsky, J (1995) *Modern Astrometry* (Springer)
Miralda-Escudé, J (1996) ApJ **470** L113
Miyamoto, M. & Yoshii, Y. (1995) AJ **110** 1427-32
Miyoshi, M., Moran, J. et al. (1995) Nature **373** 127
Peale, S.J. submitted to Icarus (astro-ph/9612062)
Peterson, D et al. "Space Interferometry Mission: Taking Measure of the Universe", final report of the Space Interferometry Science Working Group (NASA, 1996).
Reasneberg, R et al. (1988) AJ, **96**, 1731
Renzini, A. et al. (1996): ApJ **465**, L23
Shao, M. & Colavita M.M. (1992) AA **262**, 353-358
Steidel, C.C., Giavalisco, M., Pettini, M., Dickinson, M., and others. ApJL **462** L17

Part 4

THE POTENTIAL OF VLTI AND THE CONNECTION WITH RELATED TECHNIQUES

Tailoring an Interferometer to its Science and Vice Versa

T.A. ten Brummelaar, H.A. McAlister, W.G. Bagnuolo, W.I. Hartkopf, S.T. Ridgway & N. Turner

Georgia State University
Atlanta GA, 30303-3083 USA
theo@chara.gsu.edu

Abstract. In the process of planning for the CHARA Array, the science/technical team devoted significant effort to exploring possible science objectives over a broad range of astronomical fields, from solar system bodies to extragalactic objects. In this analysis, we attempted to define a scientific figure of merit for each of the various science areas based upon the optimum wavelength regime, median magnitude of the object class, representative angular size, minimum required integration time, and structural complexity as it relates to requirements for imaging rather than visibility analysis. These quantities were incorporated into a feasibility estimate that was then combined with our estimate of scientific priority to define a figure of overall merit. This rather elaborate and admittedly subjective analysis clarified the role the CHARA Array will play in high resolution astronomical imaging. It also clearly demonstrated the unlikelihood that any single interferometric array will have the breadth of capability to cover the rich scientific opportunities that lay before us.

1 Stellar Astrophysics – Single Stars

1.1 Emergent Fluxes and Effective Temperatures

These quantities are determined from angular diameter measurements using the relations:

$$F = 4f/\Theta^2 = \sigma T_{eff}^4,$$

where F is the emergent flux at the stellar surface, f is the flux integrated over the photosphere as determined from absolute photometry, Θ is the measured angular diameter, σ is the Stefan-Boltzmann constant, and T_{eff} is the effective temperature. Θ is measured through a fit to the "visibility" or correlation curve given by

$$\left|V_1(b)\right| = \frac{(I_{max} - I_{min})}{(I_{max} + I_{min})} = \frac{2J_1(\pi\Theta b/\lambda)}{\pi\Theta b/\lambda},$$

where J_1 is the first-order Bessel function, b is the interferometer baseline separating the two telescopes as projected onto the sky, and λ is the mean wavelength of the observation. The determination of Θ is thus a single parameter fit to a carefully calibrated measure of $\left|V_1(b)\right|$. T_{eff} is one of the fundamental astrophysical parameters characterizing stars, and the accurate determination of T_{eff}'s has been the classical primary goal of interferometry.

1. To be of the greatest astrophysical use, Θ should be determined to an accuracy of better than $\pm 2\%$ for significant samples of stars.
2. Establish accurate T_{eff} scales within a grid of MK types, metallicities, rotational velocities, etc.
3. All spectral types and luminosity classes are available with the most difficult being the hottest and coolest stars.
4. Because limb-darkening is less in the near-IR, diameters are less subject to uncertainties when measured from single parameter visibility observations at those λ's.
5. The calibration of flux sensitive photometric parameters such as c_o is a significant goal.
6. Wavelength dependence of apparent angular diameter for stars of various spectral types.

1.2 Photospheric Limb Darkening

This phenomenon results from the projected optical depth effects when a photosphere is viewed at increasing distances from the center. Under the simplifying assumptions of plane-parallel atmospheres, isotropic scattering and gray opacities, limb darkening follows a $cos\phi$ law, where $\mu \equiv cos\phi$. Thus, from center ($\mu = 0$) to edge ($\mu = 1$), limb darkening at optical depth 0 is given by:

$$I(\mu, 0) = I(1, 0)\frac{3}{5}\left(\mu + \frac{2}{3}\right).$$

The exact behavior of limb darkening provides insight into stellar atmospheres theory. The angular diameter Θ determined from the single parameter fit to $|V_1(b)|$ described above assumes a uniformly illuminated disk. The observational measurement of limb darkening is determined from the behavior of the visibility curve beyond the first null, and is thus a second order effect over the measurement of Θ. It is best carried out at visible λ's.

1.3 Linear Diameters

For stars of known parallax Π (in arcseconds), the linear diameter D of a star is given by:

$$D(km) = 1.5 \times 10^8 d\Theta = 1.5 \times 10^8 \Theta / \Pi,$$

where the distance d in parsecs is given by $d = 1/\Pi$. Stellar radius is related to mass by

$$R = \alpha M^\beta$$

where $\alpha = 1$ for R and M normalized to solar values and $\beta \sim 0.7$ for main sequence stars. The $R(M)$ relation has heretofore been based upon studies of eclipsing binaries in which proximity effects may bias the relation over what might be exhibited by single stars. The advent of accurate parallaxes from the HIPPARCOS astrometry satellite and from the USNO ground-based programs opens up the possibility of a reassessment of this fundamental relation in astrophysics.

1.4 Star Forming Regions

Moderate resolution in the far infrared is the most promising approach in looking for the complex structures that are expected to be associated with star formation. These regions are often heavily obscured by dust.

1.5 Pre-Main Sequence Objects

Structures associated with Pre-MS objects are of great interest, and require imaging at IR wavelengths at moderate to high resolution.

1.6 Young Stellar Objects

Dyck and Kibblewhite's list gives a typical M magnitude of 1.9 and a diameter of 16 mas.

1.7 Rotation

Rotationally induced oblateness can amount to a 3 to 2 ratio of equatorial to polar diameter before breakup, and such extremes are expected on the basis of spectroscopic analyses. Ridgway (1988) mentions specifically FU Ori stars, two of which have K magnitudes of 4.9 and 5.8 respectively, and have equatorial/ polar diameters of about 2.0/ 0.2 mas. Experiments in differential techniques to determine the axis of a rotating main sequence star can be better done in the visual.

1.8 Reddened Stars

Observations of the diameters of heavily reddened M supergiants could provide better mapping of the galactic structure. At 2 kpc, a star like α Ori would have $E_v = 3.8$ and its energy distribution would peak at 1.6 um, rather than 0.9 um. Dyck and Kibblewhite note a list of 63 M supergiant stars with a median diameter of 2.8 mas, which have a magnitude of about 2.0 in K.

1.9 Flare Stars

The physical mechanisms for this phenomenon are unknown. Because time resolution of 5 min would be needed, measurements would be in "snapshot" mode. There are some 15 UV Cet flare stars with $V \leq 10$ and $K \leq 7$, but the diameters are less than 0.1 mas, out of feasible range in the near-IR. Therefore, only fairly hot dust shells could be measured.

1.10 Calibration of Cepheid P–L Relation

The ESO/VLT Committee point out that an accuracy of 2% in the mean angular diameter of 29 Cepheids would decrease the 0.2 mag uncertainty in the P-L relation to 0.1 mag. It is not obvious what advantages near-IR would have over visual for this project, except that diameters may be better determined for the stars that are resolvable. The nearest Cepheids have diameters of 2 mas, just resolvable at 400 m (resolution at 2.2 um is 1.4 mas).

1.11 Pulsation Properties of Mira-Type Long-Period Variables

There are roughly 20 Miras (LPV's) with K brighter than 2 and diameters larger than 5 mas. In particular nodal oscillations of LPV's could be determined by spatial interferometry (in conjunction with precision spectroscopy). Dyck and Kibblewhite mention a list of 49 Miras with T_{eff} in the range 2300 - 3100 K, which have a median diameter of 5.7 mas and median magnitude of 0.0 in K [translating into K magnitude, as DK give fluxes in M.] It appears that these objects would be well observed.

1.12 Surface Inhomogeneities

1. Starspots on RS CVn stars requires $\leq$0.2 mas imaging and would be enhanced by spatially resolved Zeeman splitting observations.
2. Detailed features on giants and supergiants to answer such questions as the role of chromospheric prominences vs. photospheric convection. ("Betelgeusean Astrophysics") In the K band for a 400 m baseline, α Ori would have about 46 resolution elements across the disk, or 1700 total resolution elements. The magnitude is about - -5.0 in K, or about 3.1 per resolution element [the same for any star of this type]. One important question, given that many of these M stars have similar temperatures, is how filled does our dilute aperture have to be? If it is too dilute, we will not be able to recover useful images in a reasonable time. One major astrophysical question is how big are the turbulent eddies of stars of this type. It seems reasonable to suppose that the near-IR band is relatively better for observing stars of this class than the visual.
3. Surface structures of stars of all MK types.
4. Imaging in Zeeman sensitive lines as a first direct measure of stellar magnetic field strengths plays an important role in the "solar-stellar connection".

1.13 P-Mode Oscillations

Obtained via complementary high-resolution spectroscopy by observing in and out of selected lines.

1.14 Wolf-Rayet Stars

The diameter measurements of these hot objects could be done in the visual. IR speckle measurements of several objects [WC8-9 Ve 2-45, Dyck 1984] noted shells up to 250 mas in diameter at 2 um. Ridgway mentions a target list of 41 Pop I W-R stars with median K $\leq$ 6.

1.15 Cool Evolved Stars with Dust Shells

Dyck and Kibblewhite note 63 cool stars with dense shells having median diameter of 25 mas, median M magnitude of -0.3. (Need temp estimate for K magnitudes. These features can still be observed in K, but would be better in the thermal IR.)

1.16 Novae & Supernovae as Targets of Opportunity

The temporal evolution of the expanding structures would be the primary goal.

2 Stellar Astrophysics – Binary & Multiple Stars

2.1 Masses

The masses of stars are of fundamental significance to stellar astrophysics through "Vogt's Theorem" which states that the entire course of evolution of a star is determined *a priori* by its initial mass and its initial chemical composition. Stellar masses are determinable only from binary star orbital motion studies through the application of Kepler's Third Law

$$(M_1 + M_2) = a^3/P^2,$$

where M_1 and M_2 are the mass (in units of solar masses) of the primary and secondary stars of a binary system, a is the orbital semimajor axis (expressed in AU) and P is the orbital period of revolution (expressed in years). The orbital elements a and P, along with six additional elements, are determined from a set of measurements of the angular separation ρ between component stars and the position angle θ, which specifies the relative orientation of the components, as a function of time t. Thus the basic observational material consists of sets of data (θ,ρ,t) obtained through single parameter fits of visibility curves, i.e. a binary star is resolved essentially through the detection of the first null in visibility by a multi-parameter fit to the visibility curve given by:

$$|V(b)|^2 = \frac{1}{(1+\beta)^2}\left[\beta^2|V_1(b)|^2 + |V_2(b)|^2 + 2\beta|V_1(b)||V_2(b)|cos\left\{\frac{2\pi b}{\lambda}\rho\ cos\psi\right\}\right],$$

where β is the brightness ratio of the two stars (where $\beta \geq 1$, $|V_{1,2}(b)|$ are the visibility contributions arising from from the individually resolved disks of the two stars (the subscript 1 refers to the "primary" or brighter star while the subscript 2 refers to the "secondary" or fainter component), $\rho \cos\psi$ is the projection of the angular separation of the two stars onto the interferometer baseline with ψ being the angle in the plane of the sky between the position angle vector direction (determined by θ) and the projected interferometer baseline. The shape of the visibility curve for a binary is thus modified according to whether or not either or both of the two components are themselves resolved into disks.

1. Accurate mass determinations exist for fewer than 100 systems whereas thousands of binaries are known.
2. The $L(M)$ and $M(R)$ relations are currently completely insensitive to effects of evolution, metallicity, core rotation, etc.
3. Almost no mass data exists for Pop II stars and highly evolved disk population stars.

2.2 Double-Lined Spectroscopic Binaries

The direct resolution of DSB's results in a complete determination of the individual masses *and* in a geometrically determined *orbital parallax*. These values then lead to fundamentally determined masses and luminosities for the individual components providing that the value of Δm is also extracted from the interferometry. This is the most important class of binary for interferometry. Considered here is a set of 180 "guaranteed" objects whose angular separations are predicted to be $\geq$1 mas.

2.3 Single-Lined Spectroscopic Binaries

Some 70% of the SB's in the Batten *et al. Eighth Catalague...*, i.e. some 1,000 systems (distributed over the entire sky) have predicted $\rho \geq 0.2$ mas. The northern hemisphere is preferred for historical reasons. Many of the SSB's in this sample will be converted to DSB's through improved radial velocity techniques, and many will have accurate parallaxes from modern astrometry programs. Thus this is the second priority sample for mass determinations.

2.4 Astrophysics of Close Binaries

These objects are invariably SB's and often eclipsing binaries as well. The associated phenomena are: tidal distortion; mass exchange; circumstellar/circumsystem streams, shells, and disks; systemic mass loss; "reflection" effect; apsidal rotation; Roche lobe induced photospheric distortions, etc. Such effects will undoubtedly require very high resolution imaging. The imaging of an X-ray binary would reveal structures of accretion disks and tidal distortion mappings could lead to the definitive detection of the black hole and determination of its mass. This is an extremely challenging area.

2.5 Duplicity/Multiplicity Surveys

The frequency of this phenomenon is of fundamental significance to all areas of stellar astronomy ranging from the details of star formation to the frequency of life in the Universe. Surveys have heretofore been hampered by the limited sensitivity any particular technique has to detection, and interferometry will provide several decades of spatial resolution sensitivity. This will provide unprecedented completeness to surveys. New masses and luminosities will also result from newly discovered binaries that then become DSB's. Surveys should be conducted among broad samples of field stars delineated as to MK type, metallicity, rotation rate, abnormal spectral characteristics, etc.

2.6 Distance Calibrations

These result from resolution of DSB's (i.e. from determining orbital parallaxes) but are mentioned separately because of their specific significance. Resolution of cluster DSB's and field DSB's are possible out to distances of 10 kpc for a 400-m baseline thus offering new direct calibration of the galactic and extragalactic distance scales. Resolution of DSB's in the Magellanic Clouds might be feasible for a southern interferometer.

2.7 Low-Mass Companions

Cool stars, brown dwarfs and planets can be sought through the detection of submotions in the orbital motions of binaries if such a component exists in orbit about one component of a resolved binary. Here accuracy, rather than high resolution, is the key. The amplitude of the submotion is inversely proportional to the system's distance from the sun. Observations of high order visibility nulls may yield 100 μarcsec accuracy.

3 Star Clusters

3.1 Proper Motions Within Globular Clusters

This is an application of differential astrometry of closely spaced cluster members, completely analogous to the binary star problem.

3.2 Duplicity & Multiplicity of Cluster Members

Applicable to both galactic and globular star clusters, the problems described above in Section B would be pursued within the cluster environment. Of special interest would be the direct measurement of cluster distances through the resolution of DSB's, the relative frequency of occurrence of duplicity and multiplicity in clusters as opposed to field stars, and the comparison of cluster $L(M)$ and $R(M)$ relations compared to those of field stars.

4 Extragalactic Astronomy

4.1 Stellar Astrophysics Extrapolated to the Magellanic Clouds

This task is obviously deferred to a southern interferometric array, but could provide significant returns in many fields if sufficient magnitude sensitivity were available. At the distance of the Clouds, a limiting angular resolution of 0.2 mas corresponds to ~6 AU. The resolution of DSB's in the Clouds would result in the direct determination of the distance to the Magellanic Clouds. With a distance modulus of $m - M = +20$, the hottest main sequence stars would still be fainter than $m_v = +15$.

4.2 Imaging of Narrow and Broad Line Emission Regions in Active Galactic Nuclei

Even the brightest AGN places severe demands upon limiting sensitivity, and it is likely that K-band observations down to K = +16 would provide great returns assuming compensated imaging can be achieved. Baselines of several hundred meters would enable the imaging of core related features, including jets, for a number of AGNs. Narrow-line emission regions are thought to originate in structures several hundred parsecs in diameters while broad-line emission regions, while brighter in many cases, are associated with regions only 0.01 parsec in diameter corresponding to angular sizes of 10^{-8} arcsec at a distance of 1 megaparsec. The structure of the narrow-line regions and the multiplicity of otherwise unresolved broad-line sources are projects of high significance.

5 Solar System Astronomy

5.1 Minor Planets

Ridgway (1988) lists 585 objects with $m_v \leq 11.0$ and $m_k \leq 10.0$ with diameters ≥ 8 mas. The near-IR is not very favorable for observations, due to the coolness of these objects. Minor planets may also be observed in the 3.0 um hydration band. Visibility related observations can lead to diameters, shapes and perhaps topographically reconstructed images. Full scale imaging of low surface brightness objects is a challenge which may also stress field-of-view constraints.

5.2 Planetary Satellites

Dyke and Kibblewhite (1986) show that the spatial detection of Io's volcanos require a resolution of 12 mas, and a flux of 3.5 x $10^{-18} w/cm^2 - \mu m$ at 5 um. [i.e. $M_M = 6.6$.] These features have T = 600 K, and therefore will be difficult to observe in the near-IR.

5.3 Comets

These would be targets of opportunity with highly variable observational circumstances

5.4 Imaging of the Solar Surface

This highly specialized application superficially benefits from high photon flux, although requirements for very narrow line imaging quickly leads to photon starved situations as in nighttime interferometry. Baselines longer than ~5 m are not useful because of mean-free path considerations for photons in the solar atmosphere. New interest is arising among the solar physics community for a solar interferometer, but that instrument would be very different from the CHARA Array.

5.5 Detection of Other Planetary Systems

This application is pursuable along two lines:

1. The direct IR detection of protoplanetary disks surrounding young stellar objects, and
2. The astrometric detection of massive planets through submotions in binary systems as described in B.7.

6 Science Characterization Parameters

In the table which follows, the Science Objectives listed in Sections A – E are characterized and quantified according to the parameters described below.

N ≡ Logarithm of the Number of Objects in this Class

λ_o ≡ Optimum Wavelength Regime
- 1 = Near infrared wavelengths (0.9 – 3.0 μm)
- 2 = Visible wavelengths (0.4 – 0.9 μm)
- 3 = Far infrared wavelengths ($\geq$3.0 μm)

M ≡ Median Magnitude at λ_o (V or K magnitude) of the sample
- 1 = Brighter than +6
- 2 = Brighter than +11
- 3 = Fainter than +11

S ≡ Angular Size
- 1 = 1–5 resolution elements
- 2 = 5–25 resolution elements
- 3 = $\geq$ 25 resolution elements

T ≡ Integration Time Limitation
- 1 = Diurnal restrictions only
- 2 = $\leq$1 hour
- 3 = $\leq$1 minute

C ≡ Complexity Parameter
- 1 = Single parameter fit to visibility

2 = Multi-parameter (2-4) fit to visibility
3 = Many-parameter ($\geq$5) fit to visibility
4 = Complex extended object requiring full reconstructive imaging

F $\equiv$ Feasibility = $(\lambda_o + M + S + T + 2C)/6$
1 = Highly feasible, essentially straightforward
2 = Moderately difficult
3 = Very difficult

P $\equiv$ Priority (Significance of Scientific Return)
1 = Highest (of greatest scientific significance to many fields)
2 = Moderate (of high, but more restricted significance)
3 = Lower

Q $\equiv$ Figure of Overall Merit = $(F + 2P)/3$

7 And the Winners Are.....

1. **Single Stars**
   (a) T_{eff} and F
   (b) YSOs
   (c) Surface Structures
   (d) Novae/Supernovae
2. **Binary & Multiple Stars**
   (a) Masses
   (b) Double-lined SBs
   (c) Distance Calibrations
   (d) Low-Mass Companions
3. **Solar System Astronomy**
   (a) Extra-Solar Planets
4. **Visibilities & Imaging**
   (a) Interferometry at Visible Wavelengths
   (b) Imaging Required for most IR Problems

References

Batton Batton, A. H., Fletcher, J. M. & MacCarthy, D. G. (1989), *Eigth catalgue of the orbital elements of spectroscopic binary systems*, Vol XVII, Publications of the Dominion Astrophysical Observatory.

Beckers, J. M. (1989): *The VLT Interferometer Implementation Plan*, VLT Report No. 59b

Dyke, M. & Kibblewhite, E. (1986): *Giant Infrared Telescopes for Astronomy: A Scientific Rationale*, PASP, **98**, 260

Kippenhahn, R., Weigert, A. (1990): *Stellar Structure and Evolution* (Springer, Berlin, Heidelberg), 68–76

McAlister, H. A. (1989): *The CHARA Array: A Feasibility Study for Long-Baseline Optical Interferometry*, 1989, final report to NSF grant AST 84-21304

Ridgway, S. (1988): *Astrophysical Programs for High Angular Resolution Optical Interferometry*,in Diffraction-Limited Imaging with Very Large Telescopes, D. Alloin and J. M. Mariotti, ed. (Kluwer Academic Publishers, Dordrecht), 307-326.

Some Aspects of the VLTI's Potential for Binary Star Research

John Davis

Chatterton Astronomy Department, School of Physics, University of Sydney, NSW 2006, Australia

Abstract. The response of an interferometer to a binary system, the influence of the various parameters of the binary star on the response, and the limitations imposed by the interferometer itself are outlined. The scientific returns from the study of binary stars and, in particular, the determination of fundamental stellar properties such as masses, radii and luminosities and the distances to binary systems by the powerful combination of interferometric, photometric and spectroscopic data, are discussed. The advantages of an array and the potential of the VLTI for binary star studies at various stages of its development are summarised.

1 Introduction

Our knowledge of stellar masses comes from the study of binary stars and, in particular, from visual binaries of known distance and from eclipsing binaries that are also double-lined spectroscopic binaries. In spite of the enormous effort expended in their study, the number of well determined masses and luminosities remains relatively small. High angular resolution interferometry, in combination with spectroscopy, has the potential to revolutionise the field but it is not generally appreciated just how powerful the combination of high angular resolution interferometric data with complementary spectroscopic data is for the determination of fundamental properties including mass, radius and luminosity for individual components of binary systems.

The VLTI, even in its initial phase before an imaging capability is available, could make significant contributions to the determination of fundamental stellar properties and it is this aspect of its potential that is the primary focus of this contribution. The addition of the imaging capacity will be particularly important for the study of close, interacting binary systems but it will also increase the efficiency of the instrument for binary star studies of all types.

2 Background

Before discussing the role of interferometry, the fundamental stellar properties obtainable with the classical methods for studying binary stars will be summarised. The properties are listed in Table 1 for visual binaries, single-lined and double-lined spectroscopic binaries (SB1s and SB2s respectively), and for eclipsing binaries. For each technique a range of orbital parameters can be determined, but only those directly relevant to the determination of stellar properties are shown.

Table 1. Summary of stellar quantities obtainable from binary stars using classical techniques

Binary Type	Parameters/ Functions	Additional Data	Intermediate Functions	Fundamental Properties
Visual	a'', P	π''	$(m_1 + m_2)$	
		a_1'', a_2''	m_1/m_2	m_1, m_2
		π'', $\int_0^\infty f_\lambda \, d\lambda$		L_1, L_2
SB1	$a_1 \sin i$			
	$m_2^3 \sin^3 i/(m_1 + m_2)^2$			
SB2	$a \sin i$			
	$m_1 \sin^3 i$			
	$m_2 \sin^3 i$			
Eclipsing	i	SB2 data		m_1, m_2
	R_1/a, R_2/a			R_1, R_2
		π'', $\int_0^\infty f_\lambda \, d\lambda$		L_1, L_2

In Table 1, binary stars are grouped according to the techniques used to study them. Unfortunately, the selection effects of the various techniques result in very little overlap. In order to be observed as a visual binary the relative orbit of a binary star must be large with a corresponding long orbital period. In contrast, to be observed as a spectroscopic binary, the component stars must have high orbital velocities with smaller orbits and shorter orbital periods. Griffin (1992) has illustrated the lack of overlap between the techniques by means of histograms of visual and spectroscopic binaries against orbital period. In an attempt to rectify the situation, Griffin has carried out an extensive program of spectroscopic observations of long period binaries and McAlister and colleagues (Hartkopf et al, 1992) have conducted a comprehensive program of speckle interferometric observations to observe "visual" binaries of shorter periods than those observed with the classical techniques. The result has been only a marginal improvement in the overlap of binaries observed by both spectroscopic and spatial resolution techniques. Eclipsing binaries, that are also SB2s of known parallax, enable the major parameters of mass, radius and luminosity to be determined for the component stars but they are relatively few in number.

The addition of interferometric observations will extend the "visual" regime to short period orbits. Interferometry, like visual observations, can provide orbital parameters including the inclination of the orbit. Importantly, the angular resolution is no longer limited by the aperture size of a single telescope but by the separation of the input apertures, the baseline of the interferometer. Long-baseline interferometers, such as the Sydney University Stellar Interferometer (SUSI) (Davis, 1994), the United States Navy Prototype Optical Interferometer (NPOI) (White et al, 1994), the VLTI (von der Lühe et al, 1994), and Georgia State University's CHARA Array (McAlister et al, 1994), will provide essentially complete overlap with spectroscopic studies, thus removing the problem of non-overlapping techniques that has handicapped studies of binary stars.

A major difference between visual and interferometric observations arises when the angular resolution of the interferometer is sufficient to resolve the individual stars in a binary system. In such a case the angular diameters of the individual stars can be combined with observed absolute flux distributions to determine absolute flux distributions at the stellar surfaces (assuming that the relative flux contributions can be established as a function of wavelength - see Section 3). The effective temperatures can then be determined. For SB2s, $a \sin i$ from the spectroscopy, and a'' and i from the interferometry can be combined to give the distance to the system. In turn, this can be combined with the angular diameters of the component stars to give their radii. The masses of the components can be determined from the spectroscopic values of $m_1 \sin^3 i$ and $m_2 \sin^3 i$. For SB1s of known parallax, the addition of interferometric data permits the determination of masses and luminosities—and radii if the angular sizes of the component stars are determined. Hipparcos parallax data for SB1s will clearly be important here.

The only binary system for which a component star has been resolved to date is α Vir with the Narrabri Stellar Intensity Interferometer (Herbison-Evans et al, 1971). In addition to orbital parameters and the angular size of the semi-major axis, the angular size of the primary star was determined. Table 2 summarises the principal parameters determined for α Vir by combining the interferometric data with spectroscopic and photometric data.

Table 2. Parameters of the double-lined spectroscopic binary α Virg derived with the aid of interferometric data (Herbison-Evans et al, 1971)

Parameter	Value
Distance	84 ± 4 pc
Brightness ratio of components	6.4 ± 1.0
Mass of primary	$10.9 \pm 0.9\, m_\odot$
Mass of secondary	$6.8 \pm 0.7\, m_\odot$
Radius of primary	$8.1 \pm 0.5\, R_\odot$
log g for primary	3.7 ± 0.1 (g in cgs units)
Primary effective temperature	$22,400 \pm 1000$ K
Primary luminosity (log $L/L_\odot$)	4.17 ± 0.10

With the new interferometers and interferometric arrays mentioned above, including the VLTI, it will be possible to carry out a comparable study of a large number of systems covering a wide range of spectral types and luminosity classes. In favourable circumstances, given an adequate signal-noise ratio, it should prove possible to determine the angular size of the secondary component as well as of the primary, allowing radii and luminosities for both to be determined.

3 The Response of an Interferometer to a Binary Star

The complex degree of spatial coherence and the angular distribution of brightness across a source are a Fourier transform pair (see for example Born & Wolf, 1964). The modulus of the complex degree of coherence is equivalent to the fringe visibility and its phase is the phase of the fringes. In principle, measurement of the fringe visibility and phase enables the brightness distribution to be determined uniquely. In practice, for measurements made with a single baseline on the ground, it is not possible to determine the true phase since it is corrupted by turbulence in the earth's atmosphere. In order to determine the brightness distribution it is necessary to use three or more input apertures simultaneously, and to employ the technique of phase closure to overcome the problem of phase corruption. However, for single stars and binary systems a great deal can be achieved without measuring the phase - for single stars symmetry in the brightness distribution is assumed and, in the case of a binary system, there is a 180° ambiguity in the orientation of the system. For many purposes, these limitations have little significance, and most two aperture interferometers measure the square of the fringe visibility. This will be the type of measurement possible in the VLTI's initial phase.

The response of a two aperture interferometer that measures the square of the fringe visibility $V(d)$ at wavelength λ and baseline d is given by

$$V(d)^2 = \frac{1}{(1+R)^2}\left[\Gamma_1(d)^2 + R^2\Gamma_2(d)^2 + 2R\Gamma_1(d)\Gamma_2(d)\cos\frac{2\pi\theta_s d\cos\psi}{\lambda}\right] \quad (1)$$

$$\text{where } \Gamma_1(d) = \frac{2J_1(\pi\theta_1 d/\lambda)}{(\pi\theta_1 d/\lambda)} \text{ and } \Gamma_2(d) = \frac{2J_1(\pi\theta_2 d/\lambda)}{(\pi\theta_2 d/\lambda)},$$

θ_1 and θ_2 are the equivalent uniform disc diameters of the primary and secondary respectively, R is the brightness ratio of the secondary to the primary ($R \leqq 1$), θ_s is the angular separation of the component stars, and ψ is the angle in the uv-plane between the projection of the line joining the component stars and the baseline. In equation 1 it is assumed that $\lambda/\Delta\lambda \gg 1$ where $\Delta\lambda$ is the spectral bandwidth used for observations.

Figure 1(a) shows the instantaneous response of a two aperture interferometer to a binary star with identical components separated by a projected angle ($\theta_s \cos\psi$) equal to ten times the angular diameter of the individual stars. The envelope of the modulation is determined by the angular diameter of the individual stars. As the projected angular separation varies, due to orbital motion and varying parallactic angle, the spatial frequency of the modulation varies from a maximum, for the maximum projected angular separation, to zero when the projected angular separation is zero. Observations at a given baseline will give a variation in $V(d)^2$ with time which is dependent on the orbit of the binary system. It follows that the orbital parameters and, importantly, the inclination of the orbit can be determined from such observations.

Figure 1 (b) illustrates the responses for binary systems with different brightness ratios (R) for their components. The values of R correspond to magnitude differences of 0, 1, 2 and 3 magnitudes. The figure illustrates that, in addition to the orbital parameters, it is possible to determine the brightness ratio of the components from interferometric observations.

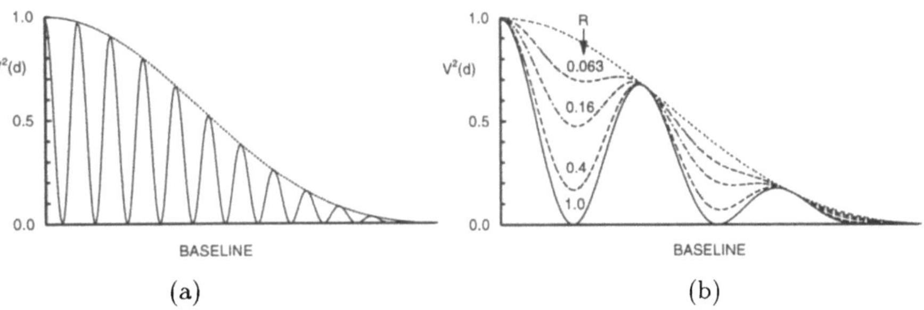

Fig. 1. The response of a two aperture interferometer for a binary star: (a) for component stars of equal brightness, separated by a projected angle equal to ten times the angular diameter of the individual stars. The full curve represents the observed square of the fringe visibility ($V(d)^2$) as a function of the baseline d. The dashed curve is determined by the angular diameter of the individual stars and represents the envelope to the observed curve. It is also the curve that would be observed when the projected angular separation of the component stars is zero - for the line joining the stars perpendicular to the baseline; (b) responses for binary systems with different brightness ratios (R) of the secondary to the primary component. In each case the projected angular separation of the components is equal to four times the angular diameters of the component stars, which are assumed equal for this figure.

A number of observing techniques have been used or proposed for measuring the fringe visibility. For example, a dynamical optical path length equalisation system is used in the SUSI and NPOI systems so that the fringe visibility is measured with the optical paths via the arms of the interferometer matched at the point of beam combination. Other schemes involve scanning the fringe pattern as the path difference between the beams changes, either due to the earth's rotation (Carleton et al, 1994), or by deliberately modulating the path difference with a triangular waveform of chosen amplitude and frequency (Baldwin et al, 1994). The fringe amplitude decreases either side of the matched path position at a rate dependent on the width and shape of the spectral bandwidth used for the observations.

For widely spaced binaries the fringe patterns due to the individual component stars will not overlap unless very narrow bandwidths are used. This is illustrated in fig. 2. The condition for non-overlapping fringe patterns is

$$\theta_s \cos\psi > \frac{1}{d}\frac{\lambda^2}{\Delta\lambda} \tag{2}$$

The projected angular separation in such cases can be determined directly from the separation of the maxima of the fringe patterns.

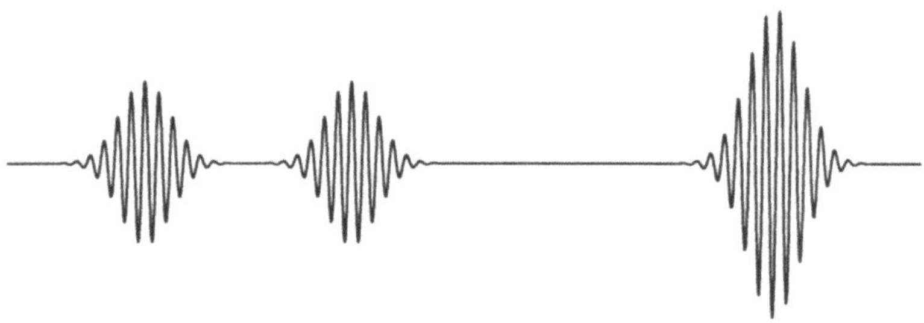

Fig. 2. Simulated scan of two binary systems illustrating the fringe pattern as the systems pass successively through the matched optical path position. The modulation of the fringes is due to the spectral bandwidth used, assumed to have a gaussian profile (in this example $\lambda/\Delta\lambda = 5$). From the left hand side of the scan the first two fringe patterns are for two equal components of a binary system which are separated by a projected angle given by $15\lambda/d$. The third fringe pattern is for a similar binary system but with a projected angular separation 12 times smaller.

In principle, projected angular separations can also be obtained from the separation of fringe peaks in the power spectra of channel fringes produced when the combined light in an interferometer is dispersed. A similar criterion to equation 2 applies except that now $\Delta\lambda$ is the total length of spectrum analysed.

The systems of greatest interest to interferometry are spectroscopic binaries. These are generally of small angular separation and do not meet the criterion expressed in equation 2. In these cases a single fringe pattern comprised of overlapping patterns due to the two component stars is produced. An example is shown in fig. 2. The fringe visibility, given by equation 1, can be determined either by tracking the zero path difference position or by scanning through the fringes.

An estimate of the angular resolving power of an interferometer with a maximum baseline d_{max} is given by

$$\theta_{min} \gtrsim \frac{\lambda}{2d_{max}} \tag{3}$$

where θ_{min} is the minimum angular separation of a binary star or the minimum angular size of a component star that can be measured. This is an approximate guide since it may be possible to measure smaller angles for observations of high signal to noise ratio.

Since the angular sizes of the component stars will be less than their angular separation it follows that, even if the angular separation can be resolved, it may not be possible to measure the angular sizes of the component stars. For example, fig. 3(a) shows the response to a binary system for baselines up to 120 m where the projected angular separation is clearly resolved but the individual stars are not. Examples of the variation of $V^2(d)$ with time for a given baseline for α Equ, a system for which the component stars were unresolved, can be found in Armstrong et al. (1971). These observations, made with the Mark III Interferometer on Mount Wilson, also show a variation of modulation depth with wavelength, corresponding to variations in the brightness ratio R with wavelength. It follows that multi-wavelength observations can be used to separate the relative flux distributions for the individual stars, as required for the determination of the emergent flux distributions and effective temperatures (Section 2).

Figure 3(b) shows the response to a binary system for baselines up to 120 m where the projected angular separation and the primary component are clearly resolved. Examples of the variation of $V^2(d)$ with time for a given baseline for α Vir, for which the primary component was resolved, can be found in Herbison-Evans et al. (1971).

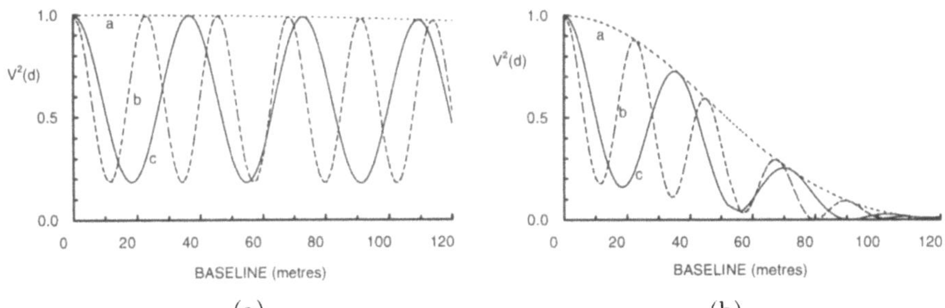

Fig. 3. (a) The response of a two aperture interferometer for a binary system with component stars unresolved at a wavelength of 442 nm with the range of baselines shown. The brightness ratio $R = 0.4$ and the projected angular separations θ_s in arcmilliseconds (mas) are (a) 0 mas; (b) 2.5 mas; (c) 4.0 mas. The angular diameters of the primary $\theta_{UD1} = 0.1$ mas and secondary $\theta_{UD2} = R\sqrt{\theta_{UD1}}$; (b) the response for a binary system with component stars resolved with the range of baselines shown. The only change from (a) is that the angular diameter of the primary $\theta_{UD1} = 1.0$ mas and θ_{UD2} has been scaled proportionately.

4 Imaging of Binary Stars

In Section 3, the need to combine light from a minimum of three apertures in order to form images with an interferometer working through the earth's turbulent atmosphere was outlined. For example, for a three-aperture array, measurements of fringe phase and amplitude are made simultaneously for three baselines to yield three fringe visibilities and one closure phase. The response for each baseline will be of the form shown in figs. 1 and 3.

Imaging with separated aperture interferometers at optical wavelengths is in its infancy. The first image, of the binary star Capella (α Aur), has been reported by the COAST group (Baldwin et al, 1996) and some binary star images have subsequently been generated by the NPOI group. The COAST images of Capella were obtained at a wavelength of 830 nm with three input apertures and demonstrated, for the first time, the feasibility of operating long-baseline optical/near-infrared interferometers for high-dynamic range, high-resolution imaging. For each image, observational data were collected over a five hour period to give sufficient coverage of the uv-plane for images to be constructed. Complex visibilities were determined for each of the three baselines for time intervals comparable with the atmospheric coherence time and closure phases were derived from them. The images were constructed for two epochs, using techniques developed in radio astronomy, and these showed the orbital motion of the binary system. In these observations the angular resolving power of the baselines used (maximum baseline $\sim$ 6.1 m) was insufficient to resolve the angular dimensions of either of the component stars.

The full potential of imaging binary systems will be achieved for observations which enable at least one of the component stars to be resolved. The reasons are the same as for observations made with single baselines (Section 2). The angular sizes of the component stars, in conjunction with spectroscopy, yield fundamental properties of the stars. However, there are several advantages of imaging with simultaneous multiple baselines compared with observing with a single baseline at a time. They include:

1. Increased efficiency as several spatial frequencies and baseline orientations are observed at the same time. This is particularly important for short period systems.
2. Higher dynamic range should be possible enabling systems with a larger magnitude difference between components to be studied.
3. The identification and study of multiple systems ($>$ 2 components).
4. The imaging of close binary systems will enable phenomena such as tidal distortion, reflection effects and mass exchange to be studied.

Currently there are two interferometric arrays capable of producing images, namely COAST and NPOI. Both are in the early stages of their development. The imaging of binary systems is a key scientific driver for the design of the CHARA Array which is now under construction. These instruments are all located in the northern hemisphere. In the southern hemisphere, the VLTI is the only imaging interferometer currently planned. Although it is not an imaging instrument, SUSI has significantly greater angular resolving power than that planned for the VLTI and the two instruments will complement one another.

5 The VLTI and Binary Stars

The capability of the VLTI for binary and multiple star studies will evolve as the instrument is developed. In its initial phase it is expected to be restricted to two aperture, single baseline observations at near and mid-infrared wavelengths. During this phase, studies of double and single lined spectroscopic binaries for the determination of fundamental stellar properties, as outlined in Section 2, will be possible. However, it is noted that unless the spectral cover is extended to visual wavelengths, the limited angular resolution will restrict the number of systems, and particularly the number of component stars, that can be resolved.

Imaging will become possible when three apertures can be combined and realistically, for binary system studies, this means when a third Auxiliary Telescope (AT) is added. The addition of the third AT will also result in a significant improvement in the efficiency of observation even without the formation of images.

An important feature of the VLTI is the ability to move the ATs to different locations. This gives a range of baseline orientations as well as a range of baseline lengths. This is important for good *uv*-plane coverage since, for a fixed baseline orientation, the range of parallactic angle variation with hour angle depends on the declination of the source and, for systems close to zero declination, this can be small or even zero depending on the baseline orientation.

Some of the potential binary star programs for the VLTI include:

1. The detection of low mass companions.
2. The determination of fundamental stellar properties and distances - mass, radius, effective temperature and hence luminosity.
3. The imaging of binary systems and, in particular, close binary systems where tidal distortion, reflection effects, mass exchange etc. are significant.
4. The discovery of new binary and multiple systems.

6 Summary

The VLTI can be expected to make significant contributions to the study of binary and multiple systems. In its initial phase, while operating as a two element interferometer, its main contribution will be the determination of fundamental stellar properties. With full imaging capability, observations will be made more efficiently and the study of more complex systems such as multiple systems and close interacting binaries will become possible.

It is important that the wavelength range is extended towards short wavelengths, to give the maximum possible angular resolving power for the fixed range of baselines, if the full potential of the VLTI for binary and multiple system studies is to be reached. While it would be difficult to justify additional ATs for binary work alone, they would significantly improve the efficiency of imaging programs. Finally, it is also important that the interferometric measurements are complemented by accurate, high-resolution spectroscopy for the determination of the fundamental parameters of stars in binary and multiple systems.

References

Armstrong, J.T., Mozurkewich, D., Vivekand, M., Simon, R.S., Denison, C.S., Johnston, K.J., Pan, X., Shao, M., Colavita, M.M., (1992): The orbit of α Equulei measured with long-baseline optical interferometry: component masses, spectral types, and evolutionary state. AJ **104**, 241–252

Baldwin, J.E., Beckett, M.G., Boysen, R.C., Burns, D., Buscher, D.F., Cox, G.C., Haniff, C.A., Mackay, C.D., Nightingale, N.S., Rogers, J., Scheuer, P.A.G., Scott, T.R., Tuthill, P.G., Warner, P.J., Wilson, D.M.A., Wilson, R.W. (1996): The first images from an optical aperture synthesis array: mapping of Capella with COAST at two epochs. A&A **306**, L13–L16

Baldwin, J.E., Boysen, R.C., Cox, G.C., Haniff, C.A., Rogers, J., Warner, P.J., Wilson, D.M.A., Mackay, C.D. (1994): Design and performance of COAST, in Breckinridge, J.B., ed, SPIE **2200**, Amplitude and Intensity Interferometry II, 118–128

Born, M., Wolf, E. (1964): Principles of Optics, 2nd Edition, Pergamon Press, Oxford.

Carleton, N.P., Traub, W.A., Lacasse, M.G., Nisenson, P., Pearlman, M.R., Reasenberg, R.D., Xu, X., Coldwell, C., Panasyuk, A., Benson, J.A., Papaliolios, C., Predmore, R., Schloerb, F.P., Dyck, H.M., Gibson, D. (1994): Current status of the IOTA interferometer, in Breckinridge, J.B., ed, SPIE **2200**, Amplitude and Intensity Interferometry II, 152–165

Davis, J. (1994): The Sydney University Stellar Interferometer (SUSI), in Robertson, J.G., Tango, W.J., eds, Proc.IAU Symp.158, Very High Angular Resolution Imaging, Kluwer, Dordrecht, 135–142

Griffin, R.F. (1992): Spectroscopic Studies of Binary and Multiple Star Systems, in McAlister, H.A. and Hartkopf, W.I., eds, IAU Colloquium 135, Complementary Approaches to Double and Multiple Star Research, 98–109

Hartkopf, W.I. (1992): Twenty Years of Speckle Interferometry, in McAlister, H.A. and Hartkopf, W.I., eds, IAU Colloquium 135, Complementary Approaches to Double and Multiple Star Research, 459–468

Herbison-Evans, D., Hanbury Brown, R., Davis, J., Allen, L.R. (1971): A study of α Virginis with an intensity interferometer. MNRAS **151**, 161–176

McAlister, H.A., Bagnuolo, Jr., W.G., ten Brummelaar, T., Hartkopf, W.I., Turner, N.H., Garrison, A.K., Robinson, W.G., Ridgway, S.T. (1994): CHARA Array, in Breckinridge, J.B., ed, SPIE **2200**, Amplitude and Intensity Interferometry II, 129-139

von der Lühe, O., Ferrand, D., Koehler, B., Zhu, N., Reinheimer, T. (1995): The Interferometric Mode of the ESO Very Large Telescope, in Breckinridge, J.B., ed, SPIE **2200**, Amplitude and Intensity Interferometry II, 168–179

White, N.M., Millis, R.L., Franz, O.G., Loven, J.M., Hutter, D.J., Johnston, K.J., Armstrong, J.T., Mozurkewich, D. (1994): Progress Report on the Construction of the Navy Prototype Optical Interferometer at the Lowell Observatory, in Breckinridge, J.B., ed, SPIE **2200**, Amplitude and Intensity Interferometry II, 242–247

Combining Mid-Infrared Lunar Occultations and VLTI Measurements

Bringfried Stecklum[1], Hans-Ulrich Käufl[2], and Andrea Richichi[3]

[1] Thüringer Landessternwarte, Sternwarte 5, D-07778 Tautenburg, Germany
[2] ESO, Karl-Schwarzschild-Straße 2, D-85748 Garching, Germany
[3] Arcetri Observatory, Largo E.Fermi, 5, I-50125 Firenze, Italy

Abstract. The angular resolution offered by both the lunar occultation (LO) technique and the VLTI in the infrared spectral range opens up the possibility to study the multiplicity of stars along with the distribution of warm dust surrounding young stellar objects as well as late-type stars in great detail. A simple analysis shows that the limiting magnitude for LOs for an 8m VLT unit telescope is very close to the limiting magnitude for a camera at the combined coherent interferometric focus (at 10μm). However, a single LO observation only allows to extract a one-dimensional brightness profile whereas the VLTI (at least at the beginning of operation) will have only limited coverage of the uv-plane. We propose to combine both types of measurements by means of a coherent data reduction effort in order to provide a unique and powerful tool to facilitate the reconstruction of interferometric images. A straightforward application of this scheme to binary stars will provide useful results for both main-sequence and pre-main-sequence objects.

1 Basics of the Lunar Occultation Technique

The measurement of lunar occultations (LO) of celestial sources offers the possibility to estimate their angular size with very high resolution (in the range of milliseconds of arc - mas). The application of the LO technique includes the measurement of angular diameters of stars (photospheres, e.g. Richichi et al. 1992) and shells (around late-type stars, e.g. Richichi et al. 1995), the detection of close binary stars (Simon et al. 1995) and the investigation of circumstellar matter around young stellar objects (YSOs, Stecklum et al. 1995a). The reader is referred to the recent review of Richichi (1994) for a more detailed description of the method and further references.
Some aspects of the new possibilities offered by the completion of 8m-class telescopes for LO work have been summarized already elsewhere (Stecklum 1994, Stecklum et al. 1995b). In this contribution, we stress the point that the application of the LO method in the mid-infrared (MIR) spectral range and its combination with VLTI measurements are very promising.
In order to make the reader familiar with the principles of the technique, a short introduction is given. The basic geometric circumstances are illustrated in Fig. 1. It is obvious that the difference in position angles between the lunar motion and

* Based on observations collected at the European Southern Observatory, La Silla, Chile, ESO No. 57-E-0321

the position of the object at the lunar limb (contact angle) influences the duration of the disappearance or reappearance. Furthermore, local deviations of the lunar limb from spherical shape will modify this effect.

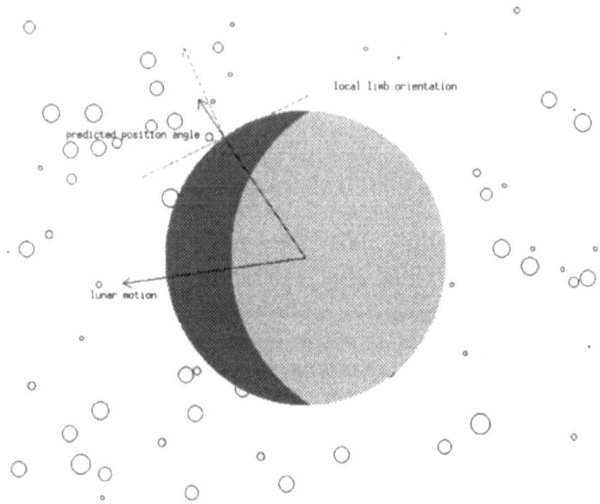

Fig. 1. Geometry of a lunar occultation. Observable events occur at the dark limb.

The diffraction at the lunar limb causes a fringe pattern which can be described by Fresnel diffraction. The angular width of the main fringe (Fresnel scale, Θ_F) depends on the square-root of wavelength λ

$$\Theta_F = \sqrt{\lambda/(2D)}, \tag{1}$$

where D is the lunar distance. It amounts to Θ_F ~25mas at 10μm (N band) which corresponds to a linear extent of ~50m at average distance of the Moon. Note that the observed fringe pattern is the convolution of the point-source pattern with the one-dimensional brightness profile of the object (integrated parallel to the lunar limb).

Due to the average lunar velocity of 0.″5/s, the fringe pattern has to be sampled at high time resolution (millisecond range).

The LO technique has been applied from the radio to the X-ray domain. With the implementation of interferometric arrays like the VLTI, the question arises whether these new facilities will render the LO method useless. In order to address this issue, the following table lists the aperture size for an interferometer needed to achieve the same angular resolution as obtained with LO in dependence on wavelength.

Table 1

Wavelength [μm]	0.5	1.0	10	350	1300	6000
Aperture size [m]	25	35	100	600	12000	26000

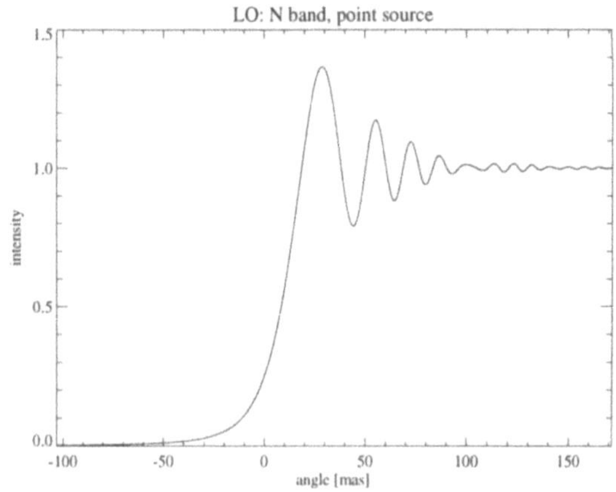

Fig. 2. Fresnel diffraction pattern of a point source for the N band.

It is obvious that in the optical and near-infrared (NIR) range, the new facilities will outperform the LO technique in terms of angular resolution. This holds already for the radio range with operating interferometers like BIMA, OVRO, or VLA. Thus, the mid- and far-infrared (FIR) might be the domain of future LO applications provided the thermal lunar background can be corrected. However, it must be noted that LO take very little time which warrants their use even if other techniques yield similar or better resolution.
One of the conditions for LO work, sampling the light curve at high speed, is met by state-of-the art MIR detectors. In order to prevent saturation from thermal background, these arrays have to be read out with frequencies up to 100Hz.

2 Lunar Occultations with TIMMI

ESO's MIR instrument TIMMI (Käufl et al. 1994) has been used to prove the feasibility of the concept outlined above. On 1996 June 3rd, the reappearance of IRAS 18020-1820 was observed at the 3.6m telescope at the integration time of 27.7ms in the N1 band. The IRAS 12μm flux density of the source is 11.9Jy which led to a signal-to-noise ratio (SNR) of about 8. Fig. 3 shows the lightcurve with the best fit superimposed.

The angular diameter of the source in the N1 band derived from the fit of a uniform disk is 26.2±5.2mas. JHK imaging using IRAC2 at the ESO 2.2m telescope was performed to complete the spectral energy distribution (SED). The IRAS-LRS spectrum of the source shows the SiO feature in absorption indicating the presence of a dense circumstellar dust shell. Fig. 4 displays the SED which can be represented by a superposition of two components. The photospheric component is characterized by $T_{eff} \sim 1250$K and angular diameter $\Theta_{phot} \sim$ 1mas and the corresponding values for the shell are 300K and 40mas. While the SED also suggests the presence of a dust shell, a more rigorous comparison of the angular sizes must be based on radiative transfer calculations.

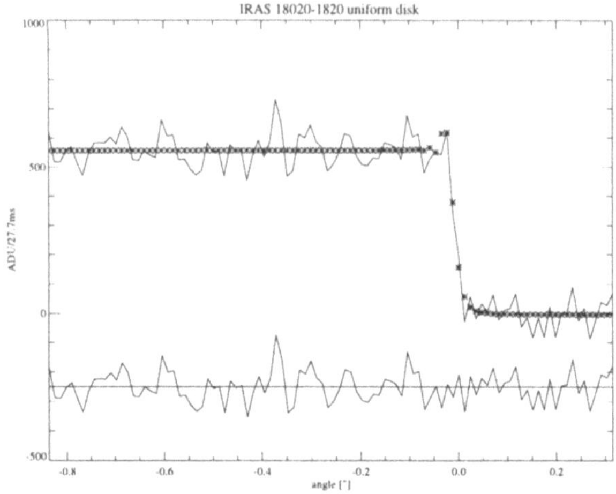

Fig. 3. N1 band occultation light curve of IRAS 18020-1820 obtained with TIMMI (solid line) with best fit (asterisks).

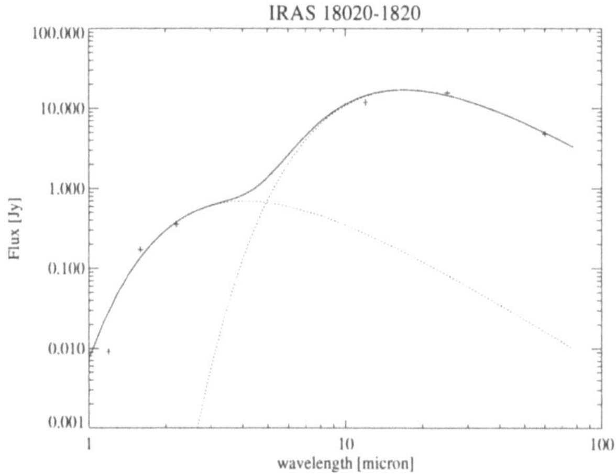

Fig. 4. Spectral energy distribution of IRAS 18020-1820 (crosses) and representation by the superposition of two modified Planck curves.

These results confirm that TIMMI is well suited for MIR LO work. However, for weak sources it is necessary to correct for the thermal emission of the lunar surface. This is obvious from Fig. 5 which is a "shift-and-add" N1 band image of the Moon with IRAS 18020-1820 being distorted at the limb. Various hot and cool areas of different emissivity at the lunar surface can be seen.

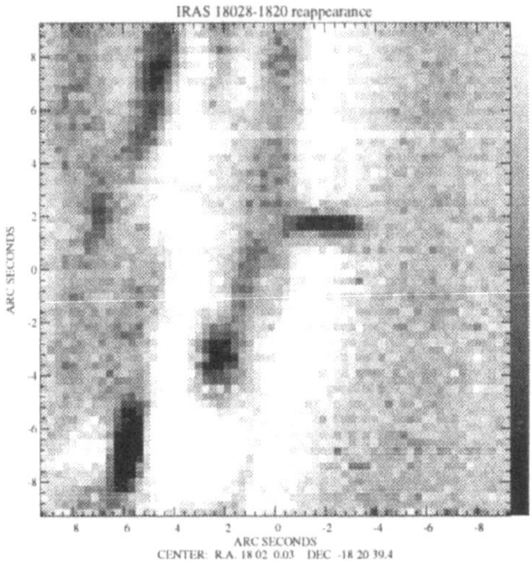

Fig. 5. Negative grey-scale N1 band "shift-and-add" image of the reappearance of IRAS 18020-1820. The image of the source at the lunar limb appears elongated since individual frames were centered on the Moon.

3 MIR Capabilities of LO and VLTI

In the following, we compare the achievable resolution and sensitivities of both techniques in the MIR under the assumption that LOs can be observed using one unit telescope (UT). In this respect, two major cases of LO work can be distinguished. High SNR ($\geq$10) is required to extract intensity profiles from the observed light curves while the search for binaries can be conducted already at low/medium SNR ($\sim$1). For the flux limits listed below, an integration time of 25ms has been adopted for the LO observations versus 10min for the interferometric measurements.

Table 2

	VISA	LO (intensity profiles)
angular resolution	$\sim$10mas	$\sim$10mas
sensitivity	400mJy	200mJy
	VIMA/Hybrid	LO (binary search)
angular resolution	$\sim$20mas	$\sim$50mas
sensitivity	700μJy	2mJy

While the VIMA/hybrid scheme of the VLTI is superior to LO for the search of binary and multiple stars, the capabilities of VISA and LO using one UT are very similar. Therefore, LO may supplement VLTI measurements (and vice versa).

The LO method has inherent problems to deliver the 2D intensity distribution from multiple observations and follow-up application of tomographic techniques

because of missing north-south "baselines" (most occultation events will be head-on). The application of VLTI results to LO measurements will remove ambiguities in the brightness distribution of the source. An example for this case which utilizes inteferometry with a non-redundant pupil mask is the investigation of M8E by Löwe et al. (1996).
On the other hand, the sparse/moderate uv-coverage of VLTI measurements is problematic for image reconstruction. In this case, intensity profiles derived from LO measurements may serve as constraints when image reconstruction is attempted. Furthermore, LO in the MIR explore targets at the level of angular resolution intended to be achieved by VLTI. This supports the preparation and interpretation of future interferometric measurements.

4 LO and the VLT

Major potential targets and scientific aims were discussed already by Stecklum et al. (1995). Here, we focus on novel aspects related to LO in the MIR and/or with regard to VLTI.
The sky coverage of the LO method resulting from the lunar orbit is about 12%. This area covers about 31 000 IRAS point sources which are, according to the above sensitivity figures, potential targets for LOs with the VLT-class telescopes. The case of the IRAS source 18020-1820 illustrates that LOs can be used for an unbiased, high-angular resolution survey of groups of various astrophysical objects. This will lead to serendipitous discoveries for which the carbon star IRAS 06088+1909 (Richichi et al. 1996) is another proof.
However, apart from the large number of potential targets, the sample comprises several objects of key character, e.g. the YSO L1551-IRS5, the ultracompact HII region G5.89-0.39, IRC+10216, and the Galactic Centre.

Binary research will remain one of the most productive areas of LO work (e.g. Simon et al. 1995). The completion of the distribution of pre-main-sequence binaries at small separations essentially relies on LO observations (Leinert et al. 1993). The application of MIR observations will enable to study close binary systems in an evolutionary state prior the T Tauri stage. Such measurements will greatly benefit from simultaneous observations at Paranal and La Silla. Application of different filters will yield color indices. Moreover, the average difference in position angle of $\sim 20°$ for the two sites enables the derivation of true binary parameters.
This argument also holds for circumstellar dust distributions. Multiple LO observations may yield hints on nonspherical/periodic nature of mass loss for stars evolving toward the asymptotic giant branch (AGB). For YSOs, such measurements in the MIR are suited to constrain the size and inclination of circumstellar disks. Both quantities are major input parameters of radiative transfer calculations (Thamm et al. 1994). Due to the optical properties of grains, circumstellar dust distributions look very compact at MIR wavelengths (Yorke and Shustov 1981). Thus, resolving the inner parts of circumstellar envelopes/disks around

both YSOs and late-type/AGB stars requires the highest attainable resolution. We note that the application of the VLTI to the study of YSOs can be complicated by the fact that nearby sources may turn out to be too extended and/or without compact component needed for fringe tracking. Also, for these targets residing in dense molecular clouds, the chance of finding a nearby field star suited for wave front sensing is minor. Thus, there might be some objects or even classes of targets for which the LO technique is the only method to achieve high-angular resolution.

Simultaneous observations using two telescopes at the VLT site enables the estimation of the local lunar limb slope in the absence of fringe pattern (extended sources) by cross-correlating the light curves. This fixes an otherwise poorly constrained parameter of fits to the light curve.
In contrast to normal MIR observations which apply chopping and nodding for background subtraction, LOs have to be performed in staring mode. Thus, slight detoriation of the sky conditions during daytime should not have no great influence on the limiting magnitude for LO in the N band. Day-time measurements of LO in the MIR effectively double the number of potential events, and should be considered in the operation scheme of the VLT.

5 Conclusions

Both, MIR LO observations utilizing one UT and interferometry using VISA yield about the same angular resolution for objects of similar limiting magnitude. Thus, results from both kinds of measurements can be very well combined. The major drawback of the LO method of being one-dimensional can be overcome by quick-look VLTI/VISA measurements. When VLTI/VISA becomes operational, LO may provide highly supplementary information for the reconstruction of brightness distributions in case of poor uv-coverage. In the near future, LO still may provide a unique angular resolution capability in the mid- and far-IR (e.g. for air- and spaceborne instruments) as well as for certain target classes. The time-saving realization of LO measurements warrants their use even when large-baseline interferometers will cover most of the spectrum.

References

Käufl., H.-U., Jouan, R., Lagage, P.O., Masse, P., Mestreau, P. and Tarrius, A., (1994): Infrared Phys. Technol. 35, 203

Leinert, Ch., Zinnecker, H., Weitzel, N., Christou, J., Ridgeway, S.T. Jameson, R. and Haas, M., (1993): A&A 278, 129

Löwe, M., Stecklum, B., von der Lühe, O. and Quirrenbach, A. (1996), in: Proceedings ESO Workshop "Science with the VLTI", Ed. F. Paresce (this volume)

Richichi, A., Di Giacomo, A., Lisi, F. and Calamai, G. (1992): A&A 265, 5 35

Richichi, A., Chandrasekhar, T., Lisi, F., Howell, R.R., Meyer, C., Rabbia, Y., Ragland, S. and Ashok, N.M. (1995): A&A 301, 439

Richichi, A. (1994): Proceedings IAU Symp. No.158 "Very High Angular Resolution Imaging", Eds. J.G. Robertson and W.J. Tango, Kluwer Acad. Publ., 71
Richichi, A., Calamai, G., Lisi., F., Stecklum, B., Herbst, T. and Thamm, E. (1996): IAU Symposium 177, "The carbon star phenomenon", Antalya 27-31 May 1996, Ed. Wing, R.F., in press
Simon, M., Ghez, A.M., Leinert, Ch., Cassar, L., Chen, W.P., Howell, R.R., Jameson, R., Matthews, K., Neugebauer, G. and Richichi, A. (1995): ApJ 443, 625
Stecklum, B., (1994): SPIE Proc. Vol. 2198, Instrumentation in Astronomy VIII, Eds. D.L. Crawford and E.R. Crane, 1425
Stecklum, B., Henning, T., Eckart, A., Howell, R.R. and Hoare, M., (1995a): ApJ 445, L153
Stecklum, B., Richichi, A. and Herbst, T.M., (1995b): Proceedings ESO Workshop "Science with the VLT", Eds. R.R. Walsh und I.J. Danziger, 99
Thamm, E., Steinacker, J. and Henning, Th. (1994): A&A 287, 493
Yorke, H.W. and Shustov, B.M. (1981): A&A, 125

Part 5

HOT STARS

Interferometric Observations of Be Stars

Andreas Quirrenbach

Max-Planck-Institut für Extraterrestrische Physik, Giessenbachstraße, Postfach 1603, D-85740 Garching, Germany

Abstract. The envelopes of seven classical Be stars have been resolved with the I2T, GI2T, and Mk III interferometers in the light of the Hα line. The Hα emission regions have characteristic sizes of 4 to 12 stellar diameters. The interferometric observations strongly support the "equatorial disk" model of Be star envelopes. It will be possible to study these disks in detail with the VLTI.

1 Introduction

Classical Be stars are defined as rapidly rotating hot stars of luminosity class III to V that have shown emission lines in the visible or near-infrared at least once. Struve (1931) recognized the correlation of rapid rotation with the presence of emission lines, and suggested a causal connection of the two phenomena. His proposal that rapidly rotating B stars "form lens-shaped bodies which eject matter at the equator, thus forming a nebulous ring" has long been the fundamental paradigm of Be star envelopes. The subsequent detection of fairly strong intrinsic linear polarization from many Be stars has provided independent evidence for a flattened geometry. Balmer line profiles, continuum flux measurements, and polarization data provided the constraints for a detailed empirical model of the circumstellar envelope of γ Cas, which was presented by Poeckert and Marlborough (1978). They concluded that the circumstellar material forms a disk, rather than a ring. The "wind-compressed disk" model by Bjorkman and Cassinelli (1993) explains the formation of a rather thin (typical opening angle $\sim 3°$) equatorial disk as a natural consequence of the geometry of stream lines in a rapidly rotating stellar wind. A competing model advocates an "onion-like" radial sequence of atmospheric regions (Doazan 1987); in this model the rotation has only a minor effect on the geometry of the envelope. The realization that these models could be tested directly, and subsequently refined, provided the motivation for interferometric observations of Be stars carried out during the past decade.

2 Observations of γ Cas with the I2T and GI2T

Early attempts to resolve γ Cas, the brightest Be star in the northern hemisphere, were carried out with the I2T interferometer of CERGA at 550 nm. The star appeared unresolved, and an upper limit to the photospheric diameter $a^\star \leq 0.9$ mas could be derived from data taken with a 48 m baseline (Vakili et al. 1984).

The first successful observations of a Be star in the light of the Hα line were carried out in 1985 with the same instrument (Thom et al. 1986). Using north-south baselines ranging from 8.0 to 25.0 m, they found an angular diameter $a = 3.25 \pm 0.4$ mas (FWHM of a Gaussian model) for the Hα-emitting region of γ Cas. Combining this value with an estimate for the stellar diameter $a^\star \sim 0.5$ mas, one obtains $a/a^\star \sim 6$, in agreement with the model of Poeckert and Marlborough (1978). Making use of the improved sensitivity afforded by the 1.5 m apertures of the GI2T, Mourard et al. (1989) observed γ Cas with a spectral resolution of 1.5 Å, corresponding to seven resolution elements across the Hα line. They found changes of visibility modulus and phase across the line and concluded that their data agreed well with a rotating envelope model.

While these results clearly demonstrated the large potential of data sets that combine high spatial and spectral resolution, it became also clear that extensive modelling is required to correctly interpret measurements obtained with a very limited sampling of the *uv* plane. Stee et al. (1995) have developed an axisymmetric model for Be star envelopes based on a radiatively driven wind. By comparing predictions from their model with line profiles and visibility data from the GI2T, they show how the free parameters of the model can be constrained by the observations. Additional information can be obtained from polarization-sensitive interferometry. Rousselet-Perraut et al. (1996) present initial results from an experiment in which the GI2T was equipped with a film-sheet polarizer. While they detected extended continuum emission, attributed to photospheric light scattered in the envelope, they did not find any polarization-dependent effects within their measurement accuracy. More exhaustive and more sensitive observations are clearly desirable to determine the structure of Be star envelopes in polarized light.

3 Results from the Mk III Interferometer

A more direct test of the different models mentioned in the introduction was performed with the Mk III interferometer on Mt. Wilson. In the fall of 1992 observations of γ Cas were carried out through a 1 nm wide filter centered on Hα; baselines ranging from 4.2 m to 23.6 m were used (Quirrenbach et al. 1993). Data were taken over the largest possible range in hour angle, in order to obtain a good coverage of the *uv* plane through Earth-rotation synthesis. A clear asymmetry was detected between the visibility amplitudes measured before and after meridian transit, demonstrating for the first time directly that the envelope of γ Cas is strongly non-spherical. Since the Mk III allows rapid switching between stars, it was possible to include a number of other Be stars in the target list, along with an equal number of (nearly) unresolved stars needed for an accurate calibration of the visibilities. For one of the program stars, namely ζ Tau, the data were of sufficient quality to allow an image of the Hα emission region to be reconstructed from the visibility amplitudes with a maximum entropy algorithm (Quirrenbach et al. 1994). Again, a very elongated structure was found, consistent with a disk seen nearly edge-on.

Table 1. The MkIII observing list. The V magnitudes are from the Bright Star Catalog, Hα excesses E_α from Coté and Waters (1987), $v \sin i$ from Slettebak (1982), the intrinsic polarization position angles χ and interferometric results from Quirrenbach et al. (1997). The central part of the table gives Gaussian model fits to the interferometer data, with formal uncertainties. The models have three free parameters: major axis a, axial ratio r, and position angle of the major axis ϕ. Point source models accounting for the photospheric emission were subtracted before model fitting. The two columns on the right list the difference between polarimetric and interferometric position angles, and minimum inclination derived from the MkIII data.

Star	V	E_α [Å]	$v \sin i$ [km s^{-1}]	χ [°]	a [mas]	r	ϕ [°]	$\|\phi - \chi\|$ [°]	i_{min} [°]
γ Cas	2.47	22.7	230	-69	3.47 ± 0.02	0.70 ± 0.02	19±2	88	46
ϕ Per	4.07	57.6	400	25	2.67 ± 0.20	0.46 ± 0.04	-62±5	87	63
ψ Per	4.23	43.2	280	45	3.26 ± 0.23	0.47 ± 0.11	-33±11	78	62
η Tau	2.87	11.9	140	–	2.65 ± 0.14	0.95 ± 0.22	(19)	–	18
48 Per	4.04	24.2	200	–	2.77 ± 0.56	0.89 ± 0.13	(68)	–	27
ζ Tau	3.00	26.9	220	33	4.53 ± 0.52	0.28 ± 0.02	-58±4	91	74
β CMi	2.85	11.5	245	–	2.65 ± 0.10	–	–	–	–

In total, seven Be stars were included in the Mk III observing list (Quirrenbach et al. 1997, see also Table 1). Depending on the declination of the star, the longest baseline used was either 23.6 m or 31.5 m. While the program stars were unresolved in the 550 nm continuum, as expected from estimates of the photospheric diameters, they were all resolved in the Hα filter. For a quantitative analysis of the data, simple models consisting of a single elliptical Gaussian were fitted to the visiblities. In the case of β CMi, however, the uv coverage was insufficient to constrain the ellipticity of the envelope; therefore a circular Gaussian model was used for this star. The contribution of the photospheric light within the 1 nm wide filter was estimated for each star, and a small uniform disk component with appropriate brightness was subtracted before the model fitting. For all stars, the major axes of the best-fit models fall in the surprisingly narrow range of $a = 2.6 \ldots 4.5$ mas. Photometric estimates of the photospheric diameters based on the Barnes-Evans relation give $a^\star = 0.34 \ldots 0.74$ mas; i.e., the Hα emission regions have sizes (FWHM of Gaussian model) of about 3.5 to 12 photospheric diameters. In the Mk III sample of seven stars, the normalized envelope size $a/a^\star$ is correlated with the strength of tracers of the circumstellar material, such as the equivalent width of Hα and the 12 μm flux. However, there is no apparent correlation of $a/a^\star$ with the equatorial rotation velocity.

The axial ratios r derived from the Gaussian model fitting span a wide range (see Table 1): ϕ Per, ψ Per, and ζ Tau are strongly elongated ($r < 0.5$), γ Cas has an intermediate ellipticity ($r = 0.7$), and η Tau and 48 Per appear nearly circular ($r \sim 1$). In the framework of disk models, this result can easily be understood as an inclination effect. The strongly elliptical envelopes must then be disks seen nearly edge-on, γ Cas is at an intermediate inclination, and the circular objects are disks seen face-on. When it is assumed that the disk lies in the equatorial plane of the star, this is consistent with rotational velocity data; the stars with edge-on disks have generally higher values of $v \sin i$ than those seen face-on. It is clear the the interferometric data do not allow a separate determination of inclination and the disk thickness. For example, the disk around ζ Tau could either be extremely thin and viewed at an inclination of $\arccos 0.28 = 74°$, or alternatively a somewhat thicker disk seen exactly edge-on. Under the assumption of axisymmetry it is possible, however, to derive a minimum inclination from $i_{\rm min} = \arccos r$ (see Table 1).

An independent estimate of the inclination can be obtained from polarimetry and infrared photometry (Coté and Waters 1987). This method is based on the idea that the infrared excess measures the amount of circumstellar material regardless of inclination, whereas the degree of linear polarization depends on both density and inclination of the disk. A very good agreement is found between i determined with this method, and $i_{\rm min}$ obtained from the interferometry. This implies that $i_{\rm min}$ is actually a good indicator of the of the inclination of the envelope, and that the disks are fairly thin. The data on ζ Tau can also be used directly to derive a maximum disk opening angle of $\sim 20°$. Furthermore, it is noteworthy that η Tau and γ Cas, which have been classified as shell stars (Bidelman 1976), are clearly viewed at fairly small inclination. It is therefore clear that a shell star classification is insufficient to conclude that a star is seen equator-on.

The combination of the lower limit on the inclination $i_{\rm min}$ with the projected rotation velocity $v \sin i$ gives an upper limit $v_{\rm max}$ to the true equatorial rotation velocity v, which is thus restricted to the range $v \sin i \leq v \leq v_{\rm max}$. It is now possible to compare v to the critical rotation velocity $v_{\rm crit}$, i.e., the rotation velocity at which the gravitational force would equal the centrifugal force at the equator. The value of $v/v_{\rm crit}$ places important constraints on the mechanism that drives the equatorial disk. With the data listed in Table 1, and theoretical estimates of $v_{\rm crit}$ from Collins (1974), it is found that γ Cas and ζ Tau rotate at a clearly subcritical rate, whereas the rotation rate of ϕ Per and ψ Per is closer to the critical rate. (For η Tau and 48 Per the allowed range of v is too large to draw any useful conclusions.) However, it should be kept in mind that both the observational uncertainty in $v \sin i$ and the theoretical uncertainty in $v_{\rm crit}$ are quite large and difficult to quantify. The situation could be improved with models of rotating stars, which have $v/v_{\rm crit}$ as a free parameter. The profiles of photospheric lines and limits on the inclination from interferometry could then be used together to constrain these models; this would give a direct estimate for $v/v_{\rm crit}$.

Additional information is contained in the position angle ϕ of the major axis of the Gaussian on the sky. In the four cases where the ellipticity is significant, ϕ is perpendicular to the position angle χ of the intrinsic linear polarization (see Table 1). This is exactly what is expected if the polarization is due to electron scattering in an optically or geometrically thin disk. Models in which the polarized flux is predominantly generated in the polar regions are ruled out by these data. The disk of ψ Per has also been resolved in the radio regime at 15 GHz with the VLA (Dougherty and Taylor 1992). The best-fitting Gaussian model at this frequency has a major axis of 111 ± 16 mas at position angle $-32^\circ \pm 10^\circ$, identical to that of the Hα disk. This means that there is no misalignment between the inner disk, which produces the optical polarization and Hα emission, and the outer disk at ~ 100 stellar radii seen at cm wavelengths.

4 Prospects for VLTI Observations

Compared to the instruments that have been used to obtain the results summarized in the previous sections, the VLTI will open new observational opportunities through significant technical improvements and enhanced capabilities. The most important characteristics of the VLTI for more detailed observations of Be stars are its large collecting area, baselines of up to 200 m, dense coverage of the *uv* plane, wide wavelength coverage, and versatile instrumentation. For a number of the brightest Be stars, it will be possible to obtain emission line images with a few resolution elements across the disk. The imaging will be facilitated by the presence of the stellar photosphere, which can serve as a reference for fringe tracking and mapping. Using referenced phases instead of closure phases retains more information and thus provides a higher dynamic range in images reconstructed from observations with few-element interferometers. The phase change across spectral lines is also an important tool to investigate the kinematics of the disk; it may be the only reliable diagnostic to determine the rotation law (Stee 1996). The good sensitivity of the VLTI will allow a spectral resolution of 10 to 30 km/s; it will thus be possible to obtain data cubes similar to those from present-day mm-wave interferometers. The sensitivity and wavelength coverage of the VLTI will play an important role for observations of weaker spectral lines, which probe different radii of the disk and carry information on the ionization structure and density distribution of the circumstellar material. In particular, a close look at the transition zone between photosphere and envelope will be needed to understand the physical mechanisms of wind acceleration and disk formation.

Continuum observations on the longest baselines of the VLTI will just resolve the photospheres of the brightest Be stars. Simple diameter measurements can be used to determine the effective temperatures. With more accurate data, the rotational flattening, the variation of the effective temperature with latitude, and limb darkening can be investigated. Together with spectroscopic data, these measurements can be used to refine models of rapidly rotating stars. Continuum images and size measurements of the envelope at visual and infrared wavelengths will complement interferometric observations in spectral lines. Many Be stars are known to be binary systems; interferometric searches for the companions will help us determine their nature and potential influence on the distribution of the circumstellar matter. In some cases, it may be possible to measure interferometric orbits and use them to determine the stellar masses.

While such comprehensive observations of individual stars can give a wealth of information on Be stars and their disks, there is also much to be learned from less detailed data on a large and sytematically selected sample of Be stars. With the VLTI it will be feasible to determine the global properties of a fair number of Be star disks; observations similar to the Mk III results discussed above should be possible for hundreds of objects. This will allow statistical correlations of disk parameters with stellar properties to be explored, and interferometric, spectroscopic, photometric and polarimetric diagnostics can be compared and cross-calibrated. By extending the sample to related stellar classes (O and A-type emission line stars, Wolf-Rayet stars, Herbig Ae/Be stars), we can learn under which conditions Be-type disks are formed. Long-term monitoring with the VLTI can also shed new light on the many types of variability observed in Be stars. Variable mass-loss rates, non-radial pulsations, disk instabilities and warps, and passages of companions through the disk plane may all play a role, and disentangling of these phenomena requires a concerted effort with all observational techniques available.

In summary, the VLTI is an extremely powerful tool to probe a wide variety of phenomena in Be stars. The following is a short list of open questions in this area that can be addressed with the VLTI:

- What is the disk opening angle?
- What is the radial and vertical structure of the disk (density laws, ionization structure)?
- Does radial motion play a significant role in the kinematics of the disk material, or is only rotation important?
- What is the rotation law in the disk?
- Is the disk axisymmetric?
- How do the properties of the disk depend on the stellar characteristics (effective temperature, rotation rate, evolutionary state)?
- What is the relation of the disk to the stellar wind?
- Can the wind-compressed-disk model explain the Be star phenomenon?
- Does binarity play an important role for the formation and structure of Be star disks?
- What is the nature of Be shell stars?
- Why are Be stars so variable on so many different time scales?
- Can we observe the consequences of disk instabilities, or of sporadic ejection of matter?

It is probably not realistic to expect that the VLTI will provide the final answers to all these questions, but we can certainly hope for significant progress in most of these areas. The physics of Be stars is thus one of the fields, on which the VLTI can have a profound impact. Interferometric observations of Be stars should also be considered in a broader context: of all types of circumstellar disks, those found around Be stars are by far the brightest, and probably the largest at visual and near-IR wavelengths (in terms of angular diameter). We can therefore use them to study phenomena of general interest for the physics of circumstellar disks, such as transport of matter through a disk, instabilities, and warps.

5 Technical Requirements for the VLTI

The implementation of the VLTI will progress step by step from a modest beginning with only two telescopes coupled for "first fringes" to a powerful and versatile instrument with a broad range of capabilities. Each phase of this development will enable new and interesting measurements of Be stars; the complexity of these observations will grow with the number of telescopes available and the sophistication of the interferometric instrumentation. In the beginning, the VLTI can build on the results from other interferometers; a single-baseline configuration with an instrument working at the wavelength of Hα and providing a spectral resolution of ~ 3000 would be adequate. In the long run, instruments covering the whole visual wavelength range as well as the near- and mid-IR will be required. The visual and near-IR instruments should give a spectral resolution in the range $10\,000 \lesssim R \lesssim 30\,000$, while $300 \lesssim R \lesssim 1\,000$ is sufficient in the mid-IR.

The development of the "standard" instruments for the VLTI will obviously require substantial effort over an extended period of time, because they have to cover large ranges in wavelength and resolving power. Polarimetric capabilities add so much complication that they have not been considered in the instrumentation plan for the VLTI. Furthermore, calibrating the instrumental polarization with so many oblique reflections is very difficult. Nevertheless, this topic should be kept in mind for future upgrades of the VLTI, so that one day we may obtain images of Be star disks in polarized light.

The spatial resolution of the VLTI is just high enough for the more ambitious observations of Be stars envisaged; the longest baselines available (up to 200 m for the auxiliary telescopes) are clearly needed. While the Mk III observations show that the overall shape of Be star disks does not change dramatically over a few months (Quirrenbach et al. 1997), short-timescale photometric and spectroscopic variability may well be an indication of structural changes on smaller spatial scales. Variability must therefore be taken into account for synthesis imaging; the possibility of reconfiguring the array several times within one night is essential to avoid artefacts in the images caused by variability. It may even turn out that Earth-rotation synthesis, which takes a full night, is not adequate. Snapshot images using the four 8 m telescopes and three or more auxiliary telescopes could solve this problem, but a sufficient number of delay lines and a multi-way beam combiner would be needed.

Since observations of Be stars rely heavily on the visible wavelength range, the sensitivity all the way down to Hβ is an important issue. Even at a superb site like Paranal, the atmospheric coherence parameter r_0 will rarely be larger than 30 cm at 500 nm. While multi-r_0 techniques can be used initially, adaptive optics will eventually be required to exploit the full sensitivity of the VLTI. In view of the high servo bandwidth required, development of affordable adaptive optics systems working in the visible will be a challenging task.

6 Conclusions

During the past decade optical and near-infrared long-baseline interferometry has advanced from the state of technology demonstration to an astronomical tool with an increasing potential of applications to "real-world" problems. Observations of Be stars with the I2T, GI2T, and Mk III instruments have confirmed the equatorial disk model for the envelopes of these objects, and a number of important parameters have been measured for a small sample of Be star disks. Building on these initial results, the VLTI will be able to address fundamental questions about the formation and structure of these disks, and their relation to the stellar photosphere and the wind. Important results in these areas can be expected during all stages of the implementation of the VLTI; both detailed studies of individual objects and long-term monitoring of a fairly large number of stars will be needed. To get the most out of the VLTI data, they must be interpreted in the context of complementary information from photometry, spectroscopy, and polarimetry; in some cases coordinated observations will be required. It is likely that future progress in our understanding of the Be star phenomenon will rely heavily on information obtained with optical interferometry. And in spite of the worldwide competition expected from a number of powerful new instruments, much of this information could come from the VLTI.

References

Bidelman, W.P. 1976, in IAU Symposium 70, Be and Shell Stars, ed. A. Slettebak (Dordrecht: Reidel), 457
Bjorkman, J.E., & Cassinelli, J.P. 1993, ApJ, 409, 429
Coté, J., & Waters, L.B.F.M. 1987, A&A, 176, 93
Doazan, V. 1987, in IAU Colloquium 92, Physics of Be Stars, ed. A. Slettebak & T.P. Snow (Cambridge: Cambridge University Press), 384
Dougherty, S.M., & Taylor, A.R. 1992, Nature, 359, 808
Mourard, D., Bosc, I., Labeyrie, A., Koechlin, L., & Saha, S. 1989, Nature, 342, 520
Poeckert, R., & Marlborough, J.M. 1978, ApJ, 220, 940
Quirrenbach, A., Bjorkman, K.S., Bjorkman, J.E., Hummel, C.A., Buscher, D.F., Armstrong, J.T., Mozurkewich, D., Elias, N.M., & Babler, B.L. 1997, ApJ, in press
Quirrenbach, A., Buscher, D.F., Hummel, C.A., Mozurkewich, D., & Armstrong, J.T. 1994, A&A, 283, L13
Quirrenbach, A., Hummel, C.A., Buscher, D.F., Armstrong, J.T., Mozurkewich, D., & Elias, N.M. 1993, ApJ, 416, L25
Rousselet-Perraut, K., Vakili, F., Mourard, D., Morand, F., Bonneau, D., & Stee, Ph. 1996, A&A, in press
Slettebak, A. 1982, ApJS, 50, 55
Stee, Ph., de Araújo, F.X., Vakili, F., Mourard, D., Arnold, L., Bonneau, D., Morand, F., & Tallon-Bosc, I. 1995, A&A, 300, 219
Stee, Ph. 1996, A&A, 311, 945
Struve, O. 1931, ApJ, 73, 94
Thom, C., Granes, P., & Vakili, F. 1986, A&A, 165, L13
Vakili, F., Granes, P., Bonneau, D., Noguchi, M., & Hirata, R. 1984, PASJ, 36, 231

Spectroscopic and Photometric Support for Interferometric Observations of Be Stars

Petr Harmanec

Astronomical Institute, Academy of Sciences of the Czech Republic, 251 65 Ondřejov, Czech Republic, Internet: hec@sunstel.asu.cas.cz

Abstract. The nature of Be stars and true physical causes of their spectral and brightness variations on several timescales remain largely unknown. It is argued that real progress in their understanding may come from systematic spectro-interferometric observations supplemented by simultaneous spectroscopy and photometry. Several unsolved problems and existing hypotheses to be addressed this way are discussed. It is argued that the present-day interferometers can mainly help to understand the geometrical structure of the circumstellar matter around Be stars and the nature of the long-term (years and longer) and (in some cases) medium-term (weeks to months) changes. Studies of rapid changes on timescales shorter than about 2 d and (in most cases) investigation of possible duplicity of Be stars will require a much higher angular resolution to be only achieved by future interferometers or by application of the promising technique of differential interferometry. It is clear, however, that Be stars are very suitable and challenging objects to be studied by means of interferometry.

1 Basic Terms and Ideas

Be stars are objects of spectral type B (T_{eff} from about 10 000 K to 30 000 K) which showed emission in at least one Balmer line on at least one occasion. I still prefer this original definition over a more restricted one, widely used now, which limits the Be phenomenon to objects of luminosity classes III-V only (for detailed arguments, see Harmanec 1983). Some authors make a compromise and call the B III-Ve stars the "classical Be stars". It is generally believed that the Balmer emission in the spectra of Be stars is a sign of the presence of circumstellar matter around them. Here, the agreement ends however: Different interpretations of the geometry of the Be envelopes range from the classical flattened equatorial disks over spheroidal envelopes, accretion disks in binary systems to helmet-type loops related to local magnetic fields or focused stellar winds (cf., e.g., contributions in Harmanec and Pavlovski 1983, Slettebak and Snow 1987, Balona et al. 1994). It is true, however, that flattened disks of one sort or another represent the most often considered model with growing observational evidence in its support. There is no general agreement on the very nature of the Be phenomenon and many investigators believe that there are several different ways how to produce the observed Balmer emission. Whatever the phenomenon is, it seems to extend to the neighbouring spectral classes: Oe stars (Frost and Conti 1970) and Ae stars (Andrillat et al. 1983) are also known.

Certainly the most challenging property of Be stars is their *variability* occurring simultaneously on several timescales from hours to decades. And it is what makes the Be stars so attractive for future interferometric studies. Because of different sensitivity of different observing techniques to geometry, velocity fields, brightness or density distribution and because of the time scales involved, the interferometric studies of Be stars require systematic observations supported by parallel spectroscopy, photometry and perhaps also polarimetry. More important than exactly simultaneous observations with different techniques are *numerous enough* observations of each kind, *well distributed in time* with respect to the time scale studied. Similar systematic studies could help to separate geometrical and physical effects and, therefore, lead to a substantial progress in the understanding of still puzzling Be stars. It is my belief that **the new pieces of information provided by the rapidly developing interferometric techniques and accurate determinations of distances to particular Be stars belong to the most promising ways how to progress in understanding the Be phenomenon.**

There are over 300 Oe, Be and Ae stars known within the $6\overset{\text{m}}{.}5$ magnitude limit of the Bright Star Catalogue. Quite a number of them are also known to have visual or at least optical companions. Interferometric studies can profit from the latter fact in some cases: bright visual companions can serve as 'point' sources which is also true about the Be stars themselves: their visibilities can be compared to visibilities of circumstellar matter around them. The first results of interferometric studies of Be stars, presented by the GI2T or Mark III groups (cf., e.g., the reports in Balona et al. 1994) are very encouraring and promising.

Below, I review most of so far recognized types of Be-star variability and their current interpretations and show where and how the interferometry could help to test the existing hypotheses. Readers interested to learn more about Be stars in general are referred to the following reviews and conference proceedings: McLaughlin (1961), Slettebak (1970), Jaschek and Groth (1982), Kogure and Hirata (1982), Hirata and Kogure (1984), Slettebak and Snow (1987), Baade (1991), Balona et al. (1994) and Harmanec (1983).

2 Long-Term Changes

2.1 Emission-to-Continuum (E/C) Changes

The strength of the Balmer and other emission lines in most of Be stars varies secularly in time. One observes a transition

B $\rightarrow$ Be $\rightarrow$ Be shell $\rightarrow$ Be $\rightarrow$ B

which typically lasts a few decades. Well-documented examples are, e.g., BU Tau (HD 23862) or γ Cas (HD 5394). These spectral changes are accompanied by two different types of light variations: Either, the development of a new Be phase is accompanied by a light decrease or by a light increase. The development of a strong shell phase, with shell lines also seen in the metallic lines, is usually reflected by a secondary light extremum. Moreover, not always the above described cycle of the phase transitions occurs. For instance, only B $\rightarrow$ Be $\rightarrow$ B changes are also known. For κ Dra, Juza et al. (1994) confirmed an earlier suspicion that its Balmer-emission strength varied periodically (P = 23 yrs).

Canonical explanation of the phase transitions is *the formation and dispersal of a circumstellar envelope*. There are, however, various interesting suggestions concerning the origin of this phenomenon and its nature. Struve (1931) argued that the Be envelopes are rotationally flattened disks and this is the basic geometry postulated even now (cf., e.g., Dachs et al. 1986, Waters 1986, Hanuschik 1995). However, Höflich (1988) was remarkably successful in modelling the observed Balmer-line profiles on the assumption of spherical envelopes. Harmanec (1983) argued that the inner parts of the Be envelopes can mimic 'pseudophotospheres', slightly cooler than the underlying stars, and suggested interpretation of the two types of the correlation between brightness and emission strength in terms of different aspect effect. Also Hirata (1995) explains the phase transitions using the concept of a variable envelope and 'extended photosphere'. Waters (1986) and Hanuschik (1996) conclude that the Be-star disks are thin near the star (in the direction perpendicular to the equatorial plane) and have a typical half-opening angle of 15° (Waters) or 13° (Hanuschik). Hanuschik also argued that the B $\rightarrow$ Be shell phase transitions *do not represent* a necessary part of the whole process of phase transitions but a phenomenon which can only occur in cases when the inclination of the rotational axis of the star in question is close to 70° ± 5°. Harmanec (1982) suggested that the re-appearance of the emission-line phases could be a phase-dependent effect of orbital motion in a wide (as yet unrecognized) binary system. No such binary has been found for κ Dra so far but it is interesting to note that binaries of suitable properties were found for three of the four stars he mentioned: BU Tau (Gies et al.1990), V832 Cyg (HD 200120; McAlister et al. 1984) and o And (HD 217675; Blazit et al. 1977, Harmanec et al. 1987).

It is clear that interferometry is ideally suited to test most of above mentioned suggestions and bring decisive answers. Many bright Be stars are found at distances smaller than some 200 pc from us. Since various available estimates indicate that the radii of their envelopes are at least 5–15 times larger than the radii of respective stars, one can assume that the angular diameters of Be envelopes may be of the order of $0\farcs001$ or larger, i.e. within the reach of already existing spectro-interferometers able to observe in Hα. Systematic observations of a group of brighter Be stars, lasting many years (even with a sampling of only 1–2 observations per month and star) could provide clear answers to questions like: How the shape and extent of the envelope varies with the phase transitions? What happens during the formation of a metallic shell spectrum? Do the envelopes of early and late B stars differ in their geometry and/or shape? Note again the need for parallel spectroscopic and photometric monitoring of the same Be stars to achieve the goal. This type of monitoring is suitable even for observatories in less ideal climatic conditions and for serious amateur astronomers as well.

2.2 Cyclic V/R and RV Variations

Slow variations of the relative intensity of the violet and red peaks of the double Balmer emission (V/R changes) and the accompanying changes in the radial velocity of the whole emission line (RV changes) can only be observed during the Be and/or Be shell phases of the E/C cycle. A typical time scale of these changes is 3–10 years. Most often, they only last over 3–4 such cycles and each cycle has a different length and amplitude. However, in some Be stars the cycles were so regular that they were initially misinterpreted for orbital motion in a binary system (cf. BK Cam = HD 20336; McLaughlin 1961).

Two basic interpretations of the slow V/R and RV variations exist: (1) Model of a gradually revolving elongated envelope (Struve 1931, McLaughlin 1961 and Huang 1973), and (2) Kato's (1983) model of one-armed global oscillation of the disk. Kříž and Harmanec (1975) and Harmanec and Kříž (1976) have shown how an elongated disk can be formed and forced to revolve if the Be star is the primary of a binary system and demonstrated a good numerical agreement for the case of ζ Tau. However, no evidence of duplicity has so far been found for a number of the long-term V/R variables. Recently, Okazaki (1994) made specific predictions for the one-armed oscillation model: The V/R and RV curves should vary in phase (which is observed) and there should also be accompanying cyclic light variation which should lag for $0\overset{\mathrm{P}}{.}25$ behind the V/R curve. The data at hand seems to speak in favour of the elongated-disk model for γ Cas (see, e.g., Telting et al. 1993) but more stars will have to be investigated.

Obviously, the systematic observing program, defined in the previous subsection, will also conclusively demonstrate whether the Be envelopes with spectroscopically observed slow V/R and RV variations are changing their spatial asymmetry or radius or both.

3 Medium-Term Changes

Most of Be stars are also known to exhibit spectral and brightness variations on time scales from several days to months. At least two types of such variations can be distinguished and are discussed here.

3.1 Periodic RV, V/R, Line-Profile and Light Changes

Such phase-locked changes were found for, e.g., Beta Lyr (HD 174638), AX Mon (HD 45910), V839 Her (HD 142926), V744 Her (HD 162732), ζ Tau (HD 37202), KX And (HD 218393), LQ And (HD 224559), CX Dra (HD 174237), RX Cas (BD+67°244), SX Cas (HD 232121) (I apologize for not crediting the original investigators due to limited space here - see references in Harmanec 1983, 1987a), more recent discoveries include V832 Cyg (HD 200120; Tarasov and Tuominen 1987), FY CMa (HD 58978; Grady et al. 1988, Grady 1989), HR 2545 (HD 50123; Sterken et al. 1994) or 3 Pup (HD 62623; Plets et al. 1995). In all these cases, the medium-term variations were found to be due to the duplicity of the Be stars in question.

A typical binary system with a Be star has an orbital period of the order of 100 d and a total mass of about 10 $M_{\odot}$, therefore a separation of the components of about 200 $R_{\odot}$. Some of such binary systems could already be resolved with the existing interferometers. In an ideal case, the geometry and spatial orbit of the binary and the structure of the circumstellar matter could be obtained. This, in turn, would help to derive more accurate masses for particular non-eclipsing Be stars. Moreover, the interferometric imaging may help to conclude whether the structure of the Be envelopes found in binaries differ from those around seemingly single stars.

3.2 Flare-Like Emission-Line and Light Changes; Quasiperiodic Changes

A sudden appearance of the Balmer emission with a subsequent slower transition back to absorption phase, taking place within one or a few weeks and accompanied by a light variation of similar shape, was reported for several stars. One of the first well documented cases of this type of behaviour was QR Vul (HD 192685) – see Pavlovski et al. (1983) and references therein – while probably the most systematic study was carried out for μ Cen (HD 120324) by Hanuschik et al. (1993). They found that the rise of the emission to its full strength always occurred during 2–5 d while the whole emission episodes lasted 20–80 d. They clearly demonstrated that this variability *is not* periodic and interpreted it as ejections of material from the stellar surface, perhaps facilitated by magnetic activity of the star.

Yet another type of variability on a time scale of weeks are quasi-periodic light variations, reported by Mennickent et al. (1994) for EW CMa (HD 56014) and ω CMa (HD 56139) to appear prior to their secular light decreases.

Interferometry could show what is really happening in the geometrical sense. Does the formation of the envelope proceed isotropically or – for instance – by ejection of gas from certain areas on the stellar surface? However, a successful monitoring of similar unpredictable events would require a very good spatial resolution and also flexibility in allocation of the observing time and a well-prepared international collaboration.

4 Rapid (Multi)periodic Changes

By "rapid variations" students of Be stars usually mean changes on timescales from about $0\overset{d}{.}1$ to $2\overset{d}{.}0$ but I find it more appropriate to speak (somewhat less strictly) about variations occurring on a time scale comparable to or shorter than the rotational periods of respective stars. Periodic or multiply periodic line-profile, RV and light changes were first reported for particular Be stars by Hill (1926) (κ Dra = HD 109387), Guthnick (1941) (o And = HD 217675) and Walker (1953) (EW Lac = HD 217050) but a real boom in this area occurred early in eighties, especially in connection with the growing use of electronic detectors in stellar spectroscopy and systematic photometric studies of Be stars (see, e.g., Walker, Yang and Fahlman 1979, Baade 1982, Bolton 1982, Percy 1983, Vogt and Penrod 1983, Harmanec 1983, 1984, Balona and Engelbrecht 1986 or Smith 1986).

Three competing hypotheses to explain these rapid changes co-exist:

• non-radial oscillations in the photosphere (Walker et al. 1981, Baade 1982, Bolton 1982, Vogt and Penrod 1983);

• corotating surface structures (Struve 1952, Walker 1953, Walker et al. 1979, Harmanec 1983, Balona and Engelbrecht 1986); and

• corotating structures in circumstellar matter slightly above the photosphere (Walker et al. 1981, Vogt and Penrod 1983, Harmanec 1989).

The decision which of the above hypotheses (or combination of them) is the correct one may be a challenging task for future interferometers, able to resolve the stellar disks. However, even now the technique of differential interferometry (Beckers 1982, Petrov in these Proceedings) may perhaps be applied.

In my defense of the model of corotating circumstellar structures (Harmanec 1989) I suggested that the equatorial surface 'spots' may act as roots of outflowing circumstellar spokes or streams, corotating with the stars. I argued that the spokes may be responsible for the formation of the so-called narrow absorption components, seen in UV resonance lines. This idea found recently a strong support in a systematic UV study of the B0.5Ib star HR 3090 (HD 64760) by Prinja, Massa and Fullerton (1995). Naturally, such structures may extend into distances of several stellar radii and may, therefore, be more easily detectable with interferometers.

5 Examples of Another Type of Problems to Be Solved

I argued several times (Harmanec 1983, 1987b) that it is conceivable that some OB stars classified by luminosity classes I-II need not be genuine supergiants in the evolutionary sense but rather *shell* stars with dense and extended envelopes in which a main-sequence object is hidden.

Baade (1986, 1988) reported rapid line-profile changes for ζ Pup, an O4Ief star, and also for several other high-luminosity emission-line stars. He showed that the time scales of these variations exclude explanation in terms of corotating surface structures and suggested they were produced by non-radial pulsations. However, if the stars observed by Baade *are not* genuine supergiants but main-sequence objects hidden in extended shells, then the rotation of the central stars may be much higher than what is measured from the observed (shell?) lines and corotation models could also be considered. It is interesting to note that Harries and Howarth (1996) concluded from their linear spectropolarimetry of the Hα line of ζ Pup that the object is not spherical but has a density enhancement at equatorial regions. They interpreted this as a compression of the stellar wind but their observation may have some relevance to the problem raised here.

Also in this case, interferometry combined with other types of observations could help to decide whether supergiants are indeed spherically symmetric and rather slowly rotating stars as expected for "well-behaved" supergiants in evolutionary phases after the core hydrogen burning or rotationally flattened configurations reminiscent of classical Be stars. More generally, it would also be interesting to see whether Ae and Oe stars differ in their geometry from Be stars or not, or whether the disks of Be stars are truncated as Waters' (1986) analysis of IRAS data indicates.

Acknowledgements

I am grateful to Drs. Dietrich Baade, Anne-Marie Hubert, Pavel Koubský, Frédéric Morand, Denis Mourard and Myron A. Smith for a number of very useful comments and suggestions on the draft version of this review. I apologize to them, however, that I remain pig-headed on some points concerning my "non-canonical" views of certain problems related to Be stars. Moreover – to cope with the page limit set up for this review – I was unable to include some of their excellent suggestions.

References

Andrillat Y., Jaschek M., Jaschek C. (1983): Hvar Obs.Bull. 7, 193
Baade D. (1982): A& A 105, 65 and A& A 110, L15
Baade D. (1986): in *Seismology of the Sun and the Distant Stars*, Ed. by D.O. Gough, Reidel, Dordrecht, 465
Baade D. (1988): in *O Stars and Wolf-Rayet Stars*, Ed. by P.S. Conti and A.B. Underhill, NASA SP-497 Publ., 137
Baade D. (Ed.) (1991): *ESO Workshop on rapid Variability of OB Stars: Nature and Diagnostic Value*, ESO Conf. Proc. No. 36
Balona L.A., Engelbrecht C.A. (1986): MNRAS 219, 131
Balona L.A., Henrichs H.F., Le Contel J.M. (Eds.) (1994): *Pulsation, Rotation and Mass Loss in Early-Type Stars*, IAU Symp. 162, Kluwer, Dordrecht
Beckers J.M. (1982): *Differential Speckle Interferometry*, Opt. Acta 29, 361
Blazit A., Bonneau D., Koechlin L., Labeyrie A. (1977): ApJ 214, L79
Bolton C.T. (1982): in *Be Stars*, IAU Symp. 98, Ed. by M. Jaschek and H.-G. Groth, D. Reidel, Dordrecht, 181
Dachs J., Hanuschik R., Kaiser D., Rohe D. (1986): A& A 159, 276
Frost S.A., Conti P.S. (1970): in *Be and Shell Stars*, IAU Symp. 70, Ed. by A. Slettebak, Reidel, Dordrecht, 139
Gies D.G., McKibben W.P., Kelton P.W., Opal C.B., Sawyer S. (1990): AJ 100, 1601
Grady C.A. (1989): Be Star Newsletter No. 20, 4
Grady C.A., Peters G.J., Bjorkman K.S., Henrichs H.F. (1988): in *A Decade of UV Astronomy with IUE*, Ed. by E.J. Rolfe, ESA SP-281, Vol. 1, 257
Guthnick P. (1941): Vierteljahrsschrift Astron. Gesell. 76, 62
Hanuschik R.W. (1995): A& A 295, 423
Hanuschik R.W. (1996): A& A 308, 170
Hanuschik R.W., Dachs J., Baudzus M., Thimm G. (1993): A& A 274, 356
Harmanec P. (1982): in *Be Stars*, IAU Symp. 98, Ed. by M. Jaschek and H.-G. Groth, D. Reidel, Dordrecht, 279
Harmanec P. (1983): Hvar Obs.Bull. 7, 55
Harmanec P. (1984): Bull.Astron.Inst.Czechosl. 35, 193
Harmanec P. (1987a): in *Physics of Be Stars*, IAU Col. 92, Ed. by A. Slettebak and T.P. Snow, Cambridge Univ. Press, 339

Harmanec P. (1987b): Bull.Astron.Inst.Czechosl. 38, 52
Harmanec P. (1989): Bull.Astron.Inst.Czechosl. 40, 201
Harmanec P., Kříž S. (1976): in *Be and Shell Stars*, IAU Symp. 70, Ed. by A. Slettebak, D. Reidel, Dordrecht, 385
Harmanec P., Pavlovski K. (Eds.) (1983): *Workshop on Rapid Variability of Early-Type Stars*, Hvar Obs.Bull. 7
Harmanec P., Hill G.M., Walker G.A.H., Dinshaw N., Yang S. (1987): in *Astrophysics*, 10th European Regional IAU Meeting, Vol. 5, Ed. by P. Harmanec, Publ. Astron. Inst. Czechosl. No. 70, 115
Harries T.J., Howarth I.D. (1996): A& A 310, 533
Hill S.N. (1926): Publ. Dom. Astrophys. Obs. 3, 349
Hirata R. (1995): PASJ 47, 195
Hirata R., Kogure T. (1984): Bull. Astron. Soc. India 12, 109
Höflich P. (1988): A& A 191, 348
Huang S.-S. (1973): ApJ 183, 541
Jaschek M., Groth H.-G. (Eds.) (1982): *Be Stars*, IAU Symp. 98, D. Reidel, Dordrecht
Juza K., Harmanec P., Božić H., Pavlovski K., Žižňovský J., Tarasov A.E., Horn J., Koubský P. (1994): A& AS 107, 403
Kato S. (1983): PASJ 43, 75
Kogure T., Hirata R. (1982): Bull. Astron. Soc. India 10, 281
Kříž S., Harmanec P. (1975): Bull.Astron.Inst.Czechosl. 26, 65
McAlister H.A., Hartkopf W.I., Gaston B.J., Hendry E.M., Fekel F.C. (1984): ApJS 54, 251
McLaughlin D.B. (1961): J.R.Astron.Soc.Can. 55, 13 & 73
Mennickent R.E., Vogt N. Sterken C. (1994): A& AS 108, 237
Okazaki A.T. (1994): in *Pulsation, Rotation and Mass Loss in Early-Type Stars*, IAU Symp. 162, Ed. by Balona L.A., Henrichs H.F., Le Contel J.M., Kluwer, Dordrecht, 380
Pavlovski K., Božić H., Harmanec P., Horn J., Koubský P. (1983): Inf.Bull.Var.Stars No.2431
Percy J.R. (1983): AJ 88, 427
Plets H., Waelkens C., Trams N.R. (1995): A& A 293, 363
Prinja R.K., Massa D., Fullerton A.W. (1995): ApJ 452, L61
Slettebak A. (Ed.) (1970): *Be and Shell Stars*, IAU Symp. 70, D. Reidel, Dordrecht
Slettebak A., Snow T.P. (Eds.) (1987): *Physics of Be Stars*, IAU Col. 92, Cambridge Univ. Press
Smith M.A. (1986) ApJ 304, 728
Sterken C., Vogt N., Mennickent R. (1994): A& A 291, 473
Struve O. (1931): ApJ 73, 94
Struve O. (1952): ApJ 115, 138
Tarasov A.E., Tuominen I. (1987): in *Astrophysics*, 10th European Regional IAU Meeting, Vol. 5 Ed. by P. Harmanec, Publ. Astron. Inst. Czechosl. No. 70, 127
Telting J.H., Waters L.B.F.M., Persi P., Dunlop S.R. (1993): A& A 270, 355
Vogt S.S., Penrod G.D. (1983): ApJ 275, 661
Walker M.F. (1953) ApJ 118, 481
Walker G.A.H., Yang S., Fahlman G.G. (1979): ApJ 233, 199
Walker G.A.H., Yang S., Fahlman G.G. (1981): in *Workshop on Pulsating B Stars*, Ed. by G.E.V.O.N. and C. Sterken, Nice Obs. Publ., 261
Waters L.B.F.M. (1986): A& A 162, 121

Imaging of Wolf-Rayet Stars with the VLTI

John Davis[1], Anthony F. J. Moffat[2], and Lindsey F. Smith[1]

[1] Chatterton Astronomy Department, School of Physics, University of Sydney, NSW 2006, Australia
[2] Départment de Physique, Université de Montréal, Montréal, P.Q., H3C 3J7, Canada

Abstract. The imaging of a Wolf-Rayet star in the light of emission lines of various ionization and excitation potentials will allow the relative distances of the layers, where the different lines are formed, to be established. Although it will not be possible to resolve an individual Wolf-Rayet star with the VLTI, its imaging capability combined with the higher angular resolution of the Sydney University Stellar Interferometer (SUSI), will enable a very detailed study to be made of at least one system.

1 Introduction

Population I Wolf-Rayet (WR) stars represent an important stage in the advanced evolution of stars with initial masses greater than ~25 solar masses, depending on the ambient metallicity. They exhibit broad emission lines superposed on a hot continuum. These prominent spectral features arise in different zones of the stratified winds surrounding the star, with hotter zones occurring closer to the star. Evolutionary models are computed in hydrostatic equilibrium but the computed radii of cores and atmospheres do not appear to agree well with the existing, very limited observations.

2 The Potential of the VLTI for Wolf-Rayet Star Studies

The imaging of a WR star in the continuum and in the light of emission lines of various ionization and excitation potentials would allow the distances of the layers, where the different lines form, to be established. By associating them with the spectroscopically determined expansion velocities, the true expansion law of the winds could be established, allowing checks to be made on models, and core radii and temperatures to be inferred. Unfortunately, the angular resolving power of the VLTI, even at wavelengths in the visual region of the spectrum combined with the longest baselines planned for the Auxiliary Telescopes (ATs), will not be sufficient to resolve an individual WR star in the continuum (the largest has an angular diameter of ~0.26 mas). However, the VLTI would be capable of imaging the emission line shells for several of the brightest WR stars if its imaging capability was extended into the visual region of the spectrum where the strong emission lines of high ionisation and excitation potential lie. In addition to probing the stratified structure of the WR winds, it would be possible to look for asymmetries in the winds, such as rotational flattening, bipolar outflows, jets and magnetic loop structures. In addition, since several of the brightest WR stars are components of spectroscopic binary systems, continuum observations combined with spectroscopic observations would provide WR masses and distances. Table 1 contains a list of the five brightest WR stars, all of which have southerly declinations, making them suitable program objects for the VLTI.

Table 1. The 5 visually brightest Wolf-Rayet stars.

WR	Name	V	Spectral Type	Notes
11	γ^2 Vel	1.74	WC8 + O9I	SB2. P=78.5d
22		6.44	WN7 + O7.5V	SB2. P=80.3d
24		6.49	WN7 + abs	Non-variable
48	θ Mus	5.69	WC6 (+O) + O9.5I	SB1. P=18.3d. Prob. triple
78		6.61	WN7	Variable

3 The Special Case of γ^2 Velorum

As can be seen from Table 1, γ^2 Velorum is by far the brightest WR star. It is a binary system with the WR component being the fainter companion to an O-type star. Observations made with the Narrabri Stellar Intensity Interferometer (NSII) (Hanbury Brown et al. 1970) demonstrated that the size of the extended region responsible for the C III-IV emission feature at a wavelength of 465 nm is of the order of five times larger than the O-type star. The sensitivity of the NSII did not permit the determination of the angular size of the fainter WR star.

The Sydney University Stellar Interferometer (SUSI) (Davis 1994) is a two aperture interferometer with much higher angular resolution than the VLTI but without an imaging capability (three or more apertures are required for imaging through the earth's turbulent atmosphere). SUSI is capable of measuring the size of the WR star in γ^2 Vel and also of measuring the dimensions of the emission regions but not imaging them. The capabilities of the VLTI and SUSI are thus complementary with the potential for a very detailed study of the γ^2 Vel system.

4 Conclusion

If the VLTI is to contribute to our understanding of WR stars it will be necessary to extend its imaging capability to the visual region of the spectrum.

References

Davis, J. (1994): The Sydney University Stellar Interferometer (SUSI), in Robertson, J.G., Tango, W.J., eds, Proc.IAU Symp.158, Very High Angular Resolution Imaging, Kluwer, Dordrecht, 135–142

Hanbury Brown, R., Davis, J., Herbison-Evans, D., Allen, L.R. (1970): A Study of γ^2 Velorum with a Stellar Intensity Interferometer. MNRAS, **148**, 103–117

Part 6

CIRCUMSTELLAR ENVIRONMENTS AND MASS LOSS

AGB and POST–AGB STARS: an Overview

Harm Habing

Sterrewacht, Leiden, the Netherlands

Abstract. In the last phase of their existence stars of low and intermediate main-sequence mass, M_{ms} below about 6 $M_{\odot}$, become very luminous red giants; in the HR-diagram they populate the "Asymptotic Giant Branch". "AGB" stars are very rare : this phase is very brief and lasts perhaps 10^5 yr. Later the star becomes a "post–AGB star" and this phase is probably even shorter, perhaps a few times 10^4 yr; during this phase the object moves towards higher effective temperatures at constant luminosity and will become a planetary nebula. AGB stars have several properties that make them special: (1) the luminosity is the highest that a star of a given main–sequence mass will ever reach; (2) the luminosity varies periodically in time with a large amplitude and a long period (Mira variables, OH/IR stars, infrared carbon stars); (3) the stars experience regularly "thermal pulses", a short episode with a duration of the order of a few years during which the luminosity is even further increased and during which new elements, produced in the stellar interior, are being dredged up; (4) mass is lost at a high rate, of the order of 1 $M_{\odot}$in 10^6 to 10^4 yr leading to the formation of a sometimes optically very thick "envelope". All these properties make the stars easily detectable especially in the infrared.

The significance of AGB and post–AGB stars for the VLTI era lies in the properties just mentioned: the objects are very luminous (when the Sun will have entered the AGB phase it will have a luminosity of 3,000 $L_{\odot}$) and have large dimensions (in some spectral lines up to 0.1 pc). Although AGB stars have the luminosity of O– and early B–type stars, they are easily distinguished by their red colours and their variability. As AGB stars are much older than OB-stars they report about star formation rates in the past.

1 Introduction

At the request of the organizers of this meeting I present a fairly general review.

The quickest way to introduce the subject is to show the observed HR diagram of the globular cluster M3 (Renzini and Fusi-Pecci (1983), see Figure 1). All stars with $V < 18$ have the same main–squence mass, $M_{\rm ms}$. Only a short while ago they were on the main sequence and had an apparent magnitude of $V = 18.5$; now they have evolved into much more luminous objects: the brightest stars are about 5 magnitudes brighter in $m_{\rm bol}$. One distinguishes three groups of evolved stars; one group is formed by the Horizontal Branch (HB) stars and there are two groups of red giants: the Red Giant Branch (RGB) and the Asymptotic Giant Branch (AGB) stars. Notice in Figure 1 that all the stars between the acronyms SGB and RGB belong to the RGB.

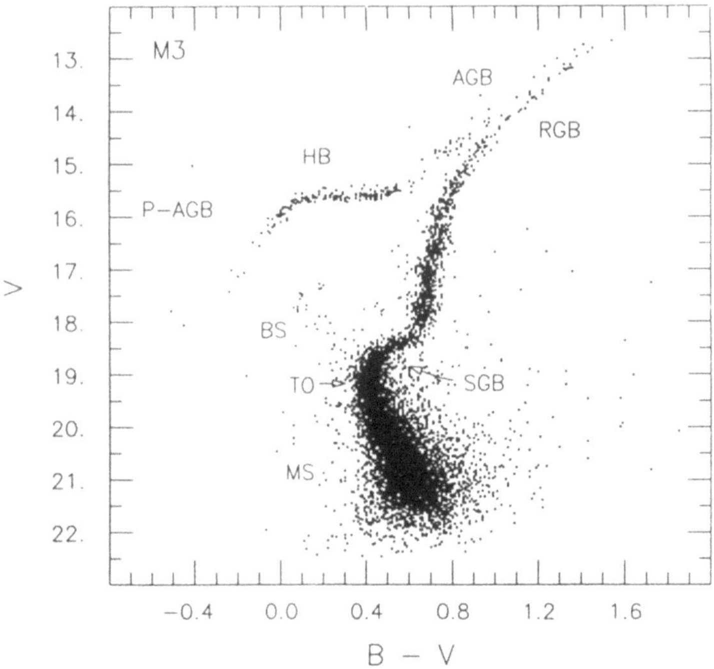

Fig. 1. The HR-diagram of the globular cluster M3 (Renzini and Fusi-Pecci (1983).

Computer simulations[1] of stellar structure show that the stellar mass determines how a star evolves after it has left the main sequence. Stars with a mass below about 6 $M_\odot$ will ultimately reach a state in which they consist of a small (about 1 Earth radius) core of gas of carbon and oxygen ions and a degenerate electron gas. The core is surrounded by a mantle of large dimensions (about 1 A.U.) and a low density. The core is covered by a thin layer of helium. The stellar energy is produced at the outside of this layer by hydrogen burning. From time to time the helium burns rapidly into carbon; this leads temporarily to a significant increase of luminosity. Such an episode, named a "thermal pulse", lasts only a short time, a few percent of the time interval between two subsequent "thermal pulses". During a pulse the layer with newly produced elements, e.g. carbon, becomes convective and the new elements may be carried to the surface ("dredge-up"). Eventually, after several dredge-ups enough new carbon atoms have been spread through the stellar mantle and the number of carbon ions now exceeds those of oxygen: the star has turned into a carbon star. Between two pulses the luminosity increases steadily, probably because of the increasing core mass. The luminosity reaches heights far beyond what the star had when it still

[1] I will use the large set of calculations published by the Padua group: Bressan et al. (1993). Other sets of calculations lead to quantitative, but not to qualitative differences.

was on the mean sequence; a star like the Sun will reach 3000 $L_\odot$. When the star reaches such high luminosities it begins to pulsate with a large amplitude and a period between 100 and 500 days (Mira-type variability); as a consequence of these pulsations and of the subsequent formation of dust particles the stars begin to loose mass at a high rate, between 10^{-7} and 10^{-4} $M_\odot$/yr. Thermal pulses, Mira-type variability and a high mass-loss rate are properties characteristic for the AGB phase.

Before entering the AGB, stars with a mass below about 2 $M_\odot$ occupy the red giant branch (RGB), see figure 1. The RGB phase lasts considerably longer than the AGB phase and there will be many more stars in the RGB phase than in the AGB phase. The structure of a star on the RGB is qualitatively similar to that of an AGB star: again it consists of a small core and a large mantle. However, the core of a RGB star consists of helium, the luminosity is lower, Mira type variability is not seen and the mass loss can probably be neglected. The RGB phase ends through the "helium flash" when the star reaches a luminosity of about 2500 $L_\odot$; the degeneracy of the electron gas disappears and the star becomes a core-helium burning star on the Horizontal Branch with a luminosity between 50 and 100 $L_\odot$. The star remains on the HB until the helium is exhausted and a transition to the AGB follows.

The hydrostatic models calculated for AGB stars and their evolution become dynamically unstable after many thermal pulses, and during such a pulse. The idea that this might happen has been around for a long time (Lucy (1987), Paczýnski and Ziólkowski (1968)) but has not been well studied. Recently Wagenhuber and Weiss (1994) have tackled the problem once again. They find the instability they look for and argue that it represents a real physical phenomenon and that it is not the consequence of numerical inaccuracies: when the increase in luminosity during a TP causes the star to expand, hydrogen in outer layers recombines and this delivers so much energy that the matter is pushed farther outward: a run-away process. Han et al. (1994) obtain similar conclusions. Wagenhuber and Weiss (1994) point out that to know what really happens one must calculate dynamical instead of static stellar models, but that this is beyond present-days techniques, because we lack an adequate theory of time-dependent convection. While this will annoy the theoreticians, the observers have less of a problem: large-amplitude, long-period variability is observed and probably reflects the limit-cycle instability described by Wagenhuber and Weiss (1994).

2 Thermal Pulses

Two aspects of the TPs are relevant here: (i) the predicted increase in luminosity during the TP; (ii) the "dredge-up", that is the convective transport of recently produced atoms from the layers where there are produced to the surface where they can be observed.

The predicted rise in luminosity has not yet been observed with certainty. The best documented case, as far as I know, is the secular variation in magnitude of a few Mira variables; see Wood and Zarro (1981). In the last few years the significant data base of magnitude estimates of long–period variables collected by the AAVSO has become accessible; this data base may contain more cases. Even better data bases will be provided as a by–product by the ongoing surveys for micro–lensing.

"Enrichment" of the stellar atmosphere has been observed in many cases. The most conspicuous examples are the carbon stars whose existence was recognized already a century ago, in the earliest years of stellar spectroscopy. Stars begin their life (and have always begun their life) with more oxygen than carbon atoms. After the triple–α process and several dredge–ups have added so many carbon atoms to the mantle that they exceed the number of oxygen atoms there is a drastic change in the chemical equilibrium of the atmosphere. The atmospheres of M–type giants are very different from those of the carbon stars, and this is also true for the circumstellar envelopes.

Several abundance changes other than that of carbon have been observed. I will quote three relevant and recent studies; many more exist.

The first is the presence of ^{99}Tc lines in all Mira variables with P> 300^d (Little et al. (1987)). The life time of this unstable element is short and one concludes that these Mira variables have undergone a thermal pulse in the recent past and that these stars must be AGB stars. The second is the promising study by Guélin et al. (1995) of isotope ratios in the circumstellar envelope of the carbon star IRC+10216. They find the relative abundances of ^{24}Mg, ^{25}Mg and ^{26}Mg. In combination with other data on N, O, Si isotopes they conclude that IRC+10216 had a progenitor with a mass of 3 to 5 $M_\odot$, but has experienced many dredge-ups. To me the possibility to measure isotope ratios in the circumstellar envelope is a break-through. The third study is the modelling of the formation of carbon stars by Groenewegen et al. (1995) and the derivation of the yields of different elements. This study underlines once again the high significance of the Magellanic Clouds as a source of stars all at the same distance. Detailed spectroscopic and photometric studies of AGB stars in the Clouds should be an important goal for the VLT.

Not every star with "peculiar abundances" is the victim of its own thermal pulses. It may have been polluted by a once more massive companion that has gone through the AGB phase long ago, has dumped its enriched material on its companion and has then turned into an invisible white dwarf. Jorissen et al. (1993) estimate that a large majority of the so–called S–stars (stars with about equal abundances of C and O) are "extrinsic" S-stars; in a volume–limited sample only a minority of S-stars is "intrinsic".

3 Mass Loss and Circumstellar Envelopes

On this topic I have written a long review that has recently appeared in print (Habing (1996)). Circumstellar envelopes are detected at all wavelengths between 400 nm and 20 cm and the study of envelopes can now be called "an industry". Dust continuum emission is detected between a few μm and 1cm; thermal line emission has been observed from many gases, although not from the dominant molecule, H_2, because its symmetries makes the molecule a poor radiator of spectral lines. A large number of maser lines from SiO, H_2O, OH have been recorded in stars as distant as the Magellanic Clouds. The ISO satellite, currently active, has already detected many H_2O lines. In NML Cyg ISO has detected the OH line at 35 μm, not in emission but in absorption: the line photons have been used to pump the strong 1612 MHz maser from NML Cyg. All these observations give a detailed description of the circumstellar envelope from almost the stellar photosphere (SiO maser lines at 1 A.U.) to distances a factor of 20,000 larger (CO lines at 0.1 pc).

Above I have already discussed the surprising detection of different isotopomers in IRC +10216. The difference between carbon-rich and oxygen-rich envelopes is very large; the dust continuum is different and also the molecular emission line spectrum (e.g. Bujarrabal et a. (1994a), Bujarrabal et a. (1994b)).

Matter flowing out of the star with a speed of 10 km/s needs 10^4 yr to reach a distance of 0.1 pc, which is roughly the outer limit of the envelope. This time interval is longer than the time between two thermal pulses, at least in stars with $M_{\rm ms}$ of, say, 1.5 $M_\odot$ or more. In several envelopes abrupt variations in the mass loss rate have now been seen on this time scale and suggest that the star recently underwent a thermal pulse (Justtanont, K. et al. (1996), Delfosse, X. et al. (1996))

A remarkable observation is the fact that a mass-loss rate higher than about $10^{-7}M_\odot$/yr is found only in long-period variables: oxygen-rich and carbon-rich Miras, OH/IR stars and carbon-rich objects like IRC+10216. There are a few OH maser stars without large amplitude variations; their overall energy distribution differs from that of variable OH/IR stars by stronger emission at long wavelengths (e.g. at 60 μm). These differences lead to the interpretation that the envelope is actually "coasting away" from the star: in non-pulsating OH/IR stars mass loss has stopped. There is further evidence in favour of the view that large-amplitude pulsations and high mass-loss rate always occur together: interferometric observations of diameters of Mira variables indicate that the atmosphere of the star is very extended and this is attributed to the pulsations. The great atmospheric height has a quite dramatic consequence: in the top layers where the temperature drops to about 1000 to 1500 K the density is still high enough for condensation of solid particles. The "smoking" of these stars provide almost all of the refractory nuclei of the interstellar dust particles. A second consequence is that because the dust particles absorb stellar photons easily they are pushed outward by the stellar light. Because dust and gas are coupled by friction also the gas is propelled outward. The fraction of momentum in the stellar light that is transfered to the gas depends on the optical depth of the envelope and can be calculated once the overal spectrum (the spectral energy distribution) is known. Knowing the final outflow velocity then allows us to estimate the mass-loss rate from the observed spectrum.

Infrared and (sub-)millimeter line observations are excellent tracers of mass loss in red giants when the rate is higher than $10^{-7}M_\odot$/yr. Mass-loss at a smaller rate is difficult to detect; a possibility to study thin circumstellar envelopes is through the detection of line photons that have been scattered in a region surrounding the star. Examples are the 589.6 nm NaI line and the 769.9 nm KI line (Mauron and Guillain (1995)).

4 The Dependence on Main-Sequence Mass and on Metallicity of the Mass–Loss Rate and of the Total Amount of Mass Loss

The mass–loss rate of a given star is not constant in time, but varies; most dramatic are the sudden interruptions that one finds in the structure of the circumstellar envelope. OH26.6+0.5 is a well–studied example (Justtanont, K. et al. (1996), Delfosse, X. et al. (1996)). In addition there is now ample evidence that the mass–loss rate is also a function of the main-sequence mass, $M_{\rm ms}$ of a star: Miras with longer periods, and thus of higher mass, show more often maser emission; and OH/IR stars at higher galactic latitudes (and therefore of lower $M_{\rm ms}$) have thinner envelopes (Likkel (1989)).

AGB stars are very luminous and very red and therefore they are easily detected elsewhere in the Galaxy and in the Magellanic Clouds. By comparing these AGB stars with those in the bulge and in the outer parts of our Galaxy one will observe similarities and differences due to lower and higher metallicity. Several pioneering studies have been made but much work remains to be done. One of the interesting observations is that in the LMC long–period variables have been detected with periods up to 1000 days as visible objects, i.e. detectable at optical wavelengths. This is in contrast to what has been found in our Galaxy where long–period variables with periods over 500 days are wrapped in optically thick shells and have thus become "invisible", but easily detected at IR wavelengths. This difference in appearance is probably caused by a higher dust–to–gas ratio in our Galaxy and not by our higher mass–loss rates. Weidemann (1987) compared the mass of white dwarfs in galactic clusters with the mass of the cores of very luminous AGB stars in the LMC and concluded that stars with the same main–sequence mass in the LMC and in the solar neighbourhood end up as white dwarfs of the same mass: the total amount of mass that is lost is about the same in the LMC and in our Galaxy.

5 The Evolution of TP–AGB Stars

To calculate the evolution of a star through the AGB phase one has to know the mass–loss rate, $\dot{M}$, as a function of time and of main–sequence mass, i.e. one needs the function $\dot{M}(t; M_{\rm ms})$. The observations have limited accuracy. Theoretical calculations ab initio are beginning to appear (Höfner et al. (1996)). In this situation courage is needed: assume some form for $\dot{M}(t; M_{\rm ms})$ with a few free parameters and calculate the evolution. Compare the statistics of the objects obtained, especially the masses of the white dwarfs that finally emerge from the AGB stars, with predicted white dwarf masses and adjust the free parameters in the mass-loss equation if necessary. Different "scenarios" along these lines are available: Vassiliadis and Wood (1993), Blöcker (1995a), Bowen and Willson (1991), Marigo et al. (1996). Figure 2 shows some results from Vassiliadis and Wood (1993). My evaluation is that these "scenarios" are the best that one may

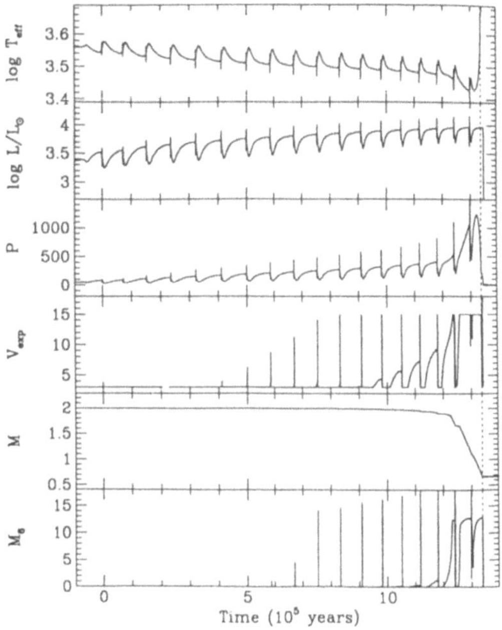

Fig. 2. The evolution of model parameters in the AGB evolution of a star with main-sequence mass of 2 $M_{\odot}$. Vassiliadis and Wood (1993).

get for some time. They should be checked by finding new tests. An example are the statistical tests by Szczerba and Blommaert (1993).

6 Post-AGB Objects

At one time these objects were called "protoplanetary nebulae" but this is a misleading term and I will use the term "Post-AGB object". Recently Kwok (1993) wrote a review, but the developments go fast and the review is already somewhat outdated; several theses on Post-AGB stars have appeared since 1993 as I will point out. Kwok states that Post-AGB objects are difficult to find and to discriminate from superficially similar objects such as young stars still inside their dust envelope, as red supergiants like NML Cyg or IRC+10420 and as symbiotic stars. Kwok lists only 28 candidate objects. A large number of new candidates (of which several may turn out not to be Post-AGB stars) and some detailed spectroscopic studies are found in the thesis by Oudmaijer (1995). If these Post-AGB objects last as long (or as short) as planetary nebulae then the total number of these Post–AGB stars within our Galaxy must be a few times ten thousand. The sample of known, well identified Post–AGB stars is still very incomplete and we may even have missed some categories of these objects altogether (Oudmaijer (1995)). Other useful theses are by Slijkhuis (1992), Bakker (1995) and Van Winckel (1995). Van Winckel has drawn attention to the fact that many of the suspected Post–AGB stars are probably double stars and that in these cases models of AGB evolution based on the evolution of single stars are not very useful.

As an example I mention the study by Riera et al. (1996) of the object IRAS 17423-1755. CCD images in various wavelength bands show the object to be a bipolar with outflow in jets that have an initial velocity of 870 km/s. The [NII]/Hα ratio is larger than usual which leads to the suggestion that the object is enriched in nitrogen and thus a post–AGB object and not a forming star. The reading of this article left me with two conclusions: many more case studies are needed and more attention should be paid to the effects of binarity on AGB star evolution.

7 AGB Stars and the VLTI: Questions and Conclusions

Many aspects of the evolution during the TP–AGB phase have been clarified in recent years, sometimes through computer simulations, sometimes from observations and in the best of cases from both sides simultaneously. We can observe by several different means the envelopes produced by the ejected stellar matter and can study them in quite some detail; convincing observations of interruptions in the mass–loss rate have been made: thermal pulses are the likely causes of these abrupt interruptions. Of equal importance is the study of the abundance anomalies. Take for example the carbon stars: whereas there are none in the bulge of the Galaxy, they appear in large numbers in the LMC and, especially, in the SMC:(Blanco et al. (1978)). Is this a consequence of abundance differences between the bulge and the MCs, or is the population of the bulge to old to produce carbon stars?

AGB stars are very luminous, as luminous as O-type stars, but they have a lower main–sequence mass, and are thus older; they trace the star formation history longer ago than OB–stars do. Turning our mind to VLTI it is easy to think of large programmes on AGB stars and Post-AGB stars: For example, mapping the detailed structure of the circumstellar envelopes of stars in the inner Galaxy; or isolating these stars from their background in the Magellanic Clouds and in other small galaxies in the Local Group and study the spectral lines. There is enough work to keep us busy for the next ten years!

References

Bakker, E. (1995): *thesis,* University of Utrecht
Blanco, B.M., Blanco, V.M., McCarthy, M.F. (1978): Nature 271, 638
Blöcker, T. (1995a): A&A 297, 727

Blöcker, T. (1995b): A&A 299, 755
Bressan, A., Fagotto, F., Bertelli, G., Chiosi, C. (1993): A&AS 100, 647
Bowen, G.H., Willson, L.A. (1991): ApJ 375, L53
Bujarrabal, V., Fuente, A., Omont, A. (1994a): ApJ 421, L47
Bujarrabal, V., Fuente, A., Omont, A. (1994b): A&A 285, 247
Busso, M, Gallino, R., Lambert, D.L., Raiteri,C.M., Smith, V.V. (1992): ApJ 399, 218
Delfosse, X., Kahane, C., Forveille, T. (1996): A&A (in press)
Groenewegen, M.A.T., van den Hoek, L.B., de Jong, T., (1995): A&A 293, 381
Guélin, M., Forestini, M.,Valiron, P., Ziurys, L.M., Anderson, M.A., Cernicharo, J., Kahane, C. (1995): A&A 297, 183
Habing, H.J. (1996): A&AR 7, 97
Han, Z., Podsiadlowski, P., Eggleton, P.P. (1994): MNRAS 270, 121
Höfner, S.,Fleischer, A.J.,Gauger, A., Feuchtinger, M.U., Dorfi, E.A., Winters, J.M., Sedlmayr, E. (1996): A&A 314, 204
Jorissen, A., Frayer, D.T., Johnson, H.R., Mayor, M., Smith, V.V. (1993): A&A 271, 463
Justtanont, K., Skinner, C.J., Tielens, A.G.G.M., Meixner, M., Baas, F. (1996): ApJ 456, 357
Kwok, S. (1993): ARA&A 31 63
Likkel, L. (1989): ApJ 344, 350
Little, S., Little-Marenin, Y., Hagen Bauer, W. (1987): AJ 94, 981
Lucy, L. (1967): AJ 72, 813
Marigo, P., Bressan, A., Chiosi, C. (1996): A&A (in press)
Mauron, N., Guillain, Ch. (1995): A&A 298, 869
Oudmaijer, R. (1995): *thesis* University of Groningen
Paczýnski, B., Ziólkowski, J.: Acta Astron. 18, 255
Renzini, A., Fusi Pecci, F (1983): ARA&A 26, 199
Renzini, A., Voli, M. (1981): A&A 94, 175
Riera, A., García-Lario, P., Manchado, A., Pottasch, S.R., Raga, A. (1996): A&A 302, 137
Slijkhuis, S. (1992): *thesis*University of Amsterdam
Szczerba, R., Blommaert, J.A.D.L. (1993): Acta Astron. 43, 397
Van Winckel, H. (1995): *thesis*, University of Leuven
Vassiliadis, E., Wood, P.R. (1993): ApJ 413, 641
Wagenhuber, J., Weiss, A. (1994): A&A 290, 807
Weidemann, V. (1987): A&A 188, 74
Wood, P.R., Zarro, D.M. (1981): ApJ 247, 247

Winds of Red Giants and Supergiants: Basic Physical Processes, Structure and Appearance

Erwin Sedlmayr, Jan Martin Winters, Axel J. Fleischer

Institut für Astronomie & Astrophysik, Technische Universität Berlin, Hardenbergstr. 36, D-10623 Berlin, Germany

Abstract. Red giants and supergiants usually exhibit three important phenomena: i) effective grain formation manifested by large infrared excesses, ii) considerable mass loss carried by slow, massive stellar winds, and iii) pronounced spectral variability induced by more or less regular stellar pulsations.

We discuss the specific relevance of these different aspects for the reliable modelling of dust forming circumstellar shells surrounding these objects. Particular emphasis is given to the strong nonlinear couplings among the basic physical processes (dust formation, thermodynamics, hydrodynamics and chemistry) which allow for a consistent modelling and a quantitative interpretation of the observed phenomena characterizing the objects in the upper right region of the Hertzsprung–Russell diagram.

1 Introduction

In the last decade considerable progress has been achieved both in observing and in modelling dust forming objects. This is due to the availability of excellent new instruments (e.g. IRAM, SEST, NTT, HHT, IRAS, ISO, HST, ...) covering the full spectral range and to the development of appropriate theoretical descriptions allowing for a reliable modelling of the physical and chemical processes involved. In the context of circumstellar dust formation, in particular a variety of effects and processes have been disclosed which provide very detailed insight into the complex interplay between the physical and chemical processes determining the structure of the objects.

In the following we confine our discussion to the extended shells of cool red giants and supergiants, Miras and LPVs, and to episodic phenomena like RCrB stars, i.e. those objects where dust formation is well known to be an important ingredient which strongly influences the structure of the shells and their temporal behavior.

The appearance of red giants and supergiants usually is dominated by a substantial reddening of the radiation field caused by effective dust formation in those regions of the shells where favorable conditions are encountered (e.g. Woolf & Ney 1969, Gillet et al. 1971). P-Cygni line profiles indicate the presence of velocity fields (e.g. Keady et al. 1988, Keady & Ridgway 1993) showing a distinct subsonic–supersonic transition and giving rise to a substantial mass loss of the order of 10^{-6} to $10^{-4} M_{\odot}$/yr (e.g. Knapp & Morris 1985). Detailed observations resolve the radial shell structure yielding spatial intensity profiles, brightness distributions and the corresponding visibility functions, which in case of pulsating variables indicate a layered structure of the shells (Ridgway & Keady 1988 , Danchi et al. 1994). Besides pronounced infrared bands and rotational line structures measuring the radially increasing complexity of the chemistry, in some cases (in particular early M giants) also emisssion lines are found which seem to be clear signatures of chromospheres located inside the dust shell. As in the outer shell region, where the chemical situation is strongly affected by the interstellar UV field, initiating a photochemistry, also chromospheres introduce substantial non-LTE effects which also may cause severe problems for a reliable modelling of the inner shell region.

With these observations, theory – attempting a reliable modelling of these objects – is confronted. Due to the complexity of the problem, a complete theoretical description proves to be a hopelessly underdetermined problem. For this reason, any reliable description has to be based on a self-consistent approach comprising hydrodynamics, thermodynamics, radiative transfer, chemistry, and dust formation and growth. Hitherto, this problem only has been solved for the (quasi)stationary situation of dust driven winds (e.g. Sedlmayr & Dominik 1995), the solution of which is fully determined by the prescription of four fundamental stellar parameters (e.g. stellar mass, stellar luminosity, effective temperature and chemical abundances ϵ_i), and for LPVs and Miras (cf. Fleischer et al. 1992), where two additional parameters (the pulsation period and its amplitude at the base of the atmosphere) have to be introduced. In contrast to these objects, where dust formation and the shell structure seem to be controlled by a self-regulating process – guarantying a secular stability of the time averaged appearance of the shell – episodic phenomena like the occultation of RCrB seem to be singular critical events, triggered by the interplay among shock waves, chemistry, and the actual cooling functions.

2 VLTI and Dust Forming Objects

Although the observations already available have provided excellent material with regard to the spectral appearance and to the spatial structure of dust forming objects, the consistent theoretical description based on the fundamental laws of physics and on appropriate theories describing the local processes needs even more detailed local, time resolved information for specifying the required input quantities and the appropriate minimum level of approach in order to arrive at a satisfactory description. This concerns in particular the local chemical structure which not only provides the transport coefficients and consequently the temperature and density structure, but also the dust formation and growth process, where the nucleating primary molecules and finally the growth species entering into the dust formation theories have to be determined. In this view, the detailed information obtained by interferometric measurements can be conceived as a necessary complement to the local processes and resulting global structures modelled by the theory. Figure 1 shows a schematic plot of this complementary view. Usually, observations focus on the more prominent features determining the appearance of the object and characterizing its general properties. In many

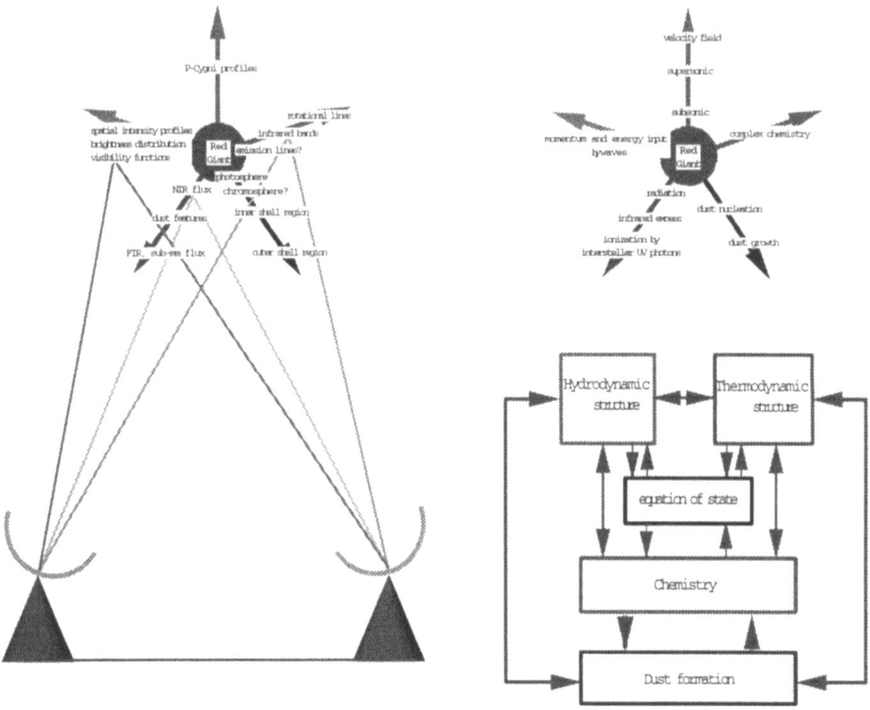

Fig. 1. Observational (l.h.s) and theoretical (r.h.s.) view of circumstellar dust shells around AGB objects. The interferometer (l.h.s.) provides very detailed information on the spatial and spectral appearance of the shell whereas the theoretical model (r.h.s.) aims at the description of the physical processes and their interrelations relevant for the structure of the shell.

cases however, these features are not primary for the understanding of the local physics which occasionally is determined by processes which are not immediately observable. Generally any reliable modelling of such a system introduces a variety of nonlinear couplings whose strength is decisive for the explanation of the local physics and the dynamical evolution of important structures. In this respect valuable information can be gained by applying interferometric methods providing a detailed local insight. As for dust forming winds the condensation process is decisive both for the local momentum transfer and for the thermal energy balance, this concerns for example the dust formation pathway, in particular the decisive critical clusters, which due to their bottleneck property hardly can be observed with other methods. In this context high resolution interferometric measurements are valuable to study the evolution of chemical complexity along the condensation pathways and to localize the transition between the nucleation and the growth regime which in all dust formation theories is crucial for the understanding of the actual condensation process.

Also detailed spatial information about the velocity field is highly needed for constraining the local conditions, in particular the concentrations of the molecular species, in order to arrive at a reliable level of theoretical modelling.

In the following, we discuss the specific value of interferometric methods for stationary dust driven winds, pulsational variables, and RCrB stars.

2.1 Stationary Dust Driven Winds

The self-consistent solution for the problem of spherically symmetric stationary dust driven winds (Gail & Sedlmayr 1987) has been applied to the prominent carbon star IRC +10216 (Winters et al. 1994a). The solution is completely determined by the four fundamental stellar parameters, which for the case of IRC +10216 yield the values $M_\star = 0.7 M_\odot$, $L_\star = 2.4\,10^4 L_\odot$, $T_\star = 2000\,\mathrm{K}$, and an overabundance of carbon to oxygen $\epsilon_C/\epsilon_O = 1.40$. From the spectral energy distribution and the visibility functions a distance of $d = 160 \pm 20$ pc is derived. In this model, classical nucleation theory is used to describe the formation of amorphous carbon grains, which leads to effective dust formation rather close to the star

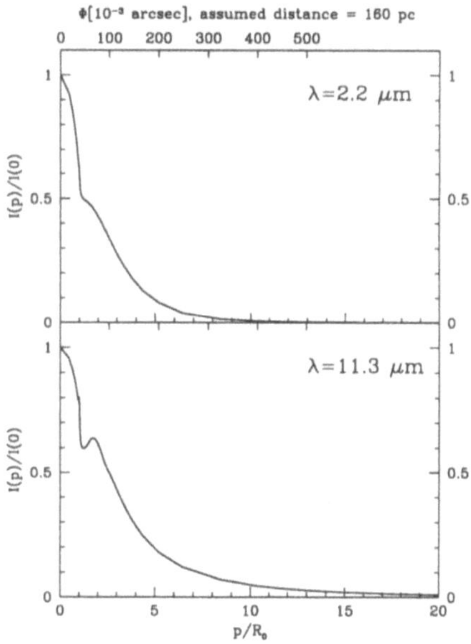

Fig. 2. Synthetic intensity profiles at $\lambda = 2.2\mu m$ and $\lambda = 11.3\mu m$ resulting from the stationary model for IRC +10216 (for details see Winters et al. 1994a).

at $\approx 1.7 R_\star$ (see Fig. 2 where two synthetic intensity profiles for this model are shown).

In order to test the applicability of classical nucleation theory in a carbon-rich circumstellar shell and, as a byproduct, to obtain a more precise distance for individual objects, it is essential to determine the actual location of the inner shell boundary e.g. of IRC +10216 by a spatially highly resolved observation. Additional high spectral resolution would provide very valuable information also on the chemical structure of the inner shell region (e.g. the radial density distribution of the C_2H_2 molecule which is of particular importance in the context of carbon dust formation).

In fact, the comparison between calculated and observed spectral energy distribution indicates the presence of inhomogeneities in the circumstellar dust shell around IRC +10216 which by now have not been modelled by means of a consistent description. From these considerations, the following observational aims for VLTI can be derived: • resolve the inner edge of the dust shell, deduce information about the nucleation process • determine the chemical structure of the inner shell region • resolve inhomogeneities in the dusty envelope.

2.2 Pulsating Atmospheres

Most objects on the Asymptotic Giant Branch (AGB) are known to be variable (e.g. Jura 1986). In order to model in detail the structure of the CDS of these objects the theoretical description has to include the effects of pulsation and thus has to be time-dependent. Since a self-consistent model of stellar pulsation which includes convection is still missing for AGB objects we apply the so-called "piston-approximation" to simulate the interior variation of the star. This approach introduces two additional parameter compared to the stationary winds (see Sect. 2.1): the pulsation period P, and the pulsation amplitude Δu. The resulting model structures show a complex interaction among the various physical components.

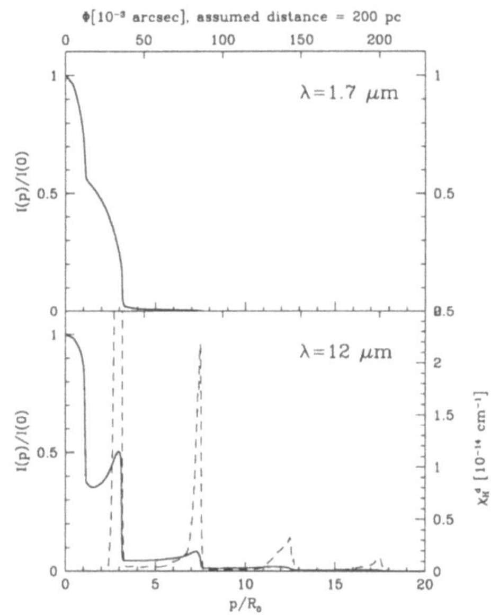

Fig. 3. Synthetic intensity profiles at $\lambda = 1.7\mu m$ and $\lambda = 12\mu m$ resulting from a dynamical dust shell model (for details see Fleischer et al. 1992).

Especially, it turns out that the shock waves travelling through the atmosphere are not caused by the interior pulsation alone but are due to a re-amplification mechanism caused by radiation pressure on dust (for details see Fleischer et al. 1992). Figure 3 shows synthetic intensity profiles for a time-dependent model calculation with the following parameter: $M_\star = 1.0 M_\odot$, $L_\star = 1.0\,10^4 L_\odot$, $T_\star = 2600\,\mathrm{K}$, overabundance of carbon to oxygen $\epsilon_C/\epsilon_O = 1.80$, $P = 650d$, and $\Delta u = 2kms^{-1}$. A general result of the the time-dependent models is that the dust is concentrated in rather narrow discrete layers (indicated by the radial distribution of the flux-mean extinction coefficient χ_H^d), leading to characteristic structures in the intensity profiles (cf. the $12\mu m$ profile in Fig. 3, a detailed discussion is given in Winters et al. 1995) and also in the calculated light curves (Winters et al. 1994b). As the strength and the shape of these structures depend on wavelength, they contain important information about the internal structure of the dust grains actually formed in the CDS and about the grain size distribution finction.

From the time-dependent models, the following observational aims for VLTI can be derived: • resolve the layered structure of the CDS and follow its temporal behavior at different wavelengths • determine the grain material, the internal structure, and the size distribution of the grains.

2.3 The Occultation Phenomenon of RCrB Stars

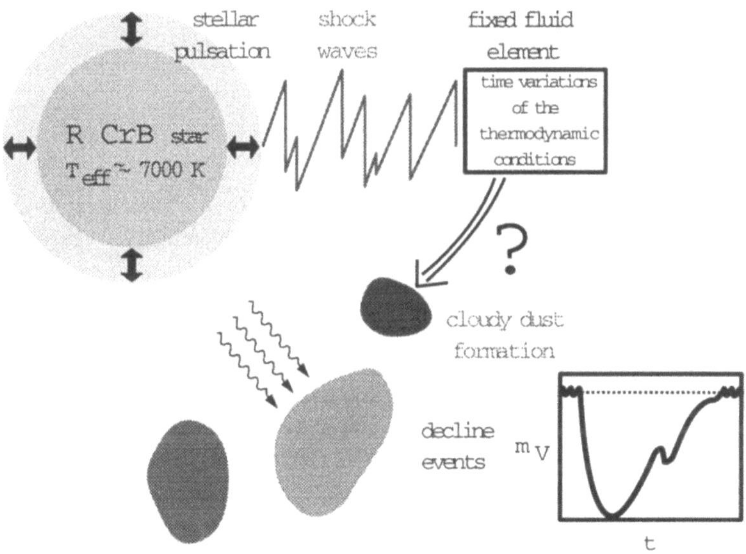

Fig. 4. Sketch of the occultation phenomenon in RCrB star atmospheres

Though being caused by instantaneous carbon grain formation, the occultation phenomenon characterizing the light curves of RCrB stars (sketched in Fig. 4) is much less understood than the more continuous dust formation process in stationary winds or pulsating atmospheres. This is due to the episodic occurrence of the phenomenon, the initial conditions of which are not really known. Several attempts have recently been made to explain the sudden dust formation either as being caused by critical conditions for the relevant thermal quantities or by a sort of episodic "blob"-event giving rise to rather rapid local cloud formation (Goeres & Sedlmayr 1992, Asplund et al. 1996). Both suggestions seem to be able to resolve the long-standing discrepancy between the observational findings (indicating a condensation radius rather close to the photosphere at only a few stellar radii) and the theoretical needs (requireing temperatures well below 1200K, which in the usual equilibrium description are encountered only far outside at about 10 stellar radii) by taking into account effective molecular cooling which for certain pressure regimes may produce very low kinetic temperatures and consequently high supersaturation ratios for the dust forming species already in the vicinity of the stellar photosphere.

In order to decide whether such type of a mechanism in fact is responsible for cloud formation and hence for the occultation phenomenon in an RCrB atmosphere it would very important to have available temporally and spatially highly resolved measurements which not only yield the evolution of the local physical and chemical conditions but also provide the critical values which have to be surmounted in order for the phenomenon to start.

Such type of observations might not only have immediate relevance for the RCrB phenomenon which may be seen as an example representing more general types of instabilities giving rise to cloudy structures in various types of dust forming objects.

3 Conclusion

The capabilities of the VLTI are ideally suited for a substantial improvement of the observations of dust–forming objects. The possibility to resolve circumstellar structures on small scales will lead to new insights in the local physics determining the evolution of these environments.

However, in order to arrive at a reliable interpretation of these new observational results, the measurements have to be complemented by (self–)consistent theoretical models. These models are already at hand for the circumstellar dust shells surrounding cool giants and supergiants. They include (time–dependent) hydrodynamics, the formation, growth and thermal evaporation of dust grains, chemistry, and radiative transfer, and at least equally important they also include the important couplings among the physical components. The theoretical models can be used to interpret the observations, and in turn, observational results will provide a test for the models and could indicate the need for improvements of the physical description of certain components, e.g. the chemistry.

A necessary prerequisite for the fruitful interplay between observation and theoretical modeling is the availability of highly time resolved observational data. Therefore, a small LPV key program should be defined for VLTI such that the objects of this program are continuously monitored, e.g. once per month.

Acknowledgements: This work was supported by the BMBF (grant 05 3BT13A 6) and by the DFG (grant Se 420/8-1).

References

Asplund M., Gustafsson B., Kiselman D., Eriksson K., 1996, A&A in press
Danchi W. C., Bester M., Degiacomi C. G., Greenhill L. J., Townes C. H., 1994, AJ 107, 1469
Fleischer A. J., Gauger A., Sedlmayr E., 1992, A&A 266, 321
Gail H.-P., Sedlmayr E., 1987, A&A 171, 197
Gillett F. C., Merrill K. M., Stein W. A., 1971, ApJ 164, 83
Goeres A., Sedlmayr E., 1992, A&A 265, 216
Jura M., 1986, Irish astr. J. 17, 322
Keady J. J., Ridgway S. T., 1993, ApJ 406, 199
Keady J. J., Hall D. N. B., Ridgway S. T., 1988, ApJ 326, 832
Knapp G. R., Morris M., 1985, ApJ 292, 640
Ridgway S. T., Keady J. J., 1988, ApJ 326, 843
Sedlmayr E., Dominik C., 1995, Space Sci. Rev. 73, 211
Winters J. M., Dominik C., Sedlmayr E., 1994, A&A 288, 255
Winters J. M., Fleischer A. J., Gauger A., Sedlmayr E., 1994, A&A 290, 623
Winters J. M., Fleischer A. J., Gauger A., Sedlmayr E., 1995, A&A 302, 483
Woolf N. J., Ney E. P., 1969, ApJ 155, L181

Study of Late Type Stars with the Infrared Spatial Interferometer of the University of California at Berkeley

Bruno Lopez[1], William C. Danchi[2], Pierre Cruzalèbes[3], Manfred Bester[2], and Charles H. Townes[2]

[1] Observatoire de la Côte d'Azur, Département Fresnel, BP4229 F-06304 Nice Cedex 4, France
[2] Space Science Laboratory, University of California at Berkeley, Berkeley, CA-94720
[3] Observatoire de la Côte d'Azur, Département Fresnel CNRS-URA 1361, Av. Copernic, F-06130 Grasse, France

Abstract. The Infrared Spatial Interferometer (ISI) of the University of California at Berkeley is a two telescopes interferometer operating in the infrared (9-12 microns). Many late type stars and their circumstellar dust shell have been observed and modelled. Several results have been obtained : location of the inner radii of the dust shell, physical conditions in the dust formation zone, optical depth of the shell, diameters of some evolved stars, location of SiO masers by regard to the dust (Danchi et al. 1994a, Greenhill et al. 1995).

We detail some of the results obtained on the carbon star IRC +10216 and on the Mira type star o Ceti. This latter study combines the ISI visibility measurements and the COMEON+ adaptive optics observations in J, H and K infrared bands (1.25, 1.65 and 2.2 microns) (Cruzalèbes et al. 1996). These measurements illustrate what kind of scientific research can be engaged in this field with the VLT Interferometer.

1 Introduction

Large mass loss toward the interstellar medium takes place at the latest stage of stellar evolution represented by red supergiants and AGB stars. The rate of matter ejection can achieve values as high as $10^{-4} M_{\odot} yr^{-1}$.

The origin of mass loss is only understood in broad outline (Lafon and Berruyer, 1991). A primary mechanism is invoked to lift the gas from the stellar photosphere to a distance large enough to allow a fraction of the gas to condense into dust particles. Then, the radiation pressure is efficient to push the dust - and the gas through collisions - up to expansion velocities which are observed.

The reason why the mechanisms of matter ejection are not yet well understood, is that, the nearby stellar environment, which contains the basic physical information, is not spatially resolved by classical observational methods. Some fundamental goals of the high resolution imaging technics (Adaptive Optics, Speckle Interferometry, Long baseline Interferometry) are the following:

- Where does the dust form in circumstellar envelopes ?

- Does it form homogeneously at a certain distance from the star or does it condense in clumps ?

- What are precisely, for the representative sources, the dust shell morphologies resulting from the dynamical process responsible for the mass loss ?

- The Mira type stars are known for their variability. How do stellar diameters change as a function of the luminosity cycle ?

- Is the mass loss rate continuous, periodic or sporadic ?

- What effects introduce a companion star onto the geometry of the mass loss?

More than twenty-five Late Type Stars have been observed up to now with the two telescope Infrared Spatial Interferometer (ISI) of the University of California at Berkeley. One of the recent results brings informations on the shell of the carbon star IRC +10216 and on the structures of the Mira type star o Ceti. This latter object has been also observed simultaneously with the ESO COMEON+ Adaptive Optics at the 3.6m Telescope. These measurements illustrate what kind of scientific research can be engaged in this field with the VLT Interferometer.

2 Scientific Background

The connection between the large scale winds and phenomena occurring at or near the photospheres of late type stars, such as pulsation and convection, is not yet well understood (Lafon and Berruyer, 1991). Some theorical models hardly begin to provide such a connection (Hearn 1990, Winters et al. 1995, Windsteig et al. 1996). During the past two decades, the envelopes which contain dust and molecules formed in the circumstellar environments of Late Type Stars have been studied by a variety of methods.

At centimeter and millimeter wavelengths, significant results were obtained at high spatial resolution, comparable to the dimension of the stars themselves. Such observations have been carried out on the Mira TX Cam and on the supergiant star VX Sgr with the Very Long Baseline Array using maser emission from SiO molecule as a tracer of material close to the star (Diamond et al. 1994, Greenhill et al. 1995).

At near-infrared and optical wavelengths high resolution works have been mainly focused on measurements of diameters of a number of these stars using the tools of speckle interferometry (Karovska et al. 1991, 1993), non-redundant aperture masking (Haniff et al. 1992, Wilson et al. 1992, Tuthill et al. 1995), and long baseline interferometry (Quirrenbach et al. 1992). Presently the diameters and shapes of two Mira type stars (R Leo and W Hyd) are also monitored with the Fine Guidance Sensor of the Hubble Space Telescope (Lattanzi et al. proposal GO-6049).

3 The Infrared Spatial Interferometer of the University of California at Berkeley

The Infrared Spatial Interferometer (ISI) is located at Mount Wilson Observatory. The interferometer operates at wavelengths ranging from 9 to 12 μm, and consists of 2 Pfund type telescopes ($\Phi = 1.65$ m for the diameter of the parabola). More details are given about the instrumental concept by Bester et al. (1994). Up to now 4, 10, 13 and 32 meters baseline length have been used.

Long baseline interferometry in the mid-infrared has provided significant observational studies of shells around Late Type Stars. A survey of thirteen stars at 11μm was recently published (Danchi et al. 1994). A monitoring of the mass loss activity of prototypical stars has also been performed. Namely the bright oxygen-rich supergiant α Orionis, the bright carbon star IRC +10216 (Danchi et al. 1994b), and on o Ceti, the prototype of Mira type stars (Bester et al. 1991, Danchi et al. 1994a, Lopez et al. 1996) were observed over a long period ~4-8 years.

4 Typical Results Obtained with the 11μ ISI Interferometer

4.1 The Carbon Star IRC +10216

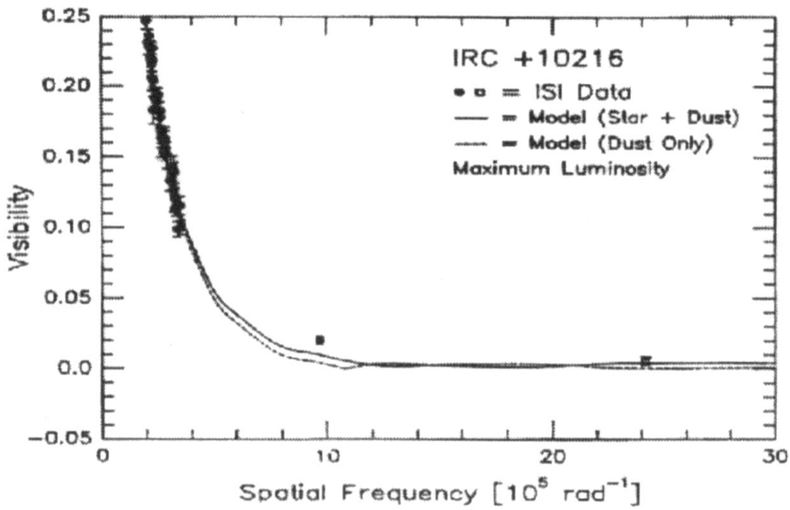

Fig. 1. The 11μm visibility curve of IRC +10216

The well known carbon star IRC+10216 is embedded in a very thick circumstellar shell. The envelope contribution to the 11μm visibility curve dominates (Figure 1) producing the steep decrease of the curve toward nearly zero when the contribution of the dust vanishes. The dust appears very close to the stellar photosphere of IRC +10216 at approximatly 2-2.5 stellar radii (depending on the luminosity cycle of the star), correponding on the sky to about 0.07-0.09″

(Danchi el al. , 1994a). With the 32 m baseline length the fringe visibility is 0.0054+-0.003 (Danchi el al., 1994b) leading to an estimate of the stellar diameter of 43.6 mas. For object belonging to the class of extreme carbon stars, OH/IR sources and post-AGB, infrared interferometry offers an unique mean of sounding stars hidden in thick envelopes.

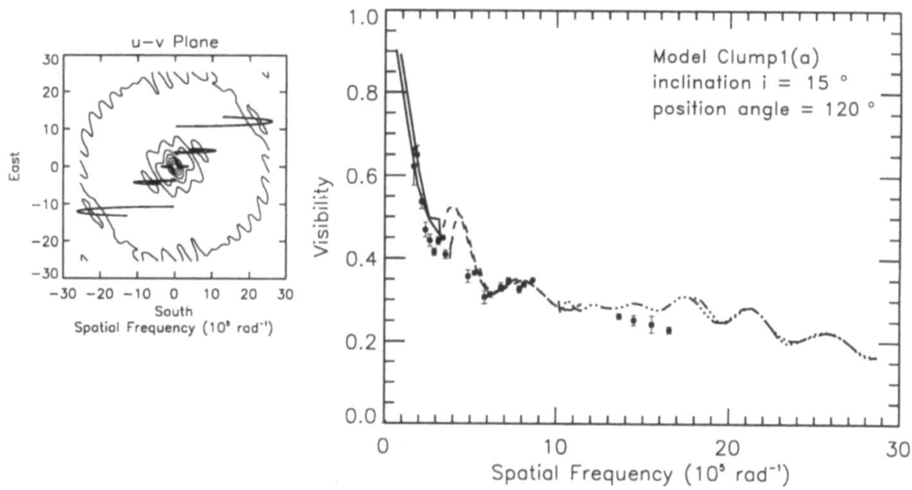

Fig. 2. Fourier plane coverage and $11\mu m$ visibility curve of o Ceti observed in 1993-1994 with ISI

4.2 The Mira Star o Ceti

Figure 2 displays the visibility curve of o Ceti as observed at $11\mu m$ in 1993-1994 with the ISI interferometer. Baselines from 4 up to 32 meters were used. The data are interpreted using radiative transfer models (Lopez et al. 1996) which involve departure from uniform mass loss, either with a disk, or multiple discrete shells or clump(s).

The monitoring of the mass loss activity was performed from 1989 up to now. Substantial variations in the $11\mu m$ visibility point out that an important dust concentration has probably occurred close to the photosphere in 1992. Dust does not seem to appear periodically following the luminosity cycle of the Mira variable but rather sporadically or multi-periodically as suggested by some theoretical models (Winters et al. 1995, Windsteig et al. 1996).

We have also observed o Ceti in September 1994 at the ESO 3.6m telescope with Adaptive Optics in the near infrared range. The dust shell is resolved in these data. At $1.25\mu m$, an extended structure in the circumstellar matter is detected (see Figure 3 also in Cruzalèbes et al. 1996) at distances approximatly

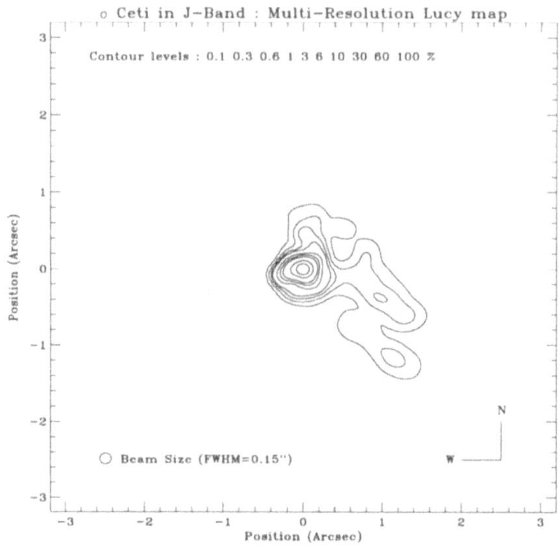

Fig. 3. o Ceti at 1.25μm with the Adaptive Optics of the ESO 3.6m Telescope.

from 0.5″ up to 2″. The observations, as well as the CO mapping made by Planesas et al. (1990), indicate that the distribution of the circumstellar matter around o Ceti is asymmetrical. Our radiative transfer models based on 11μm visibility curves remains consistent with the Adaptive Optics images.

On the other end, the o Ceti companion, a detached white dwarf discovered in 1923, is surrounded by an accretion disk (Reimers and Cassatella 1985). It has to be noticed that the structure in the 1.25μm image is elongated towards the expected position of the companion.

5 Conclusion – Performances VLTI Compared to ISI

The interest of long baseline interferometry for the study of Late Type Stars is obvious. As an illustration two examples of results obtained with ISI on the carbon star IRC +10216 and on the Mira o Ceti (results obtained on the supergiant VX Sgr were also presented during the talk - see Greenhill et al. 1995) are given. The ISI under operation provides presently some very relevant results and a lot remains to be done in this field. “Improved sensitivity”, “Spectral resolution” and “arrays of multi-telescopes for image reconstruction” are the keywords for future instrumental prospects aiming at answering the astrophysical questions of our first paragraph.

The performances of the VLTI and ISI are comparable. The telescope diameters of the VISA telescopes are identical to the ISI ones. The baselines are of comparable length. One of the goal of both interferometers is the best possible Fourier plane coverage. A third telescope in under construction for ISI, the VISA part of the VLT will be equipped with three telescope (large 8 meter telescopes should be also coupled). What is different between the two interferometers is the way the signal from each telescope is mixed together. The heterodyne detection of the ISI provide a very large spectral resolution, R about equal to 400000. This performance allows the study of various molecule locations and should allow the measurement of Doppler velocities of about $1kms^{-1}$. The knowledge of the dynamics of the mass loss should be significantly improved. Concerning the VLTI, the direct detection of the fringes gives a much larger sensitivity. At 10μm the sensitivity is improved by a difference of several magnitudes. Concerning late type stars, low brightness regions of the envelopes can be better studied and the sample of object presently observed can be also enlarged.

References

Bester, M., Danchi, W.C., Degiacomi, C.G., Townes, C.H., Geballe, T.R. 1991, ApJ 367, L27

Bester, M., Danchi, W.C., Degiacomi, C.G., Bratt, P.R.: 1994, Conference on " Amplitude and Intensity Spatial Interferometry", Proc. SPIE 2200, Paper 26.

Cruzalèbes, P., Lopez, B., Bester, M. 1996, poster of this workshop

Danchi, W.C., Bester, M., Degiacomi, C.G., Greenhill, L.J., Townes, C.H. 1994a, AJ 107, 1469

Danchi, W.C., Bester, M., Greenhill, L.J., Degiacomi, P.R., Townes, C.H.: 1994b, Conference on " Amplitude and Intensity Spatial Interferometry", Proc. SPIE 2200, Paper 27.

Diamond, P.J., Kemball, A.J., Junor, W., Zensus, A., Benson, J., Dhawan, V. 1994, ApJ 430, L61

Grennhill, L.J., Colomer, F., Moran, J.M., Backer, D.C., Danchi, W.C., Bester, M. 1995,ApJ 449, 365

Haniff, C., Ghez, A.M., Gorham, P.W., Kulkarni, S.R., Matthews, K., Neugebauer, G. 1992, AJ 103, 1662

Hearn, A. 1990, in "From Miras to PN : which path for Stellar Evolution ?" (Gif-sur-Yvette: Editions Frontières), 121

Karovska M., Nisenson, P., Papaliolios, C., Boyle, R.P. 1991, ApJ 374, L51

Karovska, M., Nisenson, P., Beletic, J. 1993, ApJ 402, L311

Lafon, J.P.J. and Berruyer, N. 1991, A&AR 2, 249

Lopez, B., Danchi, W.C., Bester, M. et al. 1996 to be submitted to ApJ

Planesas, P., Kenney, J., Bachiller, R. 1990, Apj 364, L9

Quirrenbach, A., Mozurkewich, D., Armstrong, J.T., Colavita, M.M., Shao, M. 1992, A&A 259, L19

Reimers, D. and Cassatella, A. 1985, ApJ 297, 275

Tuthill, P.G., Haniff, C.A., Baldwin, J.E. 1995, MNRAS 277, 1541

Windsteig, W., Dorfi, E.A., Hofner, S., Hron, J., Kerschbaum, F. 1996, this workshop.

Wilson, R.W., Baldwin, J., Busher, D.F., Warner, P.J. 1992, MNRAS 257, 369

Winters, J.M., Fleisher, A.J., Gauger, A., Seldmayr, E. 1995, A&A 302, 483

Interferometric Studies of Late Phases of Stellar Evolution

Gerd Weigelt[1], Yuri Balega[2], Karl-Heinz Hofmann[1], Norbert Langer[3], and Roger Osterbart[1]

[1] Max-Planck-Institut für Radioastronomie, Auf dem Hügel 69, D-53121 Bonn, Germany
[2] Special Astrophysical Observatory, Nizhnij Arkhyz, Zelenchuk region, Karachai-Cherkesia, 357147, Russia
[3] Max-Planck-Institut für Astrophysik, Postfach 1523, D-85740 Garching, Germany

1 Introduction

Objects in late evolutionary phases are very interesting targets for VLTI observations. We discuss Mira stars, carbon stars, LBVs, and Proto-Planetary Nebulae. The resolution of VLTI observations is high enough to resolve the stellar disks of many Mira stars, to reveal surface inhomogeneities, to study the strong wavelength dependence of Mira star diameters, and to study circumstellar shells and bipolar outflows of stars in late evolutionary phases. For illustration of the feasibility of the above VLTI projects we show previous speckle masking observations with the 2.2 m ESO telescope and the 6 m SAO telescope. Fortunately, most of the objects mentioned above are much brighter than the limiting magnitude of VLTI observations (which is at least about 15th magnitude in the visible and 12th magnitude in the K-band). *Multi-speckle* long-baseline interferometry techniques (modified speckle masking combined with the building block method; Reinheimer & Weigelt 1987, Reinheimer et al. 1993, and in this proceedings volume) can be applied if adaptive optics wavefront compensation is not available (for example, in the visible) or if only partial compensation is obtained.

2 Mira Stars

Mira stars show a substantial wavelength dependence of their angular size. This is caused by the wavelength dependence of the TiO opacity. Continuum and TiO band observations with suitable interference filters provide important information on the atmospheric structure and hence on the physical processes involved. Models by Bessell et al. (1989, 1996) and Scholz (1993) show that monochromatic radii observed in suitably chosen filter bandpasses are extremely sensitive diagnostic tools for studying the structure of Mira photospheres. The goal of our Mira star project is to provide the basic observations for a quantitative analysis of the photospheric structure of Mira variables and thus to test Mira star models.

Our diffraction-limited speckle masking (Weigelt 1977, Lohmann et al. 1983) observations of the Mira star R Cas with the 6 m telescope (see Figure 1 in the paper "Interferometric Observations of Mira Stars", Hofmann et al., this proceedings volume; Hofmann et al. 1995; Weigelt et al. 1996) show that much higher resolution than the resolution of single-dish telescopes is required in the future. At 700 nm wavelength a resolution of 30 milli-arcsec was achieved. The reconstructed true images are the images with the highest resolution obtained up to now at visible and infrared wavelengths. The speckle interferograms of R Cas were recorded through interference filters with only 6 nm bandwidth at center wavelengths of 671 nm (strong TiO band absorption), 700 nm (moderate TiO absorption), and 714 nm (strong TiO band absorption). The reconstructed images show that the azimuthally averaged 671, 700, and 714 nm uniform-disk (UD) diameters of R Cas are 44 mas, 36 mas, and 49 mas (± 2 mas), respectively. The images of the disks are non-uniform and elongated. For example, the two-dimensional size (elliptical uniform disk fit) of the elongated disk is 36 mas×51 mas at wavelength 671 nm (TiO absorption band filter). At wavelength 1043 nm (continuum) we measured a UD diameter of 30 mas. Mechanisms proposed for causing the non-spherical configuration of Mira stars include rotation, non-radial pulsation modes, asymmetric shock fronts, giant convective cells or bright spots, and the effects of circumstellar dust being concentrated in a disk that partially obscures the stellar surface.

3 The Circumstellar Shell Around the Carbon Star IRC+10216

IRC+10216 is the nearest and best-studied carbon star. It is a long-period variable star with a period of 650 d and spectral type C9,5. IRC+10216 is surrounded by an optically thick dust shell which is expanding at a velocity of about 15 km/s. Its mass loss rate is about 2×10^{-5} solar masses per year. Fig. 1 shows the first infrared image of IRC+10216 with 92 mas resolution (Osterbart et al. 1996). The circumstellar shell is for the first time resolved in a dominant elongated cloud (A) and three clouds (B, C, and D) with separations of 130 mas, 150 mas, and 210 mas. The observations can be used to test the time-dependent model calculations for carbon-rich circumstellar dust shells reported by Fleischer et al. (1992) and Winters et al. (1995). A general result of the calculations is the formation of discrete expanding dust layers and wavelength-dependent, step-like intensity profiles. Assuming a distance of 170 pc, the separations of the clouds B, C, and D from the dominant cloud A are 22 AU, 25 AU, and 36 AU, respectively. The radial velocity of the expanding dust shell is about 15 km/s. This corresponds to about 3 AU/yr or 18 mas/yr (at a distance of 170 pc) for motion perpendicular to the line of sight. Therefore, the tangential velocity can be determined by future observations.

4 The Proto-Planetary Nebula Red Rectangle

The Red Rectangle is an extended (~ 40") reflection nebula. Its distance is about 330 pc. The presence of carbon-rich dust and the large bolometric flux

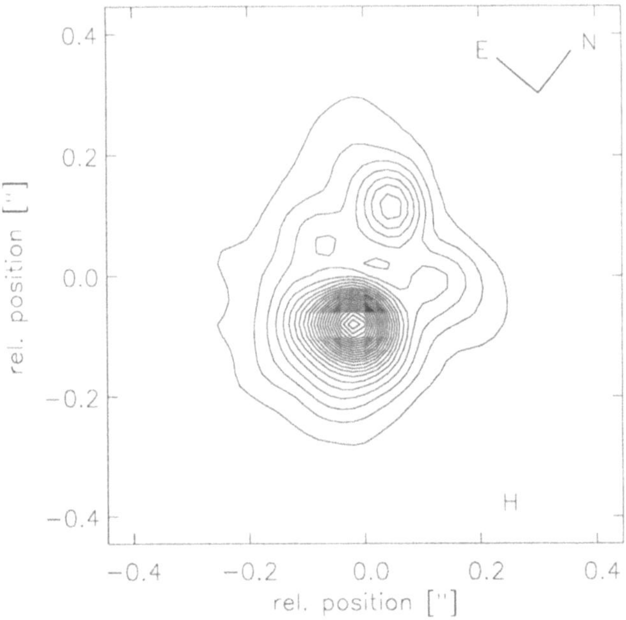

Fig. 1. Diffraction-limited H-band (1.65 μm) speckle masking reconstruction of IRC+10216. The contour levels are plotted from 4–99% of peak intensity in steps of 4%.

suggest a post-AGB nature for the central stellar object. Fig. 2 shows the first diffraction-limited speckle-masking observations of the central 0.8"×0.8" region of the Red Rectangle at optical and infrared wavelengths (Osterbart et al. 1996). The speckle data were recorded through an interference filter with center wavelength/bandwidth of 656 nm/60 nm and through a K-band filter (2.2 μm). At 656 nm wavelength the resolution of the reconstructed image is 76 mas. The reconstructed images show a bipolar structure with a dark lane between two bright lobes. The dark lane is probably produced as a result of obscuration by a dust disk slightly inclined to the line of sight, as discussed below. Recent radial velocity measurements by Van Winckel et al. (1995) show that the central object is a spectroscopic binary. Probably mass transfer has occurred in this system since AGB stars are more extended than the current binary orbit (separation $\sim$ 0.3 AU). Recent dynamical studies (Yorke et al. 1995) of this situation (common envelope evolution) show that a large fraction of the AGB envelope can be expelled from the binary system and that most of the gas remains in a disk-like configuration, confined within 20 to 30 degrees of the orbital plane. This model might explain the obscuring disk which separates the two main lobes in Fig. 2. Perhaps the Red Rectangle is the most favorable case for the formation of a stationary circum-system disk with a re-accretion of metal-deficient gas, which is proposed by Van Winckel et al. (1995) in order to explain the extremely low observed [Fe/H] ratio in the central star.

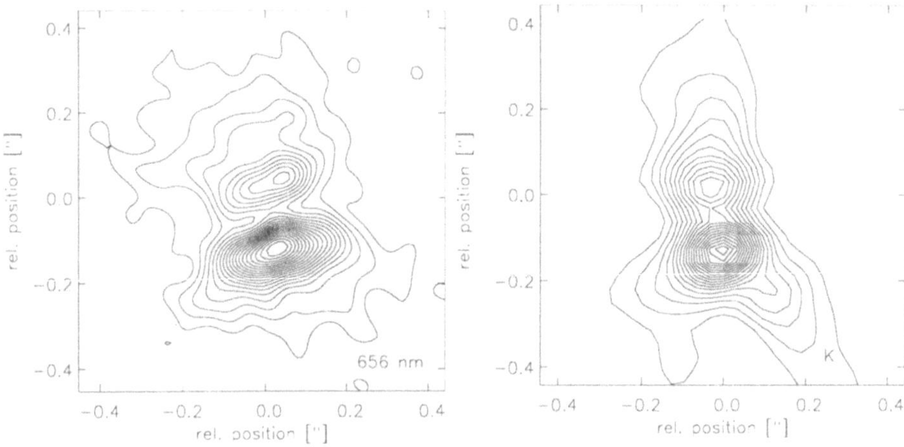

Fig. 2. Diffraction-limited speckle masking observations of the Red Rectangle: (left) 656 nm image (2.2 m ESO telescope), (right) K-band image (6 m telescope). The contour levels are plotted from 5% to 95% of peak intensity in steps of 5%. North is at the top and east to the left.

5 Eta Carinae

η Carinae is the most extreme, eruptively unstable Luminous Blue Variable, with enormous, sporadically variable mass loss. It is one of the most massive and luminous stars in the Galaxy. η Car is surrounded by several ejected objects at separations of 0.1" to 0.4" which were detected by speckle observations (Hofmann and Weigelt 1988). Fig. 3 shows a diffraction-limited Hα η Car image reconstructed from only 740 speckle interferograms recorded with the 2.2 m ESO telescope (filter 654 nm/6 nm). The tangential velocities of the ejected objects were derived from our 1985 and 1994 speckle observations. Two-dimensional objective prism spectra of these objects with diffraction-limited spatial and 0.3 Angstrom spectral resolution were obtained by objective prism speckle spectroscopy (Reinheimer et al. 1997). This resolution was high enough to measure for the first time the Doppler shift of individual ejected gas clouds at separations between 0.1" and 0.4" from the central object in η Carinae. We plan to apply similar spectroscopic methods (long-baseline speckle spectroscopy) to VLTI interferometry. Three-dimensional velocity information obtained by VLTI interferometry and VLTI spectroscopy will allow us to investigate the dynamics of the mass loss in η Carinae.

In addition to the object classes discussed above, we have investigated Seyfert galaxies using speckle techniques. These observations show that several Seyfert galaxies are bright enough for VLTI observations. For example, NGC 4151 was observed with the 6 m telescope in the K-band and diffraction-limited resolution was obtained with only a few minutes observing time. Therefore, we plan infrared VLTI imaging and imaging spectroscopy of Seyfert galaxies with a velocity reso-

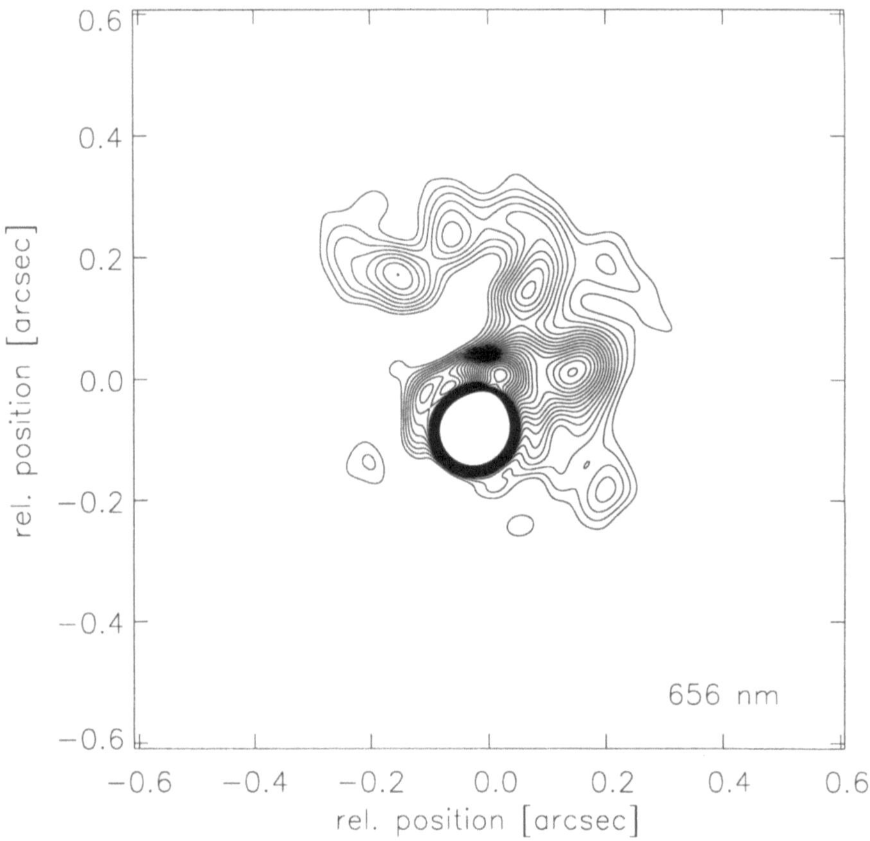

Fig. 3. Diffraction-limited Hα image of Eta Carinae reconstructed by the speckle masking method. The contour levels are plotted from 1.4% to 12% of peak intensity in steps of 0.4%. North is at the top and east to the left.

lution which is high enough to study the kinematics of the Narrow Line Region. This yields crucial input data for theoretically modeling the dynamics of this region.

References

Bessell M.S., Brett J.M., Scholz M., Wood P.R. 1989, A&A 213, 209

Bessell M.S., Scholz M., Wood P.R. 1996, A&A 307, 481 Fleischer, A.J., Gauger A., Sedlmayr E. 1992, A&A 266, 321

Hofmann K.-H., Balega Y., Scholz M., Weigelt G. 1995, in: Stellar surface structure, IAU Symp. 176, Strassmeier K.G. (ed.), Kluwer, Poster Proceedings p.45

Lohmann A.W., Weigelt G., Wirnitzer B. 1983, Appl. Opt. 22, 4028

Osterbart R., Langer N., Weigelt G. 1996, Diffraction-limited speckle-masking observations of the Red Rectangle and IRC +10216 with the 6 m telescope, Proceedings of

the IAU Symposium No. 180, Eds. H.J. Habing and H.J.G.L.M. Lamers, Groningen, 1996, Kluwer Academic Publishers, Dordrecht
Reinheimer T., Hofmann K.-H., Weigelt G. 1993, A&A 279, 322
Reinheimer T., Weigelt G. 1987, A&A 176, L17
Reinheimer T., Schöller M., Weigelt G. 1997, Objective Prism Speckle Spectroscopy of Eta Carinae, A&A submitted
Scholz M. 1993, in: New perspectives in stellar pulsation and pulsating variable stars, IAU Coll.139, Nemec J., Matthews J. (eds.), Cambridge Univ. Press, p.201
Van Winckel H., Waelkens C., Waters L.B.F.M. 1995, A&A 293, L25
Weigelt G.P. 1977, Opt. Commun. 21, 55
Weigelt G., Balega Y., Hofmann K.-H., Scholz M.. 1996, A&A submitted
Hofmann K.-H., Weigelt G., 1988 A&A 203, L21
Winters J.M., Fleischer A.J., Gauger A., Sedlmayr E. 1995, A&A 302, 483
Yorke H.W., Bodenheimer P., Taam R.E. 1995, ApJ 451, 308

Interferometric Imaging of Late-Type Stars with the VLTI

Chris A. Haniff

University of Cambridge, Mullard Radio Astronomy Observatory, Cavendish Laboratory, Madingley Road, Cambridge, CB3 0HE, UK

Abstract. First-generation interferometers have already begun to explore the detailed properties of many of the largest and nearest late-type stars. With a projected complement of 4 UTs and 3 ATs, and state-of-the-art instrumentation, the VLTIs capabilities will extend far beyond those currently available, and it will be well placed to extend studies of late-type stars both in detail and distance. In this paper, I outline a number of scientific issues well suited for interferometric study, and identify those features of the VLTIs design most relevant to its success in this role.

1 Introduction

Over the past 10 years, studies of the brightest and closest late-type stars have demonstrated the potential of interferometry for optical and infrared (ir) astronomy convincingly. Not only have they verified methods of image recovery from visibility and closure phase measurement but they have also led to a number of unexpected discoveries concerning the surface structure and morphologies of late-type giants (see e.g. Buscher et al. 1990; Tuthill et al. 1994). Although there may be a temptation to regard interferometric imaging as simply a technique looking for an application, its continued use, together with the breadth and focus of recent investigations (see e.g. Bester et al. 1996; Quirrenbach et al. 1996; Tuthill et al. 1996), suggest that a second-generation interferometer such as the VLTI will be able to address a wide range of astrophysical questions.

The broad scientific rationale for the VLTI has been presented elsewhere (see, e.e. Walsh and Danzinger 1995; Paresce et al. 1996) and so I will make no attempt here to justify observations of late-type stars in preference to the other astronomical targets reviewed in these proceedings. Rather, I shall concentrate on identifying those types of observations that will be both possible and of interest to astrophysicists, and on highlighting the implications that these may have for the design of the VLTI.

In principle there are at least two approaches to this problem: to look at theoretical models of late-type stars and predictions of the performance of the VLTI, or to extrapolate from the capabilities and successes of existing interferometric arrays. My preference here has been to choose the second of these approaches, primarily because it seems to offer a more realistic, though undoubtedly more conservative, outlook. If anything, then, the material discussed in this review should be seen as outlining a minimal, rather than maximal, set of prospects for the VLTI.

2 Present and Future Interferometric Arrays

2.1 Existing Interferometric Experiments

Traditionally, interferometric studies of late-type stars have fallen into two main categories: those using 2-element interferometers and those employing multi-element arrays. The scope of these types of instruments are outlined below:

2-Element Interferometers. Most 2-element systems, for example the MKIII (Shao et al. 1988) and IOTA (Carleton et al. 1994) interferometers, have been characterised by r_o-sized array elements, long baselines ($\gg 2 \times 10^7\,\lambda$), relatively low spectral resolution ($R \lesssim 30$), and active fringe tracking. The GI2T interferometer alone (Mourard et al. 1994) has employed larger telescopes (1.5 m) and much higher spectral resolutions ($R \sim 1500$). Observations at optical and near-ir wavelengths have been reported, though in the latter case these have used the $2.2\,\mu$m window almost exclusively. These arrays have been mainly employed for angular diameter measurements, although in a few instances sufficient Fourier data have been obtained to identify deviations from the simplest spherically symmetric models (Di Benedetto and Bonneau 1990; Quirrenbach et al. 1992).

Multi-Element Interferometers. Until very recently, multi-element interferometric observations of late-type stars have had to rely on conventional monolithic telescopes converted to operate in interferometric modes. The basic idea has been to use the telescope structure as a coherent beam combiner, with a pupil-plane mask isolating patches of the primary mirror which act as array elements. This so-called "non-redundant masking" (NRM) technique (Haniff 1989) eliminates the metrological challenges of separated-element interferometry, and allows the number, locations, and sizes of the array elements to be altered by simply changing a mask. However, the use of a conventional telescope as an optical breadboard brings with it a number of difficulties, notably a restriction of the maximum interferometer baseline and the inability to fringe-track in order to extend the basic integration time beyond the atmospheric coherence time.

Typical NRM observations of late-type stars have used r_o-sized array elements, short baselines ($\leq 10^7\,\lambda$), low spectral resolution ($R \lesssim 50$), and, because of signal-to-noise considerations associated with the lack of active fringe tracking, only small numbers of array elements (≤ 5). Nevertheless, the ability to measure closure phases and consequently recover images from such data, has meant that even rudimentary NRM observations have proved scientifically valuable.

2.2 Projections for the VLTI

The recommendations for the Phase A/B VLTI endorsed by the Science Advisory Committee clearly place it well beyond any of the interferometers that have observed late-type stars to date. In particular, the projections for the post-2003 period, which envisage 7 array elements, 5-way beam combination, operation from 1 to 20 μm, active fringe tracking, and baselines of up to 200 m exceed those of the instruments described above by orders of magnitude.

Without focusing on specific astrophysical questions it is difficult to assess which of these enhancements will be most relevant. However, it is probable that high dynamic range imaging will form the basis for future studies of late-type stars, and so issues such as Fourier plane coverage, imaging fidelity, and operation in low-visibility regimes will likely be most important.

3 Science Goals for the VLTI

For the most part interferometric observations of late-type stars have concentrated on either supergiants, typified by Betelgeuse, or AGB stars, in particular long-period variables such as Mira. Brief outlines, based on current interferometric research, of the type of science possible are given below:

3.1 Angular Sizes

The majority of interferometric observations of late-type supergiants have been angular diameter studies undertaken in an effort to better define the physical sizes and effective temperature scale of the stars (see, e.g. Dyck et al. 1996). These types of measurements have been most easily carried out in the near-ir where the effects of limb-darkening and circumstellar emission are minimised, and are sometimes quoted with formal errors smaller than a percent! At this level, however, assumptions as to the detailed surface brightness distributions of the stars have to be made – a situation that may well change when visibility measurements beyond the first null become routinely available.

The status of similar measurements for the cooler Miras has been less secure, principally because of the strong wavelength dependence of the apparent diameter (Bessell et al. 1989). Early optical observations often used spectral bandpasses far too wide to permit useful quantitative comparisons with the literature. In general, spectral resolutions of at least 100 are preferred. Better suited regions of the spectrum lie in the 1 to 2 μm window, where the advantage of the VLTIs long baselines will be extremely valuable.

3.2 Limb-Darkening

The first steps in quantitatively assessing theoretical limb-darkening models for cool giants using amplitude interferomety have been reported by Quirrenbach et al. (1996). By accurately measuring visibilities beyond the first null, these authors were able to verify the computations of Manduca et al. (1977), and simultaneously secure a precise diameter estimate, for the K1 III giant Arcturus. The success of this type of work depends critically on multiwavelength observations and, more importantly, on the ability to explore the visibility function beyond the first null, where visibilities as low as 10^{-2}–10^{-3} must be measured. Since the signal-to-noise ratio for photon-noise-limited fringe detection scales as a power of $V\sqrt{N}$ (V is the fringe visibility and N the number of photons detected per integration) this represents a significant challenge for low-contrast sources, such as Arcturus, where V is low. Even for Betelgeuse, recent data for which are shown in Fig. 1, the long baseline visibilities have a maximum of only 6%, and so a key design goal for the VLTI will be to guarantee reliable operation in the low-visibility regime.

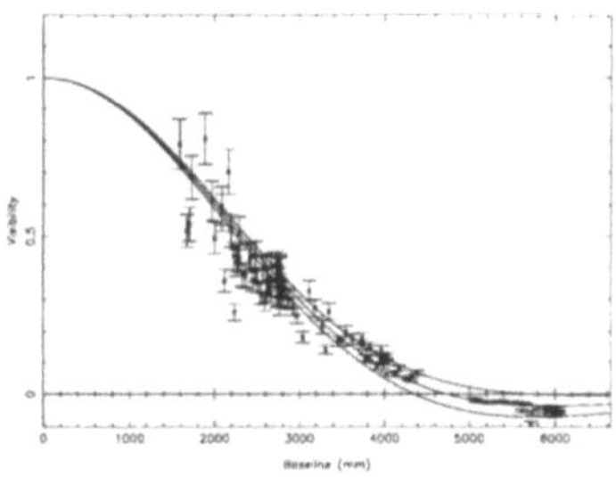

Fig. 1. Visibility measurements for Betelgeuse, obtained at 830 nm with the COAST interferometer in 1995 September. The solid lines show the best-fit and $\pm 3\sigma$ limb-darkened models. At the time the star showed no evidence for surface structure.

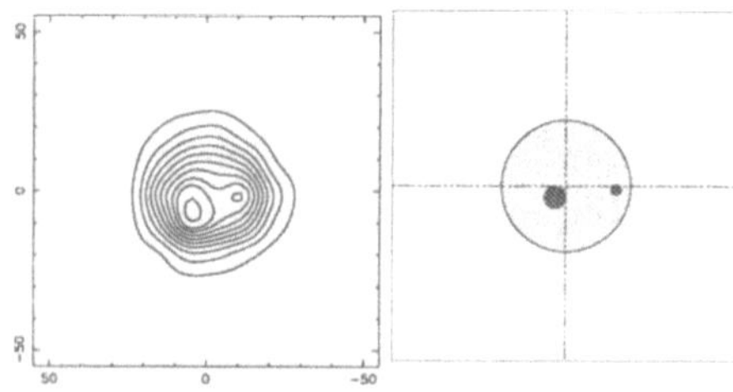

Fig. 2. Diffraction-limited image of α Her obtained at 633 nm in 1993 June using NRM methods. The right-hand panel shows a schematic of the 3-component model, comprising a uniform disk and 2 unresolved spots. The relative flux contributions of the 2 spots are denoted by their angular sizes.

Because of the difficulties of properly modelling the atmospheres of Miras, confrontation between limb-darkening models and interferometric data has only begun quite recently (Haniff et al. 1995). As for their warmer counterparts, significant progress will have to wait for long-baseline observations, albeit secured in more numerous narrow bandpasses so as to properly sample the molecular and atomic features in their spectra. These data should provide a new diagnostic tool for cool stellar atmospheres and offer new guidelines beyond those currently available from photometry and spectroscopy.

3.3 Radial Pulsation

Evidence for radial pulsation in a number of M-supergiants has come mainly from photometric and radial velocity studies which show periods of several hundreds of days (Smith et al. 1989). Yet interferometric campaigns to monitor stellar diameters have not been regularly undertaken, despite the fact that in some instances variations as large as 15% are predicted (Dupree et al. 1987). Recently however, Bester et al. (1996) have analysed several years of 11 μm interferometer data, together with optical and mid-ir photometry, for Betelgeuse and concluded that diameter changes inferred in the past may have been significantly overestimated. Dedicated interferometers like the VLTI will be well placed to resolve this ambiguity, especially if co-ordinated spectro-photometric data can be secured.

Unambiguous interferometric evidence for radial pulsation in Miras has similarly not been secured. Even for these stars, for which pulsation is a fundamental property, theoretical models provide differing predictions as to the amplitude and phase of the motion of the atmosphere, the range of expected variations spanning 5% to 50%. NRM observations of Mira between 1991 and 1993 (Tuthill et al. 1995) were unable to detect any periodic size variations, but did reveal evidence for a pulsation-induced shock propagating through the stellar atmosphere in late 1993. The detection and characterization of such activity, at both continuum and hydrogen line wavelengths, will be a challenging programme for the VLTI, but has the potential to validate current dynamical models for these stars and to clarify the long-standing debate concerning their pulsation modes.

3.4 Surface Features

Interferometric data, in particular those that retain visibility phase information, have consistently shown unequivocal evidence for asymmetric structure in the brightness distributions of M-supergiants (see, e.g. Buscher et al. 1990; Wilson et al. 1992; Tuthill et al. 1996; Fig. 2). The nature of these asymmetries, however, has remained unclear mainly because existing data have been unable to fully constrain the sizes, temperatures and locations of the features seen. A favoured hypothesis is that they represent the effects of giant convective cells, but other explanations involving patchy extinction or the effects of close companions have also been suggested. Even moderate ($\times 5$) improvements in spatial resolution would limit the set of viable explanations, as would contemporaneous measurements over a wide wavelength range. Such imaging observations would, in practice, be difficult because of the weak visibility signals expected and the need to obtain sufficiently complete Fourier coverage to ensure unambiguous mapping. They will surely though remain a priority for the VLTI.

Due to a lack of closure phase data, the evidence for surface structure on Miras is less clear. Existing measurements do show indications of non-circularity in the brightness distributions of several Miras (Quirrenbach et al. 1992; Tuthill et al. 1995; Weigelt et al. 1996), but its origin remains obscure. The most recent data for R Cas suggest that in near-ir continuum bandpasses the asymmetries disappear (Weigelt, private communication), but whether this is solely an effect of the poorer spatial resolution currently available at these wavelengths will have to be decided by instruments like the VLTI.

3.5 Mass Loss and Dust Shell Properties

Despite considerable theoretical and observational study, the physics of mass loss from late-type stars remains controversional. The presence of pulsation, convection, dust-formation and a large pressure scale height suggest a variety of mechanisms. These include radiation pressure, acoustic waves, and Alfvén waves. Imaging at near and mid-ir wavelengths on stellar scales will be one method to locate the regions of mass acceleration directly. The observed morphologies of planetary nebulae (PN) suggest strong bipolarity in the pre-PN phase: high resolution imaging offers the additional possibility of identifying the location and epoch at which gas and dust collimation can occur.

Considerable success has been achieved by the mid-ir 2-element Berkeley Infrared Spatial Interferometer (Bester et al. 1990). Despite its limited sensitivity and Fourier coverage, it has pioneered studies of the dust shells of late-type stars (Danchi et al. 1994), and has allowed not only the determination of their properties, but has also been used to identify individual mass-loss events (Bester et al. 1996). Attempts to use near-ir NRM methods to examine hotter dust closer to the stellar surface have also been successful (Fig. 3), although the longer baselines of the VLTI will be needed to extend this to a larger sample.

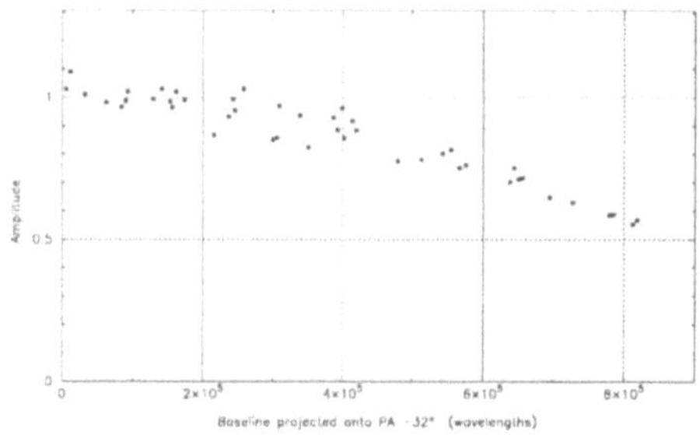

Fig. 3. L-band visibility function of RX Boo projected along PA $-32°$. In the orthogonal direction the source is unresolved at the same baseline lengths, implying a gross departure from circular symmetry for the star and its dust shell at 3.5 μm.

With larger array elements and homodyne beam combination, the VLTI has the potential to revolutionize studies in this area. Two-dimensional Fourier plane sampling, tunable spectral coverage, and better sensitivity all represent significant enhancements and should lead to much higher quality observational data.

4 Implications for the VLTI

The successful applications of interferometry to studies of late-type stars suggests that they may be among the most attractive targets during the initial operational phases of the VLTI. When planning the VLTI, then, it may be valuable to consider the following issues:

1. **Telescope Size.** Fortunately the stars we have discussed are intrinsically bright. Davis (1985) estimates that several hundred candidates later than M0 and with V magnitudes brighter than 6 will be visible from the southern hemisphere. At near and mid-ir wavelengths these stars will be brighter still, and so array elements the size of the ATs will be more than adequate. A consequence of this is that Adaptive Optics will not be required for these studies, unless observations at wavelengths below 1 μm are envisaged.

2. **Number of Telescopes.** Most interferometric observations of late-type stars have shown signs of incipient asymmetry and complexity, and so the ability to image with, e.g. 10×10 resolution elements will be crucial. Experience suggests that at least 4, and preferably more (6/7) array elements will be needed for high-quality phase recovery. Successful imaging will thus likely need all 3 ATs as well as a number of UTs operating simultaneously.
3. **Spectral Capability.** Given their low effective temperatures, studies of late-type stars are likely to be most efficient at near and mid-ir wavelengths. Operation in the J/H windows will give enhanced spatial resolution, whereas for studies of circumstellar material capabilities in the mid-ir are desirable. Throughout this range, spectral resolutions of order a hundred should suffice in the near to moderate term.
4. **Fringe Tracking.** Perhaps the most important requirement for the VLTI design, arising from the need to image, will be the ability to fringe track on resolved targets. This implies careful design of the array so that individual long baselines can be reached by combinations of shorter baselines on which fringe tracking is possible. Colavita (1994) has given approximate sensitivity figures for imaging interferometers operating in this regime. It is difficult to convert his figures directly to those appropriate for the VLTI: however it seems unlikely that imaging with large numbers of pixels ($> 10 \times 10$) will be realistic with only a small number of array elements.
5. **Baseline Length.** Dyck and Kibblewhite (1986) have examined the characteristic sizes of M and C stars, and their dust shells, in the solar neighbourhood. They quote median sizes between 5 and 25 millarcseconds for several hundred targets, implying baselines of between 40 and 200 m at 5 μm. These figures are extremely well matched to the VLTI site.
6. **Temporal Resolution.** Given the small number of VLTI elements, it is fortunate that the characteristic timescales associated with changes on and around late-type stars are months/years. Rapid mobility would clearly be desirable, but nighttime relocation of the antennae should not be required.
7. **Theoretical Framework.** Exisiting interferometric data are already beginning to challenge theoretical models of late-type stars. If the VLTI is to be successful in an astrophysical context then the appropriate astrophysical models for interpreting its data must be developed now. Urging the theoretical community to prepare for the VLTI is a crucial and urgent task.

5 Summary

The VLTI represents such a significant step beyond the interferometers that have been used to study late-type stars to date, that it is inconceivable that it will not engender a whole new class of inquiry into this type of source. The goal to image resolved low-contrast targets in narrow wavelength channels will not be easy to achieve, but at least the magnitudes, numbers and sizes of the expected population of interesting targets are favourable. Operation in a coherent mode, with long effective integration times, may be the key to success.

References

Bessell, M.S., Brett, J.M., Scholz, M., Wood, P.R. (1989) A&A **213**, 209–225

Bester, M., Danchi, W.C., Townes, C.H. (1990) Proc SPIE **1237**, 40–48

Bester, M., Danchi, W.C., Hale, D., Townes, C.H., Degiacomi, C.G., Mékarnia, D., Geballe, T.R. (1996) ApJ **463**, 336–343

Buscher, D.F., Haniff, C.A., Baldwin, J.E., Warner, P.J. (1990) MNRAS **245**, 7p–11p

Carleton, N.P., Traub, W.A., Lacasse, M.G., Nisenson, P., Pearlman, M.R., Reasenberg, R.D., Xu, X., Coldwell, C., Panasyuk, A., Benson, J.A., Papaliolios, C., Predmore, R., Schloerb, F.P., Dyck, H.M., Gibson, D. (1994) Proc SPIE **2200**, 152–165

Colavita, M.M. (1994) *Proc. IAU Symposium 158 – Very High Angular Resolution Imaging* (Kluwer, Dordrecht), 469–476

Danchi, W.C., Bester, M., Degiacomi, C.G.., Greenhill, L.J., Townes, C.H. (1994) AJ **107**, 1467–1513

Davis, J. (1985) *Proc. IAU Symposium 111 – Calibration of Fundamental Stellar Quantities* (Reidel, Dordrecht), 193–208

Di Benedetto, G.P., Bonneau, D. (1990) ApJ **358**, 617–621

Dupree, A.K., Baliunas, S.L., Hartmann, L., Nassiopoulis, G.E., Guinan, E.F., Sonneborn, G. (1987) ApJ **317**, L85–L89

Dyck, H.M., Kibblewhite, E.J. (1986) PASP **98**, 260–267

Dyck, H.M., Benson, J.A., Van Belle, G.T., Ridgway, S.T. (1996) AJ **111**, 1705–1712

Haniff, C.A. (1989) *Proc. NATO ASI Diffraction-Limited Imaging with Very Large Telescopes* (Kluwer, Dordrecht), 171–190

Haniff, C.A., Scholz, M., Tuthill, P.G. (1995) MNRAS **276**, 640–650

Manduca, A., Bell, R.A., Gustafsson, B. (1977) A&A **61**, 809–813

Mourard, D., Tallon-Bosc, I., Blazit, A., Bonneau, D., Merlin, G., Morand, F., Vakili, F., Labeyrie, A. (1994) A&A **283**, 705–713

Paresce, F., Mourard, D., Bedding, T., Beletic, J., Haniff, C., Leinert, C., Malbet, F., Mariotti, J.-M., Mozurkewich, D., Mundt, R., Petitjean, P., Quirrenbach, A., Reinheimer, T., Richichi, A., Röttgering, H., von der Lühe, O., Waters, R. (1996) ESO Messenger **83**, 14–21

Quirrenbach, A., Mozurkewich, D., Armstrong, J.T., Johnston, K.J., Colavita, M.M., Shao, M. (1992) A&A **259**, L19–L22

Quirrenbach, A., Mozurkewich, D., Buscher, D.F., Hummel, C.A., Armstrong, J.T. (1996) A&A **312**, 160–166

Shao, M., Colavita, M.M., Hines, B.E., Staelin, D.H., Hutter, D.J. (1988) A&A **193**, 357–371

Smith, M.A., Patten, B.M., Goldberg, L. (1989) AJ **98**, 2233–2248

Tuthill, P.G., Haniff, C.A., Baldwin, J.E. (1994) *Proc. IAU Symposium 158 – Very High Angular Resolution Imaging* (Kluwer, Dordrecht), 395–397

Tuthill, P.G., Haniff, C.A., Baldwin, J.E. (1995) MNRAS **277**, 1541–1546

Tuthill, P.G., Haniff, C.A., Baldwin, J.E. (1996) MNRAS In press.

Walsh, J.R., Danziger, I.J. Eds. (1995): *Science with the VLT* (Springer-Verlag, Berlin, Heidelberg)

Weigelt, G., Balega, Y., Hofmann, K.-H., Scholz, M. (1996) A&A In press.

Wilson, R.W., Baldwin, J.E., Buscher, D.F., Warner, P.J. (1992) MNRAS **257**, 369–376

Geometry of Post-AGB Shells

Christoffel Waelkens[1], Rens Waters[2,3], and Hans Van Winckel[1],

[1] Instituut voor Sterrenkunde, Celestijnenlaan 200 B, B-3001 Leuven, Belgium
[2] Astronomisch Instituut, Universiteit van Amsterdam, Kruislaan 403, 1098 SJ Amsterdam, The Netherlands
[3] SRON Groningen, P.O. Box 800, 9700 AV Groningen, The Netherlands

Abstract. Post-AGB stars are low- and intermediate-mass objects in the rapid evolutionary stage between the asymptotic giant branch (AGB) and the planetary-nebula stage. Their study is important both because it concerns a relatively poorly documented phase of stellar evolution and because during this phase the circumstellar dust ejected on the AGB is processed before being restored to the interstellar medium. The angular resolution and the mid-infrared capability of VLTI will help us to build a picture about these objects that is much more detailed and much less affected by observational bias than the view we have at present. The basic expectations from the VLTI are outlined and illustrated with three different objects.

1 Post-AGB Stars

Post-AGB stars are defined as transition objects between the asymptotic giant branch (AGB) and the planetary nebulae (PN). The alternative denomination of proto-planetary nebulae is now less in use, because of possible confusion with circumstellar disks in young objects, and also because in the objects of lowest mass the circumstellar shell may already be dispersed before the central object reaches the effective temperature necessary to ionize the nebula.

This terminal evolutionary stage of stars of low and intermediate masses is one of those that are least understood, since because of its short duration it is relatively scarcely populated. The study of post-AGB stars is of importance to verify the chemical alteration of stellar photospheres as the result of nucleosynthetic and dredge-up processes at the end of the AGB; while some objects indeed are conform to theoretical expectation (e.g. HD 56126, Klochkova 1995; HD 187885, Van Winckel et al. 1996), the chemical composition of others is puzzling (Luck, Bond and Lambert 1990, Van Winckel 1996). The study of the circumstellar shells of post-AGB stars is important to study the mass loss history and also the early processing of the dust; it is thought that more than 90% of the dust in the interstellar medium is processed from dust ejected from AGB stars.

It is highly probable that our present census of post-AGB stars suffers from important observational biases. It is likely that a 'typical' post-AGB star is much less conspicuous than its progenitors and its products: in the former the dense circumstellar shell and the high IR flux of the AGB star produce intense OH masing sources, in the latter the UV flux of the central star is able to ionize the nebula. The 'constant OH/IR stars' and 'young PNe' then respectively represent early and late stages, while objects with intermediate effective temperatures remain inconspicuous mid-IR sources which are waiting for the VLTI to be studied in detail. An important subclass of post-AGB candidates are the high-galactic-latitude supergiants; the low-mass and post-AGB nature of these stars is inferred from the combination of their spectral type, galactic location and kinematics and IR excess; but the low circumstellar extinction and the longer post-AGB timescale of low-mass objects that favor their detection suggests that these objects describe the low-mass tail of the distribution and so are not fully representative for post-AGB stars as a class.

Finally, a certain number of well studied sources such as the Red Rectangle owe much of their spectacular behavior to the fact that they are binaries. The conjecture that most of the objects which display a near-IR excess in addition to the mid-IR excess due to the shell ejected on the AGB are binaries (Waters et al. 1991, 1992) has been confirmed by several observations we have made since. It is then to be expected that the systematic study of mid-IR sources by VLTI will help to build a much more balanced view on the properties of post-AGB stars.

2 Basic VLTI expectations

The VLT and VLTI will greatly enhance our knowledge about post-AGB stars. High-resolution optical spectroscopy of faint, embedded, stars will enable us to complete the picture about the chemical composition of the photospheres, and so to determine observationally the chemical yield from AGB stars in the whole relevant mass range. The VLTI will offer us both the spatial resolution and the infrared spectral capability needed for unraveling the mass loss history and the dust processing in the expanding circumstellar envelope.

So far, studies of the circumstellar envelopes of post-AGB stars relied heavily on the spectral energy distribution obtained from ground-based infrared photometry and IRAS data. Significant progress has come from the advances of IR spectroscopy in the 10 and 20 μm windows, and is now also emerging from the spectrometers of ISO. However, it is well known that an unambiguous interpretation of such data in terms of a geometric model of the circumstellar shell cannot be achieved without spatially resolved information. Such information will be obtained for the most nearby sources with the first VLT unit, and will then be significantly extended with the VLTI.

Models of post-AGB stars (e.g. Blöcker 1995) show that the post-AGB luminosity and evolutionary timescale are critically dependent on the core mass and so on the initial mass of the star. Once the spatial extent of the shell and its expansion rate, derived from molecular lines, are known, these models can be checked for every star. Disagreements with model estimates would enable us then to obtain more insight on post-AGB mass loss, a parameter which is poorly constrained by present observations, but which again could be probed from spatially resolved data.

A major unknown is the deviation from spherical symmetry of post-AGB shells. Planetary nebulae, and especially those with high-mass progenitors, are well known to deviate often markedly from spherical symmetry. Recent high-spatial-resolution observations of pulsating AGB stars (e.g. Cruzalèbes, this conference) show that non-spherical effects are present since the early stages of mass loss. An exciting prospect for VLTI observations is the study of the evolution of non-sphericity of post-AGB shells and its dependence on the development of a fast wind once the UV flux of the star has become significant.

3 Examples

3.1 HD 56126 (IRAS 07134+1005)

HD 56126 may be considered as a prototype of a single post-AGB star of rather low mass. The low mass is attested by the high radial velocity (86 km/s), the metal deficiency ([Fe/H] = -1), and the relatively low infrared-to-optical luminosity ratio; the post-AGB nature is clear from the supergiant spectral type combined with high carbon and s-process abundances (Klochkova 1995) that point to a 3th dredge-up phase. Lines of C-rich molecules have been detected in this star and indicate a shell expansion velocity of 8.5 km/s (Bakker et al. 1996), a value which is similar to those found in the shells of AGB stars of rather low masses.

Assuming a luminosity $log\ L/L_{\odot} = 3.86$, typical for a 0.605 $M_{\odot}$ post-AGB star (Blöcker 1995), Bakker et al. (1996) derive that the star left the AGB 440 years ago; the inner radius of the dust shell should then be at some 20 A.U. from the star. Assuming E(B-V) = 0.40 (Bakker et al. 1996), we derive a distance of 2.3 kpc to this 8.2-magnitude star, implying an inner hole in the shell less than 10 mas in radius. However, a post-AGB mass of 0.605 $M_{\odot}$ means an initial mass of 3 $M_{\odot}$, which is too much for a star that metal deficient. If the initial mass were only 0.8 solar masses, Schönberner's (1983) models for a 0.546 $M_{\odot}$ post-AGB star would imply an age of some 5000 yrs and a distance of about 1 kpc, so that the inner radius of the shell would be at more than 0.2" from the star.

From a confrontation of spatially resolved data with state-of-the-art models the basic parameters of post-AGB stars can thus in principle be derived or, at least, be used as a check for the models. It is probable that the detailed picture will be affected by uncertain aspects of the models, e.g. post-AGB mass loss, and by deviations from spherical symmetry. In Figure 1 we show a TIMMI image of HD 56126 on which it can be seen that a perfect spherical symmetry does not hold for this object.

Besides essential basic parameters such as distances, ages and evolutionary masses, VLTI observations of the shells of this and similar stars will trace the spatial distribution of the circumstellar dust and so the mass loss history. HD 56126 is one of the few known sources with the emission feature at 21 μm, that is probably due to some carbon-rich dust feature. Spatially resolved observations at this and other wavelengths will teach us the typical temperatures at

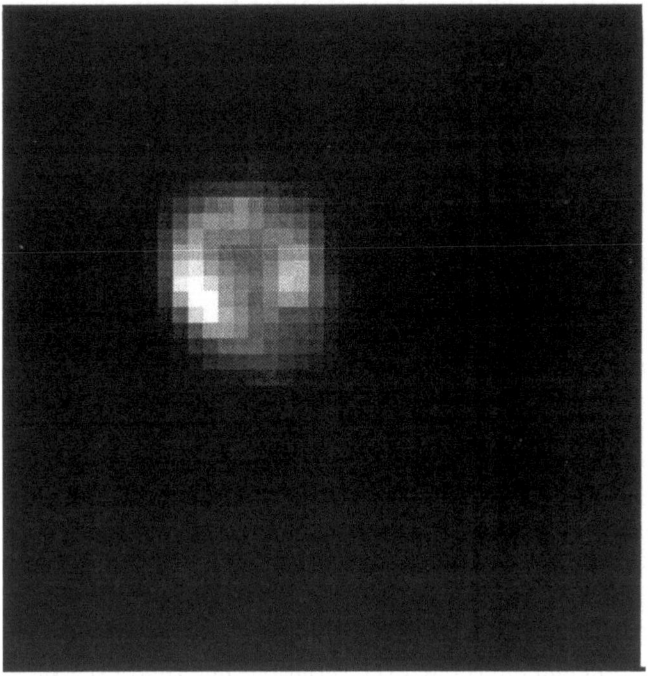

Fig. 1. A 10 μm (N-band) deconvolved image of HD 56126 taken with TIMMI at the 3.6m telescope. Note the evidence for a torus-like density enhancement.

which this feature is formed. For this kind of research, high-spatial-resolution observations in the 20 μm window are eagerly awaited.

3.2 He 3-1475 (IRAS 17423-1755)

This object, discussed independently by Bobrowsky et al. (1995) and Riera et al. (1995), has characteristics of both an expanding OH maser source and a young PN. Its IRAS colors are intermediate between those of AGB stars and PNe. Optical CCD images and red and near-IR NTT images show a central source and a bipolar outflow that extends some 15" on the sky. Short optical HST images (Bobrowsky et al. 1995) resolve the central source in a highly asymmetric structure a few arceseonds accross.

Bobrowsky et al. (1995) and Riera et al. (1995) interpret their data in terms of a central star surrounded by a thick torus seen at a certain inclination, on the far side of which the light seen by HST is reflected. The outflow consists of a fast bipolar jet and a relatively slow detached shell traced by the OH. From the present observations it is not yet possible to exclude the possibility that the object is young, but the probable N enrichment of the outflow (Riera et al. 1995) suggests that the post-AGB interpretation is more likely.

Clearly the VLTI will be instrumental for the study of such sources. High spatial resolution in the mid-infrared is important for mapping the torus. The 25 km/s expansion velocity of the 'slow' wind and the N enrichment both suggest a rather massive central star, for which observable changes can be expected within a decade.

3.3 HD 44179 (The Red Rectangle)

The Red Rectangle is one of the most popular post-AGB sources in the sky. From ground-based optical speckle data (Weigelt, this conference) and near-IR adaptive-optics observations (Roddier et al. 1995) it has become clear that the central star is not directly seen, but that its light is scattered from two lobes above and under a thick disk which is viewed nearly edge-on; this picture is beautifully confirmed by a HST snapshot observation (Bond 1996).

The central star is the peculiar A-supergiant HD 44179 orbiting a low-mass object, probably a main-sequence star (Van Winckel et al. 1995); the orbital period is 319 days. Waelkens et al. (1996) observed photometric variability with the orbital period and interpreted the variations in terms of the variability of the scattering angle around the orbit. In this picture, which is schematically shown in Figure 2, the two scattering lobes that are observed are not situated on the orbital axis (in which case no velocity variations would be seen), but on the parts of the bicone that are located in the direction towards the observer.

The central star of the Red Rectangle poses a challenge to stellar evolution, in the sense that the orbit is too close to accomodate an AGB star; yet the carbon richness of the dust underscores that a third dredge-up must have occurred in this star. A picture with Roche-lobe overflow appears unlikely, since it would probably not have led to the detached eccentric system we observe. It therefore seems that the thick circumbinary disk has been created in a process of sub-Roche-lobe mass transfer. In turn, selective reaccretion by gas from this disk must have occurred in order to explain the remarkable photospheric abundance pattern, which is characterized by an extreme depletion of the refractory elements (Waelkens et al. 1992, 1996).

The rich and diverse circumstellar environment of HD 44179 certainly deserves detailed studies at high angular resolution; at the distance of this object, 0.005" corresponds to only 2 A.U.. Knowledge of the spatial distribution of the observed CH^+ lines, the red emission and the near-IR PAH-features could yield important clues on the formation and processing of carbon-rich dust particles of different sizes, especially since the radiation field of the central star is known with a fair accuracy. Part of this can be achieved with HST, but in particular in the infrared VLTI is needed. For a detailed study of the scattering processes in the nebula, a high-spatial-resolution polarimetric mode at the VLTI would be extremely useful.

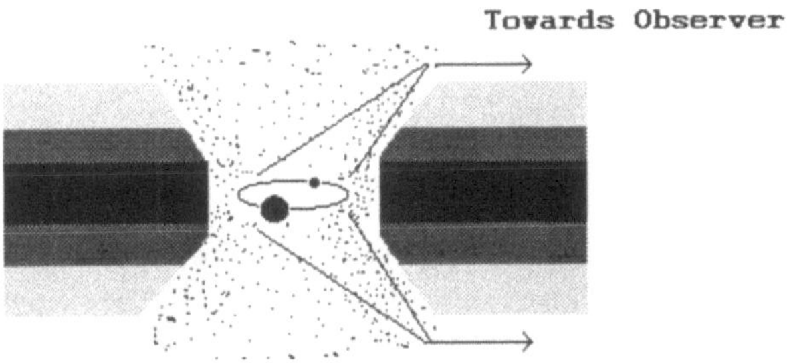

Fig. 2. A geometric model for the inner part of the Red Rectangle. Note that the thickness of the disk is two orders of magnitude larger than the size of the binary system. During the orbit, the scattering angle toward the observer varies, being largest at inferior conjunction, when minimum brightness is observed.

4 Conclusions

Post-AGB stars form an attractive research subject, both from the point of view of stellar evolution as from that of the interaction of an evolving star with its circumstellar environment, the processing of which is essential for our understanding of the dust component in the interstellar medium. The VLTI seems to be ideally suited for enabling important progress in this field, since the source sizes (between 0.1 and a few arcsesonds) and the peak of their emission (mid-IR) fall in the ranges that will be covered by the VLTI. In this sense, not only the interferometric capability but also the superb infrared conditions of Cerro Paranal will prove of extreme value. As far as the specification of the instruments goes, this research will benefit from medium-resolution spectroscopic capability in both the 10 and 20 μm windows and from polarimetric options. High angular and spectral resolution in the optical and near IR will be of great benefit for studying the photospheres of the central stars, the morphology and composition of the circumsystem disks, and also the bipolar outflows and fast winds that develop in these sources. The high angular resolution even offers the exciting prospect of witnessing directly stellar evolution in some objects on a timescale of decades.

5 Acknowledgements

We thank Frank Molster for processing the TIMMI image of HD 56126. CW and HVW gratefully acknowledge financial support from the Flemisch Fund for Scientific Research. Grant nr. 6.2145.94. LBFMW gratefully acknowledges support from the Royal Netherlands Academy of Arts and Sciences.

References

Bakker, E.J., Waters, L.B.F.M., Lamers, H.J.G.L.M., Trams, N.R., van der Wolf, F.L.A. (1996): A&A **310**, 893

Blöcker, T. (1995): A&A **299**, 755

Bobrowsky, M., Zijlstra, A.A., Grebel, E.K., Tinney, C.G., Te Lintel Hekkert, P., Van De Steene, G.C., Likkel, L., Bedding, T.R. (1995): ApJ **446**, 89

Bond, H.E. (1996): IAU Symposium 180, in press

Klochkova, V.G. (1995): MNRAS **272**, 710

Luck, R.E., Bond, H.E., Lambert, D.L. (1990): ApJ **357**, 188

Roddier, F., Roddier, C., Graves, J.E., Northcott, M.J. (1995): ApJ **443**, 249

Schönberner, D. (1983): ApJ **272**, 708

Van Winckel, H. (1996): A&A, in press

Van Winckel, H., Waelkens, C., Waters, L.B.F.M. (1995): A&A **293**, L25

Van Winckel, H., Waelkens, C., Waters, L.B.F.M. (1996): A&A **357**, L37

Waelkens, C., Van Winckel, H., Trams, N.A., Waters, L.B.F.M. (1992): A&A **256**, L16

Waelkens, C., Van Winckel, H., Waters, L.B.F.M. (1996): A&A, in press

Waters, L.B.F.M., Waelkens, C., Trams, N.R. (1993): in the Second ESO-CTIO Workshop *Mass loss from the AGB and beyond*, edr. H.E. Schwarz, p. 298.

New Insights in Close Circumstellar Physics Using Interferometry

Jean-Pierre J. Lafon[1,2] and Nicole Berruyer [1,3]

[1] Groupement de Recherches "Milieux Circumstellaires" du CNRS
[2] Observatoire de Paris - DASGAL, URA CNRS 335, 92195 Meudon Cedex, France
[3] Observatoire de la Cote d'Azur - Dpt Cassini, URA CNRS 1362, BP 229, 06304 Nice Cedex, France

1 Introduction

Circumstellar media of young and evolved stars are a wide and new field of research. They are governed by specific physics involving a lot of phenomena out of usual terrestrial or astrophysical conditions, in particular concerning physico-chemistry of gas and solids far from global and even local thermodynamic or dynamic equilibria. Radiative and collisional processes are important. The size scales belong to wide ranges: from a fraction of the stellar radius to several thousands of stellar radii. The temperatures range from several thousands of K close to the star to a few K in the outer parts, but can reach several tenths of millions of K in hard shocks crossing them.

Now, due to the distances and the sizes of the concerned stellar objects (young stars with or without protoplanetary disks or evolved stars, red giants, super giants, AGB stars, planetary nebulae, Be stars, ...), angular resolutions of the order of the milliarcsecond are crucial for breakthroughs in understanding of their physics, in particular in the inner layers of the circumstellar envelopes, within a few hundred stellar radii, which are very important at both ends of the star life. The inner layers are those where matter is exchanged between the star and the close environment and where angular momentum is transferred. For young stars, dust destruction and planetary formation occurs there. For evolved stars, the mass loss is generated and structured there, with formation of dust and complex physics and chemistry.

However improving our knowledge of these regions implies specific requirements. These regions are very inhomogeneous so that angular resolution is necessary to observe variations in density, but the motions are also complicated so that an accurate determination of the velocity fields is necessary: high spectral resolution is required (say resolutions between a few 1000 and 30000). At spectral resolution of 30000 and beyond, the Doppler imaging technique may be extended to spatial interferometry for studying fine structures or velocity fields. However, Doppler imaging techniques are often limited by particular properties of the target such as the inclination of the stellar rotation axis with respect to the line of sight, and can produce ambiguous "star images". The addition of spatial information to spectroscopy completely elinminates ambiguities. Moreover, the spectral signatures of the physical processes also appear in wide wavelengths ranges: coordinated observations in a large part of the spectrum are necessary for correctly interpreting the results obtained in each wavelength band. Finally, all the concerned objects are variable so that a good knowledge of the variation cycle, with datation of the various observations in relation with it, is needed.

All these conditions make it difficult to obtain accurate constraints for the models. However, the challenge is such that it is worth noting that, first, the circumstellar media are fine laboratories of extreme media, some of which can now be mimicked in the laboratory with interesting terrestrial analogues, and then recent developments in theory provide new input physics available for models. Hereafter we shall analyse which physical phenomena should be fruitfully observed with the VLTI, in which parts of circumstellar envelopes, at which wavelengths and with concourse of which other means.

2 Properties of the Wind of Evolved Stars

2.1 Circumstellar Environments and Driving Mechanisms

Roughly speaking, circumstellar envelopes of evolved stars can be divided into four zones with fairly different sizes and located at different radial distances from the center of the star. The first one is a thin layer usually modelled as a surface separating "the star" from "the circumstellar medium" and through which the star fixes the inner boundary conditions for the envelope. Beyond the star surface, a zone covering a few stellar radii is that in which matter begins to be accelerated by several mechanisms concerned with the gas. Then dust appears and the momentum transferred from photons to dust grains by radiation pressure is transmitted to the gas by collisional friction. However this ideal scheme is somewhat perturbed by the interpenetration of the above mentioned zones and the competition of the mechanisms.

The main driving mechanisms seem well identified but, since they are in competition, the main problem is always the determination of the ranges of parameters, in other words (depending on the type and mass of the stars) the stage of the star evolution, in which one or several of them are dominant. Authors tried to delimit regions of the HR diagram bounded by "dividing lines" linked to some spectral features expected to be sensitive to physical parameters like the magnetic field strength, the condensation threshold, the temperature (characterized by emission lines), chromospheric features, ... and so on. The topology of some of these lines has been discussed by Lafon and Berruyer (1991). However, many of these lines cross each other, clearly indicating that, though in competition, the various phenomena are interacting to build non linear and often self consistent structures. One must thus carefully look for features characteristic of the concerned processes.

In any case, the most obvious process opposed to gravity are are thermal pressure, Alfven waves, acoustic waves, shock waves and radiation pressure (which should in fact be better called radiation force).

2.2 Magnetic Field and Alfven Waves

It is well known that Alfven waves propagate in the atmosphere of the sun. Thus, they are expected to propagate through the inner atmosphere of at least low mass stars not so far from the main sequence. Nevertheless, it is also well known that evolved stars are highly convective so that they may generate magnetic fields, present at least in their inner circumstellar layers. Alfven waves can carry momentum at high rates and, if they can dissipate, can transfer a lot of momentum able to initiate an outerward motion of the stellar matter. However, the magnetic field of (hot or cool) late type stars is still largely controversial. Magnetic spots cover in average 2% of the sun surface. Thus, for stars like o Ceti (Mira) or Alpha Ori, spots should be seen. Recently using the Hubble Space Telescope A. Dupree observed the surface of Betelgeuse with an angular resolution better than the size of the earth's orbit at the distance of the star, exhibiting a hot spot. This may indicate the presence of a magnetic field, but, to determine the origin of a magnetic field, it would be necessary to follow the evolution of this spot on the star surface to determine whether it exactly follows the rotation of the star or whether there is a drift velocity. This emphasizes the requirement of regular observations of the same object at different phases. Now, at spectral resolution of 30000 and beyond the Doppler imaging technique could be extended to spatial interferometry for studying fine structures or velocity fields. However, Doppler imaging techniques are often limited by particular properties of the target, such as the inclination of the stellar rotation axis with respect to the line of sight, and can produce ambiguous "star images". The coordination of spatial information and spectroscopy completely eliminates ambiguities.

Models of Alfven driven stellar winds have been built starting from solar models, usually including the strength of a magnetic field as a free parameter. These models assume different field geometries corresponding to different assumptions on the structure of field lines (Lafon and Berruyer, 1991; see references therein). For media with high conductivities, Alfven waves are transverse waves coupling field lines and matter and cannot be dissipated without some coupling with longitudinal oscillations which can dissipated more easily. Thus, the different models dissipate waves in different ways. Even for monochromatic waves, this introduces a new scale size (which is often a free parameter): the dissipation length, characterizing the distance through which the momentum is transferred from the waves to the medium. Various attempts have been made to increase this length up to the order of a few stellar radii and to minimize the sensitivity of the model to variations of this parameter.

Some of the models described in the review by Lafon and Berruyer envisaged the presence of dust close enough to the star to reduce the ionization and so make the dissipation easier by ambipolar diffusion. Now propagation of Alfven waves has been investigated in partially ionized media including or not electrically charged dust, showing that ambipolar diffusion through ions and through charged grains can play significant roles and alter the dissipation rates by orders of magnitude. The role of ambipolar diffusion is also emphasized for star and planet formation (Ciolek and Mouschovias, 1996, and references therein). All these results indicate that the role of the magnetic field is still controversial, and that observations of star spots, Ca and K lines, ... should provide useful indicators about its magnitude and the resulting effects. However, with an angular resolution of the order of the milliarcsecond, observation of spots will be possible only for the giant evolved stars.

2.3 Thermal Pressure, Acoustic Waves

It is well known that the thermal pressure plays a role to counterbalance the gravity, even if this effect is much weaker than in the case of the sun, which possesses a hot corona (however research of coronas or chromospheres in star environments should be more systematic).

Now the contribution of acoustic waves to wind mechanism is less clear. The main argument against acoustic waves was the their too fast dissipation in an hydrostatically stable atmosphere (the dissipation length, of the order of the atmospheric height scale, is of the order of 0.01 R*). It has been shown early that in expanding atmospheres, the scale height and so the dissipation length becomes of the order of the stellar radius. Then, several models have been built (see Lafon and Berruyer, 1991, references therein and later papers by same authors on the subject). The question is then the multiplicity of the solutions for same outer and inner conditions and their stability, if one considers that the solutions cross several singular (critical) points!. Moreover the interaction of dust grains with acoustic waves was not envisaged in the models for which self consistent solutions were obtained up to several tenths of stellar radii, though it is well known that dust alters significantly the properties of acoustic waves. Acoustic waves can be observed through density and velocity variations as shock waves.

2.4 Pulsations

Evolved stars are highly convective and pulsate. Their atmospheres are also very unstable. One can first observe the variations of the "stellar radius", defined in spherical symmetry, for the cold giants and supergiants by the radius for which the atmosphere becomes opaque: practically this is the radius at which the optical depth of the TiO line (712 nm) is equal to 1. For the closest objects the so defined "stellar radius" corresponds to 0.01 to 1 milliarcsecond. It is interesting to compare the results of these measurements with those of observations of SiO masers, which, in addition, give informations about chemistry in the inner layers of the envelopes in relation with dust condensation. It will also be very interesting to couple such observations with a mapping of the abundances which can be obtained using spectroscopy (Lafon, 1995). Note also that the main contribution in CO lines between 1 and 3 microns comes from mass elements of layers located where the optical depth is of the order of 0.7 . These lines provide a lagrangian measurements (following the motion of the layers) of the velocity; indeed, it is worth noting that the velocity of these layers is maximum at the maximum of contraction of the stars (but this is interpretation is based on the assumption of spherical symmetry).

2.5 Shock Waves

The instabilities of evolved stars also favour propagation of already formed shock waves or breaking waves forming shock waves with abrupt changes in the gas velocity, following the periodic variations of the star luminosity. Many spectral features in a large (from optical to infrared) range of wavelengths are currently interpreted as evidences of the propagation of radiative hypersonic shock waves (Huguet and Lafon, 1996 and references therein). CO lines between 1 and 3 microns or hydrogen (Balmer) and helium lines allow observation of the motion of post-shock layers. These give estimations of the velocities of the layers following the shock front, but the velocity of the shock must be deduced from models. Beyond some velocity threshold, located about 40 km/s in Mira-like circumstellar envelopes, the shock structure changes strongly and becomes hypersonic (or radiative) in the sense that energetic radiations emitted beyond the hydrodynamic discontinuity in the wake of the shock, at distances of several 100 kms, cross the shock front and generate a radiative precursor. The global structure is highly non linear and self consistent. Unfortunately, in spite of huge efforts, there is still no fully self consistent shock model accounting for all the spectral features because they occur in collisional gases strongly far from any equilibrium (even far from ETL or Maxwell equilibrium), a physics unusual in laboratory and astrophysics (see Huguet and Lafon, 1996 for the most comprehensive self consistent available model). The profiles of density, velocity or temperature through such structures are highly sensitive to the approximations: in particular a description of the wake omitting the thermal separation of electrons from other (heavier) particles beyond the front leads to significantly different results, which could explain controversies about the shock velocity and the still unsolved strong discrepancies between the observations and previous shock models (Lafon and Huguet, in preparation). This could also have consequences for the formation of the Halpha line, which is formed over a thick layer of the wake, including one and two temperature regions. Recently, another type of (weaker) non radiative shock waves was invoked to provide a driving mechanism by interaction of the waves with dust, if any present close to the star. This may also be a mechanism for self consistent enhancement of dust production and momentum transfer (see Sedlmayr, 1996).

2.6 Dust

Dust grains have been considered for a long time as the component driving the winds at radial distances greater than some "condensation radius" beyond which dust appears. Indeed, dust is observed in a wide range of wavelengths, from optical to radio wavelengths.

Now, the near infrared is a window onto warm circumstellar media: it corresponds to blackbody temperatures ranging from 300 K to 3000 K i.e. to the thermal emission of warm dust and gas. Shorter wavelengths exhibit the behaviour of the star light scattered by dust. Moreover, the controversial features due to PAHs or other dust components are found between 3.3 and 12 microns

(3.3, 6.2, 7.7, 8.6, 11.2). The windows at 10 and 20 microns contain the signatures of carbon and silicate dust compounds. The region from 2.5 to 25 microns is called "the fingerprint region", because it contains all the features corresponding to the stretching, bending and rocking motions of molecules in a solid. Ice features are found at 3 microns. Molecules like SiO, SiS, SiC, NH3, PH3, CH4, SiH4, C2H2, C2H4, C2H6, which are efficient contributors to the chemistry of formation of dust grains, have transitions in the 10 micron window.

These examples emphasize the importance of high resolution imaging and spectroscopy in this wavelength range. Indeed, even if dust grains are no longer considered as the only driving component of the wind. it is well established that they play a key role in the global dynamics and thermodynamics. Nevertheless, there remains many uncertainties concerning the ways in which they condense and evolve later on, both for "oxygen" and "carbon" stars. High resolution imaging strongly suggest two salient facts:

1/ observing in a narrow window gives indications on the presence of dust made of matter exhibiting features in the adequate wavelength range. However, dust formation is the result of a complex chemistry; heterogeneous or catalytic condensation is most probable, and this implies the existence of precursors observable in other windows not only dependent on the temperature

2/ it is not so clear and probably false that dust forms very quickly, in a thin layer located at some well determined distance from the star. Of course, all models of nucleation show small typical scale sizes, of the order of a fraction of the stellar radius. Moreover the "inner shell radius" is located between 2 and 10 stellar radii. However, these results are obtained under the assumption of spherical symmetry. The medium is probably inhomogeneous not spherically symmetric: there are "blobs" containing more dust than elsewhere.

This may explain apparent paradoxical phenomena like those emphasized by Danchi et al. (1994): presence of dust closer to the photosphere than SiO masers. Finally, there are two important unsolved problems:

1/ nucleation theories are still in infancy and include rough assumptions (though the situation is quickly evolving). They should be improved, in particular to take into account chemistry (Lafon, 1991). Multi-wavelength spectroscopy and imaging would provide useful constraints

2/ dust driving do not clearly follow other driving processes at larger distances. Observations suggest that dust can be present in regions close to the star surface where it is not expected (Danchi et al., 1995). In such cases all possibilities of interaction with waves and other mechanisms have not been explored.

3 Variability

The star mass loss is not always a steady state phenomenon. It may even not be continuous. It may be stronger and faster in some periods, weaker and slower in others. This can produce inhomogeneities under the form of dense or diluted zones in the expanding envelopes. Of course this is observable with the resolution of the VLTI (consequences for outer parts of the envelopes seem already observable at the resolution of the HST), if constrasts of luminosities higher than say 1000 can be observed. This can provide informations on the "history" of the star in parts of its life in which it evolves quickly and also constraints on the limit of validity of the steady state models.

However, one should be careful in distinguishing true spatial inhomogeneities from inhomogeneities due to time variations and convected with the wind. The periods of the stars are of the order of years. Observations are truly useful only if following the phase, and if correlations can be established with the photometric variations and evidence of characteristic features of the different mechanisms.

4 Conclusion

To conclude, the construction of the VLTI is strongly wished by people working on problems of circumstellar physics, as a well fitted tool to solve several specific questions.

The spatial resolutions of the VLTI range in quite adequate values, in adequate wavelengths, for solving the above listed problems. In particular the VLTI will give much stronger constraints on the models than those imposed by the measurements of the "mass loss rate" and the "terminal velocity", which are actually not sufficient for the new models, most of which are always in agreement with the observed values. Moreover, the sensitivity will allow observations of a larger number of fainter objects than the available interferometers.

However, it appears essential for circumstellar physics that the science instrument packages include spectroscopic capabilities with spectral resolution ranging fron 10 to 30000 (and beyond if possible). It is also essential that capabilitiy of observation and imaging exist in the red part of the spectrum to allow correct interpretations of observation in the 10 micron range. The main problem is now to locate the regions where the various mechanisms are dominant and to evidence the characteristic scale sizes at which they are at work. Availability of the facility at regular intervals of the periods of the objects is also a crucial point.

References

Ciolek and Mouschovias (1996): this book

Danchi, W.C., Bester, M., Degiacomi, C.G., Greenhill, L.J., Townes, C.H. (1995) : Astron. J., 107, 1469

Huguet, E. and Lafon, J.-P. J. (1996): A & A, to appear

Lafon, J.-P. J. (1991): *Dust formation and evolution in circumstellar media, in "The infrared spectral region of stars"*Int. Coll. held in Montpellier (France), 16-19 oct. 1990, C. Jaschek and Y. Andrillat eds, Cambridge Univ. Press, 193

Lafon J.-P. J. (1995): *Prospective on circumstellar physics I Observations and models* Proc. ESO workshop on Science with the VLT, Garching bei Muenchen, 28 june - 01 july 1994, J.R. Walsh and I.J.Danziger eds, Springer, 119

Lafon, J.-P. J. and Berruyer, N. (1991): A & A Reviews, 2, 249

Lafon J.-P. J. (1995): *Prospective on circumstellar physics II Early and late type stellar objects* Proc. ESO workshop on Science with the VLT, Garching bei Muenchen,28 june - 01 july 1994, J.R. Walsh and I.J.Danziger eds, ESO publ., 7

Sedlmayr E. (1996): this book

Part 7

STAR FORMATION AND EARLY STELLAR EVOLUTION

Circumstellar Disks in the Early Phases of Star Formation

Antonella Natta

Osservatorio Astrofisico di Arcetri, Largo Fermi 5, I–50125 Firenze, Italia

1 Introduction

The current picture of the process of star formation requires the existence of circumstellar disks over a large fraction of the life of young stars, from the very early stages of the formation of a protostar to the time when the star reaches the main sequence. The disk properties and evolution are crucial for our understanding of problems such as the formation of outflows, the star formation efficiency, the termination of the accretion phase, the activity during the pre-main–sequence phase, the formation of planets and the disk dissipation.

This summary discusses the unique possibility offered by high-spatial resolution observations in the mid-infared to study the nature of disks in embedded objects and the formation of binary systems. The same material has been presented at the workshop on Infrared Space Interferometry, which was held in Toledo, Spain, in March 1996 (Natta & Butner, 1996) and the reader is referred to the proceedings of that workshop for details. This paper is just a brief summary of what published there.

2 Disk Properties in Very Embedded Objects

Let's consider the system of a star+circumstellar disk still embedded in a dense, accreting cloud of dust and gas. The effect of such an optically thick envelope on the the disk is to warm up the outer parts of the disk to temperatures significantly above the values expected for standard circumstellar disks (Butner et al. 1994). As a consequence, the disk brightness profile is significantly broader, especially at long wavelengths. The envelope also absorbs the disk radiation, and, even more important, emits at all infrared and mm wavelengths. In fact, the envelope emission dominates over that of the disk at all infrared wavelengths, and "hides" the disk. However, the envelope emission is significantly more extended than that of the disk, and with very high spatial resolution it is, in fact, possible to disentangle the disk emission from that of the envelope.

Fig.1 shows, as an illustration, the predicted brightness profiles for L 1551 (see Natta & Butner 1996 for a description of the model and the values of the parameters). In this particular case, the envelope emission overwhelms the disk at scales larger than $\sim 3\times10^{-3}$ arcsec at 5 μm, 10^{-3} arcsec at 11 μm (where the disk emission is strongly absorbed by the silicate feature), 10^{-2} arcsec at 20 μm.

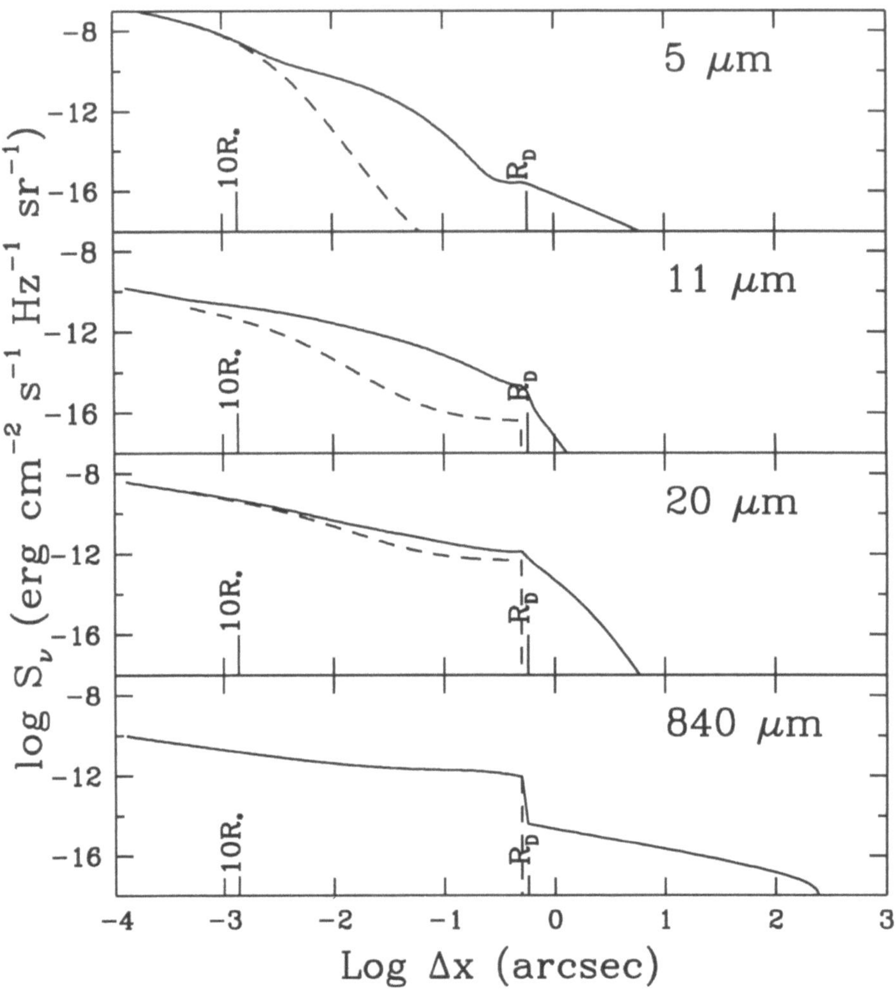

Fig. 1. Predicted surface brightness profiles for L1551 at 5, 11, 20 and 840 μm(from top to bottom). The solid lines show the total surface brightness,the dashed lines the contribution of the disk. Δx is the angular offset from the central star in arcsec.

At this wavelength, the integrated envelope emission is at most twice as large as that of the disk. At sub-mm (and millimeter) wavelengths, the disk emission still dominates at scales comparable to the disk physical size; the envelope surface brightness is about two order of magnitude lower, resulting in a sharp drop of the brightness profile coincident with the physical radius of the disk. This behaviour explains why Lay et al. (1994) were able to resolve the L1551 disk at 840 μm with a resolution of about 0.3 arcsec.

These results depend on the adopted model and how the effect of the envelope warming of the disk is treated. However, they should be indicative of the actual situation in very embedded objects.

3 The Origin of Binary Systems

The formation mechanism of binaries is still under discussion. It is extremely important to extend the statistical work performed so far on visible objects to the very early phases of star formation, i.e., to detect and study binaries in the embedded phase.

Millimeter observations have detected a few examples. However, mm observations can succeed only if the binary has a separation large enough that each star retains an extended circumstellar disk, which emits strongly in the mm. If, as is currently believed, the disk is disrupted by the companion on scales comparable to the binary separation, one can only detect at 1 mm binaries wider than 50 AU. For closer binaries, which among the T Tauri stars in Taurus account for 50% of the binaries, the circumstellar disks will be truncated at smaller radii and will emit strongly only at shorter wavelengths. The difficulties in detecting such systems are similar to those just discussed for single objects. One needs mid-IR observations with high sensitivity (since the circumstellar disk emission is strongly attenuated by the envelope) and with milliarcsecond resolution, so that the envelope emission can be separated from that of the disks.

References

Butner, H. M., Natta, A., & Evans, N. J.: 1994, ApJ 420, 326.

Lay, O. P., Carlstrom, J. E., Hills, R. E., & Phillips, T. G.: 1994, ApJ 434, 75L.

Natta, A., & Butner, H.M.: 1996, in *Infrared Space Interferometry*, A. Alberdi, T. de Graauw, C. Eiroa, C.J. Schalinski & H. Thronson eds., Astr.Space Sci., in press.

Outflows from Young Stars

Jochen Eislöffel[1] and Catherine Dougados[2]

[1] Thüringer Landessternwarte Tautenburg, Sternwarte 5, D-07778 Tautenburg, Germany
[2] Laboratoire d'Astrophysique, Observatoire de Grenoble, Université Joseph Fourier, B.P. 53X, F-38041 Grenoble Cedex, France

Abstract. Outflow activity seems to be intimately connected with the star formation process. However, the physical mechanism responsible for the mass loss is still essentially unknown. We briefly review some direct and indirect observations of outflows from young stellar objects on length-scales of parsecs down to a few stellar radii. We then present the current observational knowledge on the origin and collimation of the flows, as well as the theoretical framework in which they are explained, and review the main open questions, where considerable breakthrough is expected with the high angular resolution provided by an optical interferometer. Finally, we discuss in more detail the feasibility of such observations with the VLTI. Availability of the optical domain as well as long baselines (> 100 m) will be essential for this project.

1 Signatures of Outflows from Young Stellar Objects

Over the past decade it has become evident that mass loss is intimately connected to star formation. Outflow activity from young stellar objects (YSOs) manifests itself in many ways at all stages of stellar evolution. In the early phases of star formation, highly-collimated jets and arc-shaped Herbig-Haro (HH) objects – the bow shocks of the jets – are seen at optical wavelengths on scales of $0\farcs1$ to $1000''$ (or 15 AU to 1 pc at a distance of about 150 pc to the nearest star forming regions). Both jets and bow shocks contain shock-excited gas which emits in the Balmer lines of hydrogen and in forbidden atomic lines, e.g. of [SII], [NII], and [OI]. Discussions of the jet parameters, like tangential and radial velocities, photometric variability, line profiles and velocity dispersion, line ratios and shock velocities can be found, e.g., in recent reviews by Mundt (1993), Ray (1996), and Eislöffel (1996). In the near-infrared around 2 μm the lines of molecular hydrogen dominate the spectra. Because of the strongly reduced extinction compared to the optical they are especially useful for the study of embedded, and hence very young, objects (see, e.g. Davis and Eislöffel 1995). At mm-wavelengths the outflows are detected in CO, but also in other molecules like SiO, HCO^+, and HCN, to name but a few (for details see the recent review by Bachiller (1996)).

With the help of HST (and potentially also with adaptive optics) the flows can now be imaged to distances of about $0\farcs1$ from the source, or 15 AU at the distance of the nearest star forming regions (e.g., Ray et al. 1996, Burrows et al. 1996). Long-slit spectroscopy and integral field spectroscopy can also reach that close to the source, because they allow for the accurate subtraction of the strong stellar continuum emission. In optically visible T Tauri stars, which are believed

Fig. 1. The HH 30 jet in the [SII]$\lambda\lambda$6717,6731 lines + Continuum (Ray et al. 1996).

to be in the latest phases of accretion, forbidden line emission regions of a few arcseconds extension have been revealed by these techniques. Their morphology and kinematics can now be studied in detail down to a few tenths of an arcsecond from the bright continuum source (Solf and Böhm 1993; Hirth, Mundt, and Solf 1994; Lavalley et al. 1996).

At scales of a few stellar radii so far only indirect observations of the flows are possible. The spectra of some T Tauri stars show the Balmer lines of hydrogen and resonance lines like Na D or Ca II with quite deep and broad blue-shifted absorption throughs reaching out to several hundred $km\,s^{-1}$ (e.g., Mundt 1984). This indicates that already very close to the star gas is moving at rather high velocity.

2 Origin and Collimation of Outflows: the Need for High Angular Resolution

Observations are showing that outflow activity seems to go in hand with accretion. This association has prompted the suggestion that magnetic winds could be a likely ejection process in accreting systems (see the contribution by F. Shu). Indeed, magneto-centrifugal ejection processes provide a mechanism both to extract angular momentum from the accreting disk and focus the flow on large scales. However, both the regions for the origin ($R < 10\ R_\star$) and focusing ($100 - 1000\ R_\star$) of the flow proposed in these models are not directly accessible to observations. Thus the questions where the flow originates from (stellar wind, innermost part of the accretion disk, or larger scale disk wind?) and whether MHD winds altogether are a dominant angular momentum transport mechanism in accreting disks remain unanswered.

The resolution of an interferometer with a 100 m baseline is approximately $\lambda/B = 1$ mas at 0.5 μm, or 4 mas at 2 μm. This translates to linear distances of 10 – 50 stellar radii (assuming $R_\star \simeq 2\,R_\odot$) for young stellar objects located in the Taurus cloud (d=150 pc). The VLTI thus provides the spatial resolution necessary to study the innermost regions of T Tauri star winds, between 10 and $1000\,R_\star$, so far inaccessible to direct observations. *This will enable us to study the region of the formation and collimation of outflows from young stars directly.*

We review below the current status of observations regarding the origin and collimation of outflows, and some of the open questions where real breakthrough is expected with the VLTI. We come back to the feasibility of such observations in more detail in the following section.

2.1 The Opening Angle and the Collimation of Outflows

In order to shed some light onto the collimation process, Mundt, Ray, and Raga (1991) measured the jet width as a function of distance from the source for a number of outflows, and derived the jet opening angles. They found average opening angles of 10° more than 2″ away from the sources. Observations with HST down to 0.″1 from the source show the same result (Ray et al. 1996; Burrows et al. 1996). The jet diameters at these distances are still of the order of 0.″2.

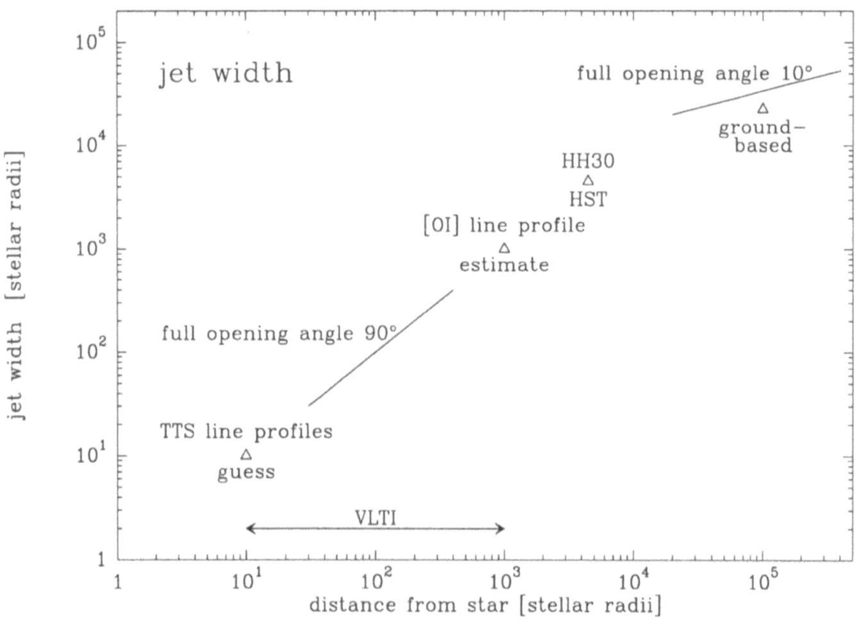

Fig. 2. Jet diameter with distance from the source. See text for further explanation on the "estimated" and "guessed" values.

So, either the opening angle of the flow must become very wide close to the central source or the ejection must originate in an extended region around the star (most likely from the surface of the circumstellar disk). Rough estimates of the opening angle for distances less than 0.″2 can be obtained from line widths of the [OI]$\lambda\lambda$6300,6363 lines and Hα line profiles observed in T Tauri stars, assuming that they originate from diverging stream lines in the outflowing wind. The measurements and these estimates seem to indicate that the jet opening angle remains rather constant at about 10° further away from the source, but bends over and increases to values of around 90° close to the source. The variation of the opening angle on scales ≤ 10 AU is a critical parameter to constrain the origin of the ejection process. As we saw above, different MHD models predict that the mass loss originates from varying radial distances at the surface of the disk. However, no direct observational constraint is yet available.

2.2 Jet Morphology

On scales larger than 10 AU, the emission along the jet is clearly knotty. Some of the outermost knots are resolved and display bow-shock like geometries. Tangential velocities for these outer knots are of the order of 100 - 200 km s^{-1}. The origin of the knotty structure is still a matter of debate. The most favoured mechanisms are working surfaces in a time-dependent ejection process and crossing shocks due to reconfinement of the jet. Determining the geometry of the jet emission on scales < 10 AU as well as the nature of the innermost knots will be essential to test both hypotheses.

2.3 A Two-Component Wind in T Tauri Stars?

The optical forbidden lines in T Tauri stars show at sufficient spectral resolution two blueshifted velocity components. The high velocity component ($\simeq 100$ km s^{-1}) arises in a collimated outflow with typical extensions of a few arcseconds, which is believed to be the analogue of the larger scale jet observed in younger stars. As seen above, it can be traced to 0.″1 ($\simeq 15$ AU) from the central source. The origin of the second lower velocity component ($\simeq 10$ km s^{-1}) is poorly understood in the context of the current paradigm for accretion disks and outflows. The emission is unresolved at present resolution, but sometimes displaced from the continuum peak by a few 0.″1. One possibility is that this velocity component arises from an intrinsically slower wind originating from the disk surface. Establishing its exact geometry will be essential to test this hypothesis and investigate whether disk winds are an important part of accretion disk systems.

2.4 Probing the Basis of the Outflow

The most common interpretation for the optical Balmer emission lines in classical T Tauri stars is their formation in energetic winds originating within a few stellar radii from the star. Spectroscopic evidence for mass outflows at velocities up to several hundred km s^{-1} comes from the observation of blueshifted absorption reversals most frequently seen in Hα and Na D. However, high spectral resolution observations also reveal, in a significant number of cases, the presence of redshifted absorption minima, mostly visible in Balmer lines of highest excitation, challenging the above interpretation. Infall of material originating at about five stellar radii along accretion columns has been proposed to account for these signatures (Calvet & Hartmann, 1992; Edwards et al. 1994). Putting a constraint on the geometry and size scale of the emitting region would thus be essential to discriminate between the different proposed mechanisms.

3 Probing YSO Mass Loss with the VLTI

The angular resolution provided by an optical interferometer with a 100 m baseline appears quite appealing for the study of the formation and collimation region in YSOs. In addition, the star itself is usually bright enough in the optical and the near-infrared to be used for wavefront sensing and fringe tracking. Forbidden line transitions occuring in the optical and near-infrared domain have critical densities ranging from 10^8 cm^{-3} ([O I] 5577 Å, [S II] 1.04 μm) to 10^4 cm^{-3} ([N II] 6583 Å) and thus help probe the jet structure over a large range of distances, typically from a few AU to a 100 AU. Hydrogen and sodium lines are expected to form within a few stellar radii at the basis of the outflow. Bearing this in mind, we now investigate the feasibility of some key observations regarding the origin and collimation of outflows that could be attempted with the VLTI.

3.1 The Collimated Jet

The exact geometry of the emission of the jet gas, as well as excitation mechanism of the collimated jet still remain unknown on scales ≤ 10 AU. As we saw before, there is some indication of a rapid change of the jet width towards the central source, with an opening angle close to 90° .Determining this variation is essential to investigate the origin of the ejection.HST observations of HH30 showed that the line emission along the jet rises as $r^{-2} - r^{-3}$, thus ensuring detection of the jet close to the central source (Burrows et al. 1996). We test in Fig. 3 our ability to constrain the jet opening angle with the VLTI by simulating expected visibility functions. For an opening angle of 90° the jet is resolved in both directions for baselines ≥ 100 m. Differences in the visibility curves between the cases of large and small opening angles (10°) amount to 4-5%. Discriminating between these two cases seems possible, but will require high accuracy calibrations.

Determining the geometry of the emission along the jet, in particular if it is still dominated by knots on scales ≤ 10 AU, will require imaging capabilities and could be attepmted only in the later stages of the VLTI. Also proper motion measurements may be possible. Indeed, for tangential velocities of 100 km s^{-1} expected displacements are of the order of 10 mas/month for sources located at the distance of the Taurus-Auriga region. Determining their proper motions would help to precise the nature of the different knots. In particular, a stationary collimation shock is predicted to occur at distances above the disk plane z_{foc} varying between 1 and 100 R_{jet} according to the different model parameters. Determining velocity gradients both across and along the jet would also be essential to discriminate between the different ejection models. Medium spectral resolution ($\simeq 10000$) would be required in order to study the kinematics of the jet (expected gradients a few $10 - 100$ km s^{-1}). The tracers most adapted to this study would be forbidden lines with highest critical densities: [OI]$\lambda\lambda$5577,6300Å. Hα emission is also expected to occur in shocked regions within the jet. With these observations of different lines it will also be possible to measure the excitation of the cooling gas at high spatial resolution.

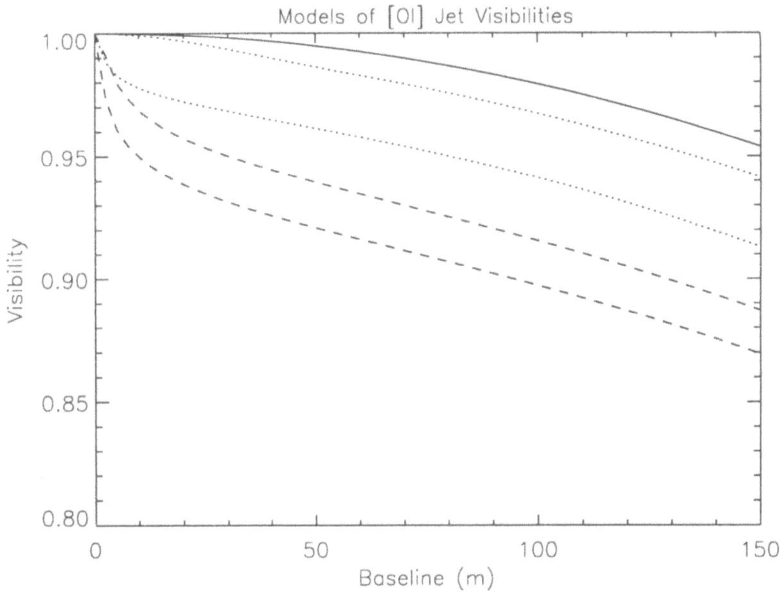

Fig. 3. Model visibilities of a jet observed through a narrow-band [O I] filter, assuming EW([O I]) = 1Å. The continuum (full line) has been approximated by a gaussian of FWHM=0.1 mas, and the surface brightness of the jet falls off as r^{-2}. Two opening angles have been considered: 10° (dotted lines) and 90° (dashed lines). For a given model, the two lines correspond to two perpendicular cuts in the visibility function.

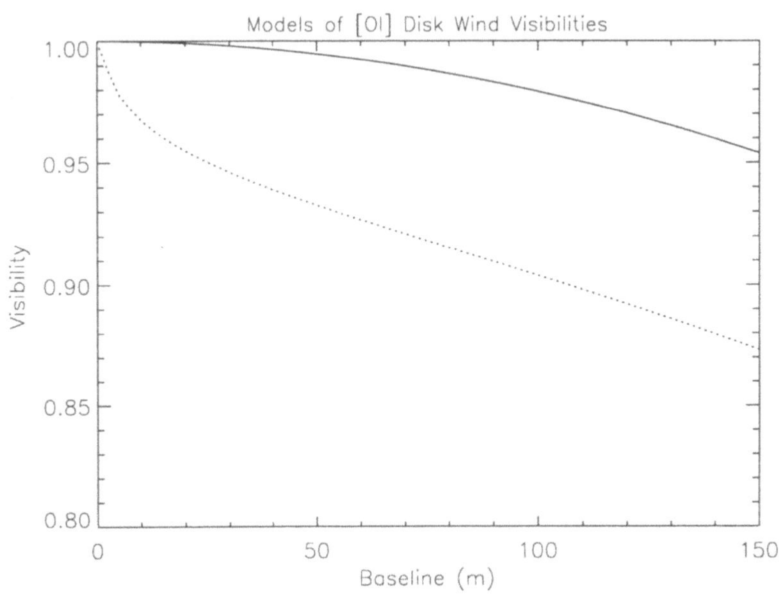

Fig. 4. Model visibility for the disk wind component in a narrow band [O I]λ 6300 filter (assuming EW([O I] = 1 Å, dashed line) compared to the continuum (full line). The surface brightness of the line component falls off as $r^{-2.2}$ for $r > 5\,R_{\star}$, but is constant below. The continuum has been approximated by a gaussian of FWHM=0.1 mas.

3.2 The Origin of the Low Velocity Component

One suggested possibility is that this velocity component arises from an intrinsically slower wind originating from the disk surface. Hartigan et al. (1995) derived from the observed velocity profile of the [OI]λ6300 line an expected surface brightness between 0.1 AU and 10 AU of such a disk wind to follow $S(r) \propto r^{-2.2}$. We compare in Fig.4 the visibility curve of such a component observed through a narrow band [OI] filter with that of the continuum alone. The visibility curve falls below 0.9 for a baseline of 120 m. Long baselines (> 100 m) and high accuracy ($< 5\%$) will thus be necessary in order to detect such a structure with the VLTI.

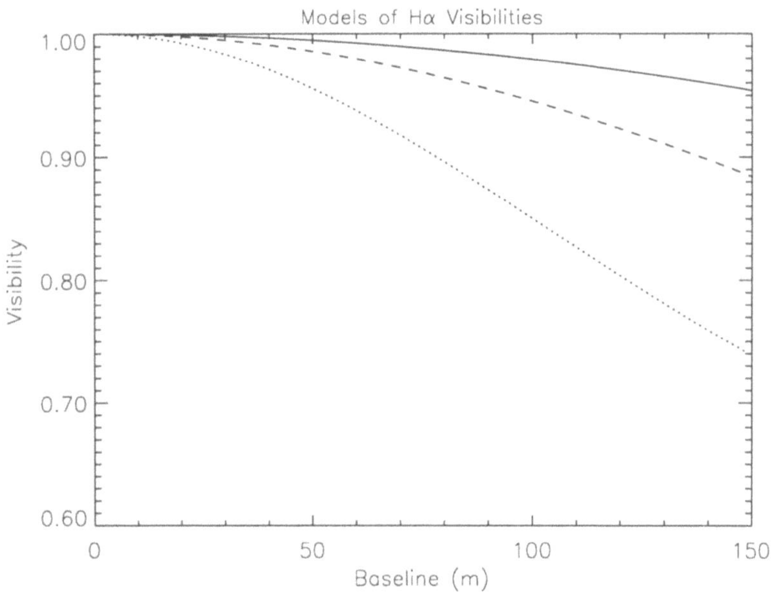

Fig. 5. Model visibilities for the Hα emission through a narrow band filter, assuming EW(Hα) = 10 Å. The continuum (full line) is approximated as a gaussian function of FWHM=0.1 mas, and the line emitting region is assumed to be gaussian-shaped with FWHM=5 $R_\star$ (dashed line), or 10 $R_\star$ (dotted line).

3.3 Probing the Basis of the Outflow

The large Balmer decrement observed in Hα and Hβ suggests that the former linw originates from a significantly more extended region than the Balmer lines of higher excitation. Hα thus appears as the most promising candidate to trace the basis of the outflow. Fig.5 shows that if the emitting region is larger than 5 stellar radii(FWHM), it will be resolvable with a 100 m baseline interferometer. Indeed, from the observed Brγ line centre flux, Najita et al. (1996) estimate a radius of the emitting region in the range 2 – 7 $R_\odot$ for an isothermal gas in LTE at T = 4000 – 12000 K. Brγ emission should then also be resolved. Putting such a size constraint on the emitting region could even be attempted during the earliest phase of the VLTI.

We are grateful to Reinhard Mundt for many interesting discussions on the origin and collimation of stellar jets, and to Fabien Malbet for making his software to calculate object visibilities available to us.

References

Bachiller, R. 1996: ARA&A, in press
Burrows, C.J., et al. 1996: ApJ, in press
Calvet, N., Hartmann, L.: ApJ 386, 239
Davis, C.J., Eislöffel, J. 1995: A&A 300, 851
Edwards, S., Hartigan, P., Ghandour, L., Andrulis, C. 1994: AJ 108, 1056
Eislöffel, J. 1996: in: "Jets from Stars and Galactic Nuclei", ed. W. Kundt, Lecture Notes in Physics 471, 104
Hartigan, P., Edwards, S., Ghandour, L., 1995: ApJ 452, 736
Hirth, G.A., Mundt, R., Solf, J. 1994: ApJ 427, L99
Lavalley, C., Cabrit, S., Dougados, C., Ferruit, P., Bacon, R., 1996: A&A, submitted.
Mundt, R. 1984: ApJ 280, 749
Mundt, R. 1993: in "Stellar Jets and Bipolar Outflows", eds. L. Errico, A.A. Vittone (Kluwer, Dordrecht), p.91
Mundt, R., Ray, T.P., Raga, A.C. 1991: A&A 252, 740
Najita, J., Carr, J. S., Tokunaga, A. T., 1996: ApJ 456, 292
Ray, T.P. 1996: in "Jets from Young Stars", ed. K. Tsinganos (Kluwer, Dordrecht), p.539
Ray, T.P., Mundt, R., Dyson, J., Falle, S.A.E.G., Raga, A. 1996: ApJ Letters, 10. Sept. 1996 issue
Solf, J., Böhm, K.H. 1993: ApJ Lett. 410, L3

Disks Around Young Stars

François Ménard[1], Fabien Malbet[1], Jean-Louis Monin[1], Jean-Philippe Berger[1], Stéphane Guilloteau[2], Anne Dutrey[2] and Gilles Duvert[1]

[1] Laboratoire d'Astrophysique, Observatoire de Grenoble, BP 53, F-38041 Grenoble CEDEX 9, France
[2] Institut de Radio Astronomie Millimétrique, 300 rue de la Piscine, F-38406 Saint Martin d'Hères, France

Abstract. The direct detection of accretion disks surrounding young stellar objects has been the "holy grail" of observational star formation research over the last ten years or so. At stakes, there is a tremendous reward that includes the comprehension of the jet/outflow driving mechanism, an understanding of the viscous and turbulent dissipation processes within the disks, and important clues regarding the formation of our own solar system. In this contribution we will present recent observational *results* that proved, at last, the existence of the accretion disks and allowed a direct assessment of their physical parameters. We also present numerical simulations of circumstellar and circumbinary disk visibility curves and demonstrate that such disks are detectable with an instrument like the VLT Interferometer.

1 What Do We Know About Disks so Far?

The idea that a large fraction of a T Tauri star's luminosity, and especially its near-infrared and ultraviolet excesses, might come from energy dissipation in a viscous disk, itself in accretion onto the central star, is not new and was first suggested by Lynden-Bell and Pringle (1974). Since then, various adaptations of this model have been proposed. Observationally, many indirect pieces of evidence exist to support the accretion disk model (see reviews by Bertout (1989) and Appenzeller (1990) for example) but until very recently, its central element, i.e., the rotating accretion disk, still eluded direct detection.

This section offers a quick glance at the most recent and spectacular results concerning the *direct* detection of disks surrounding single and binary low-mass pre-main sequence stars.

1.1 Results from Millimeter Observations

Binary Stars. Direct, unambiguous detection of rotationnally supported disks surrounding T Tauri stars are still sparse and only a few convincing cases exist. One of them, GG Tau, actually a binary, was observed with the IRAM-Plateau de Bure Interferometer (Dutrey et al. 1994). These authors showed that the rotation curve of the disk is Keplerian. The results are presented in Fig.1. Since then, rotating circumbinary disks have been imaged around a few other objects by various groups in Japan, France and the USA (e.g., Koerner et al. 1993; Duvert et al. 1997).

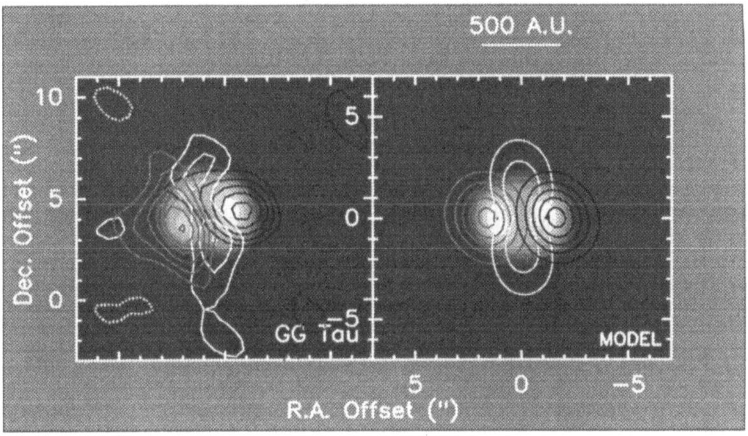

Fig. 1. 2.7mm aperture synthesis map of the GG Tau circumbinary ring. The left panel shows the observations, the right panel is a model of a rotating circumbinary ring. The grey, black and white contours are respectively to blueshifted, redshifted and systemic $^{13}CO(j = 1-0)$ velocity channels of GG Tau. These contours are superimposed on the grey scale 2.7mm thermal dust emission continuum map.

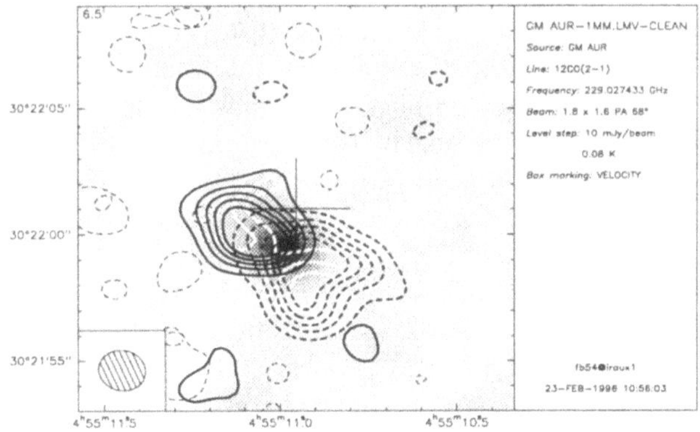

Fig. 2. 1.3mm aperture synthesis map of the single T Tauri star GM Aur. The black, dashed, and white contours are respectively to blueshifted, redshifted and systemic velocity channels. These contours are superimposed on the grey scale 1.3mm thermal dust continuum emission map.

Single Stars. The improvement of the large mm interferometer's resolution, made possible by extending the baselines and/or reducing the wavelength, lead to the rapid discovery of disks orbiting single stars too. Fig.2 shows IRAM-Plateau de Bure observations of GM Aur (Dutrey et al 1997).

It is more than likely that the various ongoing mm surveys will reveal many more objects surrounded by rotating disks. But there remains no doubt now, thanks to the first results from these surveys, that centrifugally supported disk really exist around both double and single T Tauri stars.

The Future. New millimeter arrays are currently under study and will be available early in the next century. The Large Southern Array (LSA) for example, studied by an ESO-IRAM-OSO-NFRA consortium, will have 30 times the sensitivity of current arrays. It should routinely provide $0\farcs05$ resolution images of the dust continuum emission and $\sim 0\farcs1 - 0\farcs2$ resolution for spectral lines. When these large arrays become operational, their results will complete very efficiently those of the VLTI.

1.2 Results from Optical/Near-Infrared Observations

We are quite secure in believing that the mm range is efficient at probing the outer, cold part of the disks in the continuum and that line observations have given us indications about the kinematics in these disks. But the direct detection of the thermal emission of disks at shorter optical/near-infrared wavelengths is also a goal that is very actively reached for because it diagnoses the inner, warmer part of the disk. The impact that such detections now have on our understanding of the star formation mechanism, by hinting at the accretion/ejection processes taking place early on for example, is so fundamental that is was and it still is well worth the effort. This is a key problem in the star formation research field, but it is technically challenging because it requires imaging with angular resolutions that approach the diffraction limit of larger and larger telescopes, a limit that cannot be reached without compensating for the aberrations induced by the Earth's unstable atmosphere. T Tauri stars have therefore been looked at again and again everytime a new step in increasing the angular resolution became available: from plain focal plane imagers, to tip-tilt cameras, to full adaptive optics systems and now to optical/near-infrared interferometers. The first clear images from disks came from Adaptive Optics at ESO, La Silla (Malbet et al. 1993). On the other hand, imaging in space of the scattered light by HST has also proven to be quite efficient recently! And groups are now excessively active with both the Hubble Space Telescope (e.g., Burrows et al. 1996; Stapelfeldt et al., this proceedings) and Adaptive Optics at ESO, La Silla or at CFHT.

Binary Stars. Roddier et al. (1996) successfully imaged the GG Tau ring using a Near-infrared Adaptive Optics camera at the CFHT. Modelling of the data lead to disk parameters that are strikingly similar to those deduced from the mm data of Dutrey et al. (1994). A disk-like structure surrounding Z CMa have also been imaged at 3.6μm and 4.8μm using the ESO adaptive optics systems Come-On (Malbet et al. 1993; see Fig. 3). Lately, Malbet et al. (1997) have successfully imaged that disk at shorter wavelengths, 2.2μm, confirming the previous results.

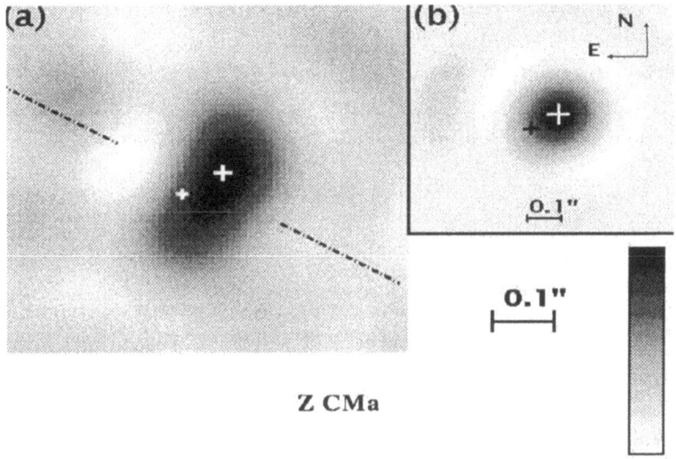

Fig. 3. Near-IR L-band image of the PMS binary star Z CMa obtain with the Come-On Adaptive Optics instrument (from Malbet et al. 1993). The "plusses" indicate the position of the binary. The left panel shows the emission from the disk after subtraction of the stellar contributions.

Single Stars. The most strinking results so far come from HST's observation of the HH 30 disk in scattered light (Burrows et al. 1996). The optical images have been presented almost everywhere and are also available on the web. Recently, the disk surrounding GM Aur has been detected by Stapelfeldt et al. (1997) also with HST (see these proceedings).

With the detectability of the disks now proven in that wavelength range also, many surveys are under way and results should start to flood the literature in the coming years.

2 What Will VLTI Give Us?

2.1 Accretion Disks Around Single T Tauri Stars

In the previous section arguments showing the very existence and detectability of equatorial disks around low-mass PMS stars have been presented. In this section we explore, with the help of numerical simulations, what we expect will be the outcome of observations carried out with the VLTI during Phase A, that is with a configuration that includes 1 UT and 2 or 3 AT's. In the following we therefore assume that VLTI will provide us with visibilities obtained with baselines up to 100m long, but that no mapping will be available.

Two approaches are possible to try and detect disks around T Tauri stars. One can either look for thermal emission or search for scattered light. The latter occurs mostly at short optical/NIR wavelengths, say shorward of 2μm while the former could possibly be detected at all wavelengths, from the UV to the mm, given one has enough sensitivity and angular resolution.

To evaluate the scientific contribution of VLTI to the T Tauri disk problem we consider, following Malbet & Bertout (1995), the *thermal emission* from 3 different models of young stellar objects. In the next subsection we will address the scattered light issue.

Model Description. For all models we assume a 45° inclination of the disk plane to the line-of-sight.

1. Model 1 is a classical T Tauri star (CTTS) with a standard accretion disk. The temperature law in the disk is $T_D \propto r^{-3/4}$. Accretion rate is 10^{-7} $M_\odot$/yr. Outer disk radius is 50 AU. The object is located at the closest distance for young stars, i.e., 150 pc.
2. Model 2 is the same model as the previous one, except for the temperature law which is $T_D \propto r^{-1/2}$. This temperature law can be derived from flared passive disks (cf. Adams et al. 1990, Beckwith et al. 1990). Accretion rate is 10^{-6} $M_\odot$/yr. Outer disk radius is also 50 AU and distance 150 pc. The spectral energy distribution can then fit a spectra flat in the infrared as the ones observed in many objects like HL Tau, T Tau. We call this model the flat IR T Tauri star (FTTS).
3. Model 3 matches the case of FU Orionis objects (FUOr). It contains a massive standard accretion disk ($\dot{M} = 10^{-4}$ $M_\odot$/yr) which overwhelms the stellar emission. Outer disk radius is 150 AU and distance is 450 pc.

Results. Fig. 4 shows the spectral energy distributions as well as the corresponding visibility curves expected for models 1,2 and 3, from top to bottom respectively. The corresponding synthetic intensity maps can be found in Malbet & Bertout (1995).

If we assume that visibilities will be measured to an accuracy of a few percents by VLTI, say 5%, then the *thermal emission* of a disk surrounding a Classical T Tauri star (model 1, top panel) will not be detected at 2.2μm even with baselines of 100m. It might be detectable at 10μm, but with difficulty at best. This can be understood since the contribution of these disks to the total luminosity of the system dominates the photosphere's only longward of 5-10μm.

More massive, more active disks, will be easier to detect however. The disks of flat-spectrum T Tauri's (model 2, middle panel) and FU Orionis objects (model 3, bottom panel) will be detectable at 2.2μm and longward for baselines of about 50m. The FU Ori disk might even be visible at 1μm although the model assumes a distance of 450pc!

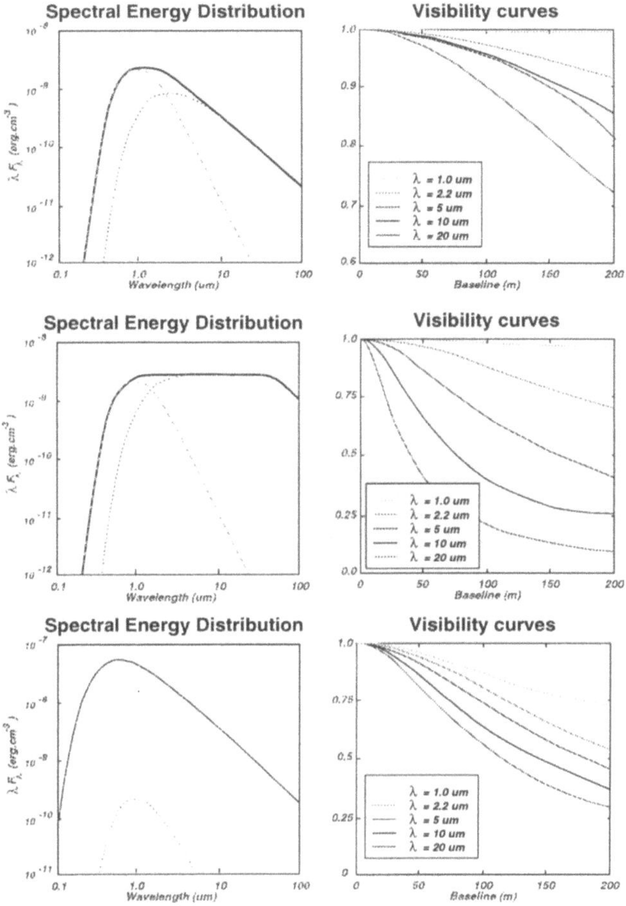

Fig. 4. Top Panel: Spectral energy distribution (left) and visibility curves (right) for Model 1, a Classical T Tauri star disk located at d=150pc. Middle panel: Idem as top panel for Model 2, a Flat spectrum T Tauri star disk located at d=150pc. Bottom Panel: Idem as top panel for Model 3, a FU Orionis star disk located at d=450pc.

2.2 Disks Around Binary T Tauri Stars

In the previous section we considered *thermal emission* to show that disks should be detected in the NIR by VLTI. However, scattered light (at shorter wavelengths) can also be used very advantageously to detect disks, as shown by HST.

In this section, we show models for scattered light only and we apply our results to binary systems. We will consider more specifically the possibility to detect circumbinary disks, eventually with large gaps. The calculations include multiple Mie scattering.

Fig. 5 presents the results of two simulations of circumbinary disks. The binary system has a separation of 3AU and the disk an outer radius of 100AU. In this first attempt to calculate visibilities for such disks, we have simulated the gaps by assuming central clearings (R_{int}) of different sizes: a small one with R_{int} = 6AU and a large one with R_{int} = 25AU. We have voluntarily placed the system quasi edge-on in order to lower the need for a large dynamic range. By doing this, the central stars are occulted by the disk and the contribution of the scattered light to what is observed is maximized. This result is general and does not apply to interferometry only.

Fig. 5 clearly shows that such large "gaps", because they effectively remove a significant fraction of scattered light when they are present, will affect the visibility, and hence be detectable, for relatively short baselines (left panel, for baselines between 10 and 30m). For now, our model is probably too simple and finer calculations need to be done before the exact visibility losses induced by a gap can be predicted accurately as many parameters affect the result. But it can be said already that VLTI will have the capacity to detect large clearings in equatorial disks quite effectively if there is enough scattered light. More details will be presented in Berger et al. (1997).

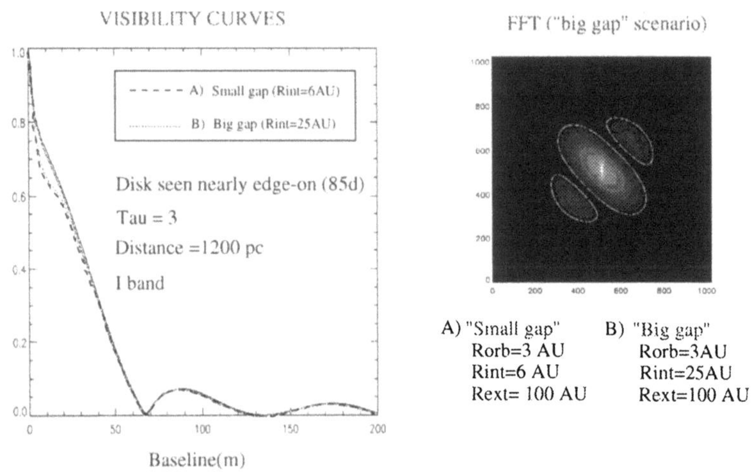

Fig. 5. Circumbinary disk model. The left panel shows the visibilities expected for a circumbinary disk with and without a large gap. The expected losses, of order 10% in that case, should be detectable by VLTI if enough light is scattered, or else, if the detectors are sensitive enough! The right panel shows the FFT map corresponding to the "large gap" model.

3 Summary

Observations in the millimeter, NIR and optical ranges are now providing us with direct evidence that disks around T Tauri stars exist. The geometrical parameters of these disks, like their radius and inclination to the line-of-sight, can now be evaluated in a few cases and their mass can also be estimated quite accurately sometimes from millimeter observations.

Many questions like the shape of their temperature and density structures still remain open however. Because VLTI will allow us to detect the disks, both in thermal emission in the NIR for the more massive ones, and in scattered light in the optical for the others, given their orientation is favorable, these questions will for the first time find answers from direct observations. Indeed, we believe from numerical simulations that VLTI, as soon as Phase A, will give us visibility curves that will differ significantly according to the exact density and temperature profiles in the disks. That possibility to study the disks in great details will represent a long awaited for breakthrough.

References

Adams, F.C., Emerson, J.P., Fuller, G.A. (1990): ApJ, **357**,, 606.
Appenzeller, I., Mundt, R. (1989): A&A Rev., **1**, 291.
Beckwith, S.V.W., Sargent, A.I., Chini, R.S., Güsten, R. (1990): AJ, **99**, 924.
Berger, J.-P., Ménard, F., Malbet F. (1997): in preparation.
Bertout, C. (1989):ARA&A, **27**, 351.
Burrows, C.J., Stapelfeldt, K.R., Watson, A.M. et al. (1996): ApJ, in press.
Dutrey, A., Guilloteau, S., Simon, M. (1994): A&A, **286**, 149.
Dutrey, A., Guilloteau, S., Duvert, G. et al. (1997): in preparation.
Duvert, G., et al. (1997): in preparation.
Koerner, D.W., Sargent, A.I., Beckwith, S.V.W. (1993): Icarus, **106**, 2
Lynden-Bell, D., Pringle, J.E. (1974): MNRAS, **168**, 603.
Malbet, F., Bertout, C. (1995): A&AS, **113**, 369.
Malbet, F., Rigaut, F., Bertout, C., Léna, P. (1993): A&A, **271**, L9.
Roddier, C., Roddier, F., Northcott, M.J., Graves, J.E., Jim, K. (1996): ApJ, **463**, 326.
Stapelfeldt, K.R. et al. (1997): these proceedings.

Part 8

STELLAR POPULATIONS AND GALACTIC CENTER

Observing the Galactic Center with the VLTI

A. Eckart, R. Genzel, R. Hofmann, S. Drapatz, R. Katterloher,
A. Quirrenbach, L. Tacconi-Garman

Max-Planck Institut für extraterrestrische Physik, 85740 Garching, Germany

Abstract. Recent observations of the Galactic Center have resulted in the detection of stellar proper motions in the innermost core of the Galaxy. After 4 years of high resolution near-infrared imaging we have determined proper motions for 39 stars between 0.03 and 0.3 pc from the compact radio source SgrA*. Proper motion and radial velocity dispersions are in very good agreement indicating that the stellar velocity field on average is close to isotropic. We have detected significant changes in the structure of the innermost complex of stars in the vicinity of SgrA*, suggesting stellar motions of $>10^3$ km/s within Taking radial and proper motion data together we find a $2.45(\pm 0.4)\text{x}10^6$ $M_\odot$ central dark mass within < 0.015 pc of SgrA*. The density of this central dark mass therefore is in excess of 10^{12} $M_\odot pc^{-3}$, implying that the central mass concentration is likely a single massive black hole.
Future interferometric measurements with the VLTI will allow us to investigate the proper motions of stars with an higher accuracy than with current single dish observations. In order to exploit the possibilities one would have with an imaging beam combiner we have constructed a beam combiner simulator COSI. First results from COSI have been published (Katterloher et al. 1994, Böker et al. 1994, and Eckart et al. 1994) and are summarized in this contribution.

1 Introduction

The Galactic center is a unique target for searching for a central massive black hole as it is the closest center of a galaxy (8 kpc, 1"≈0.039 pc, Reid 1993). The best way to find out whether or not galactic nuclei contain central massive black holes is from dynamical measurements (see Kormendy and Richstone 1995 for a review). The result of these investigations is of considerable importance for the understanding of active galactic nuclei and the evolution of galaxies. Future measurements with the VLT and VLTI will allow us to pursue these investigations in great detail. In a recent study of the radial velocities of 222 stars within 1 pc from the dynamic center, Genzel et al. (1996 see also references therein) found that a central compact (core radius <0.06pc) dark mass of 2.2 to $3.2\text{x}10^6$ $M_\odot$ if the stellar motions are isotropic. However, for determining the mass distribution from stellar dynamics without the a priori assumption of isotropic motions, it is necessary to determine both radial and proper motions of the stars. In contrast to external galaxies stellar proper motions in the Galactic center can realistically be measured in a few years time if relative stellar motions can be determined to about 10 milliarcseconds (mas). Observations with the VLTI will carry this precision into the mas and sub-mas domain. Since 1991 we have been carrying out a program of high resolution 2.2μm (K-band) imaging at the

3.5 m New Technology Telescope (NTT) of the European Southern Observatory (ESO) in La Silla, Chile. The main goal of this program is the detection of proper motions in the innermost parsec of the Galactic Center cluster. The first results of based on 7 epochs of images spanning four years have been summarized in detail in Eckart & Genzel (1996, 1997).

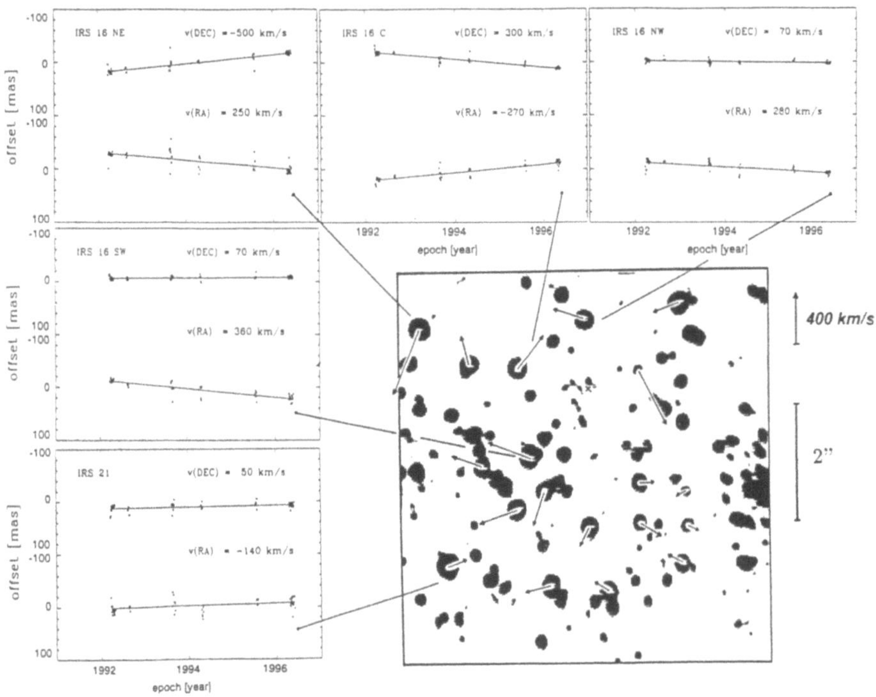

Fig. 1. Proper motion of stars in the central few arcseconds.

2 Results from the NTT

In the present analysis of our NTT speckle data we have determined positional changes as a function of time for 39 stars between 0.9" and 8.8´" and 11 stars <0.9" from SgrA*. Fig.1 shows the proper motion vectors of stars in the central few arcseconds, superposed on a 2μm image. Our data show that the relative positions of many of the stars in the central 0.1pc appear to change monotonically with time. Of the 39 stars motions at the >5σ level are detected in at least one coordinate for 10 cases and at the >4σ level for a total of 19 cases.
A simple consideration shows that the motions we have detected cannot be due to orbital motion in multiple star systems that are not resolved at 0.15" resolution. It can also be shown that the motions in the very center are not due

to gravitational lensing (Eckart & Genzel 1997). We therefore conclude that we have detected orbital proper motions of the stars in the central gravitational field. An immediate important conclusion is that the velocity dispersion is very similar in all three coordinates and hence any anisotropy of the stellar motions must be small. This can be seen from Fig.2 where we have plotted proper motion and radial velocity dispersions (from Rieke and Rieke 1988, McGinn et al. 1989, Sellgren et al. 1990, Lindqvist et al. 1992, Genzel et al. 1996, Haller et al. 1996) as a function of projected radius p from SgrA*. Taken together, the measurements for p>0.053 pc now show a very significant Keplerian ($s(p) \propto p^{-0.5}$) falloff of the stellar velocities with distance from the dynamic center (assumed to be at SgrA*), demonstrating that there must be a central, compact and dark (cf. Genzel et al. 1996) mass concentration.

The proper motion data thus strongly suggest that the dark mass concentration is centered within <0.009 pc on the compact radio source SgrA*. The SgrA* location relative to the near-infrared stellar distribution has recently been determined by Menten et al.(1996) to ±30 mas. SgrA* lies approximately at the center of a hammer-shaped concentration of fainter stars (SgrA*(IR), Eckart et al. 1995). In all our best images we obtain between 1994 and 1996 we were able to resolve the central SgrA*(IR) cluster into about a dozen components. The same structure has also been confirmed with the Keck 10m telescope (Klein et al. 1996) and via adaptive optics with the CFHT (R. Doyon, priv.com.). A close inspection of our NTT images shows that those from the two epochs in 1996 are very similar while there are fairly obvious structural changes between 1994 and 1996. The morphology of the 1995 epoch maps fits between the 1994 and 1996 data. Several of the stars in SgrA*(IR) appear to move with total proper motions of 1000 to 1600 km/s (Fig.3). The most convincing case is S1 (relative to S2/S3) which is within 0.2" (0.008 pc) of SgrA*. S1 appears to move with 1600±400 km/s with respect to the overall base position and with 1950±500 km/s relative to the centroid of S1/S2. For the SgrA*(IR) cluster we derived a velocity dispersion of 560±90 km/s at p=0.01 pc (corrected for measurement bias).

We fit to the projected velocity dispersion data in Fig.2 a model with a central point mass plus an extended isothermal cluster of dispersion at large radii of 50 km/s (see Genzel et al. 1996 , and Eckart & Genzel 1996, 1997 for details). The best fit (Fig.5 in Eckart & Genzel 1997) is obtained as a combination of a 2.45±0.25 (statistical) ±0.4 (total) $\times\ 10^6\ M_\odot$0.38 pc and a core stellar density of $4 \times 10^6\ M_\odot pc^{-3}$, representing the stellar cluster radiating in the near-infrared with a $M/L(2\mu m)$ ratio of about 2 (Genzel et al. 1996). Mass modelling with a dark cluster shows that its density then would have to be $>10^{12}\ M_\odot pc^{-3}$ and its core radius <0.0062 pc (Fig.5 in Eckart & Genzel 1997). The second one is the combination of the same visible stellar cluster with a second, dark cluster. To fit the data, the central dark cluster must have a central density of at least $6.5 \times 10^9\ M_\odot pc^{-3}$ and a core radius less than 0.035pc.

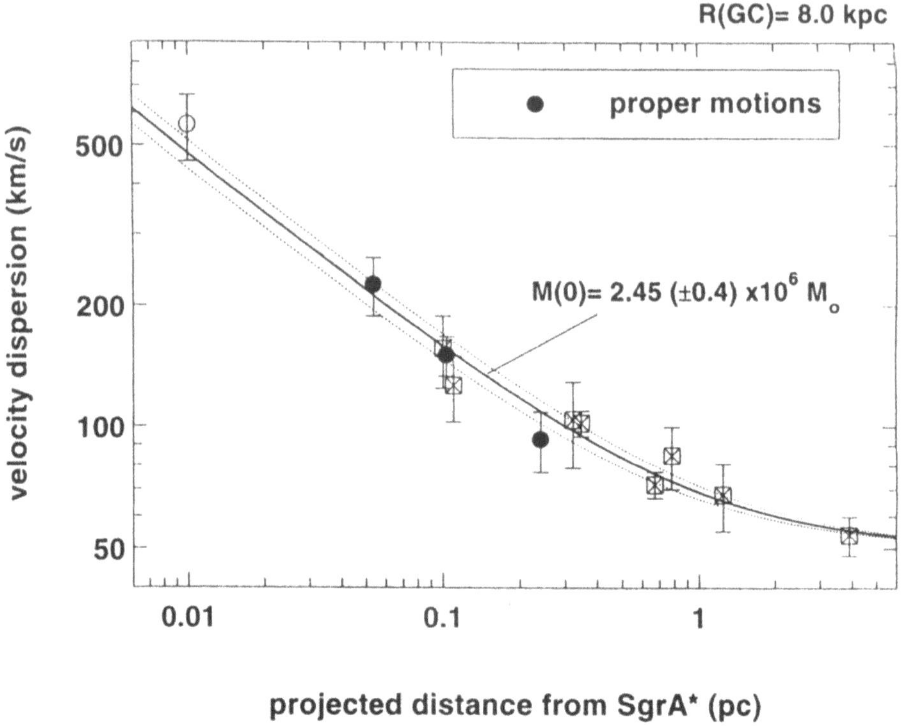

Fig. 2. Proper motion and radial velocity dispersions as a function of projected radius p from SgrA*.

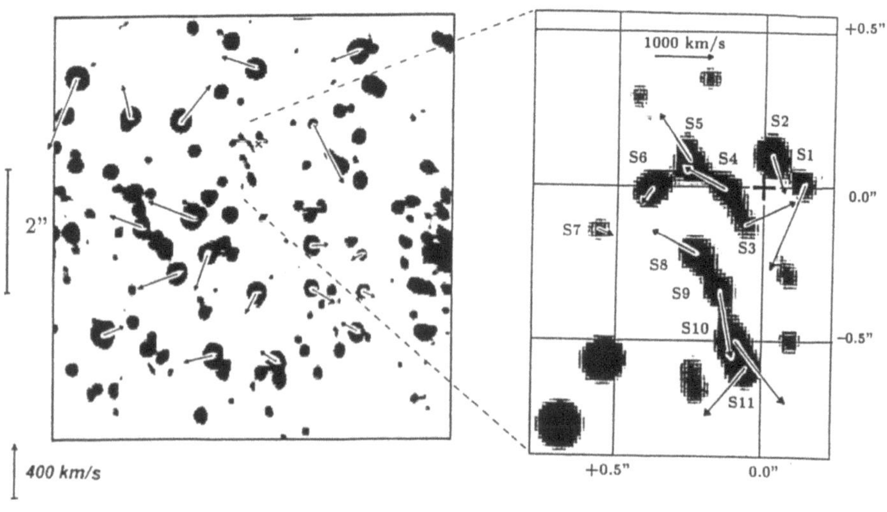

Fig. 3. Proper motions in the SgrA*(IR) cluster.

3 The VLTI Imaging Beam Combiner

An essential part of the Very Large Telescope Interferometer will be the imaging beam combiner (ESO/VLT Report No. 59b and Nr. 65). In its most sophisticated version, it will allow interferometric imaging over an 8 arcsec cophased field-of-view. This will be made possible by relaying the subpupils of the individual interferometer elements to a location where at any given time the interferometer exit pupil is arranged to be a scaled down replica of the input pupil as seen by the source. This procedure will properly correct differential delays between individual subpupils which is required for wide field-of-view (significantly larger than the primary beam of a single phased interferometer element) homothetic mapping (Faucherre et al. 1992). To investigate hardware and software requirements and to perform first experiments we have built the imaging beam combiner COSI. A detailed description of the experimental setup is given in Katterloher et al. (1994) and first interferometric imaging results have been presented in Böker et al. (1994) and Eckart et al. (1994).

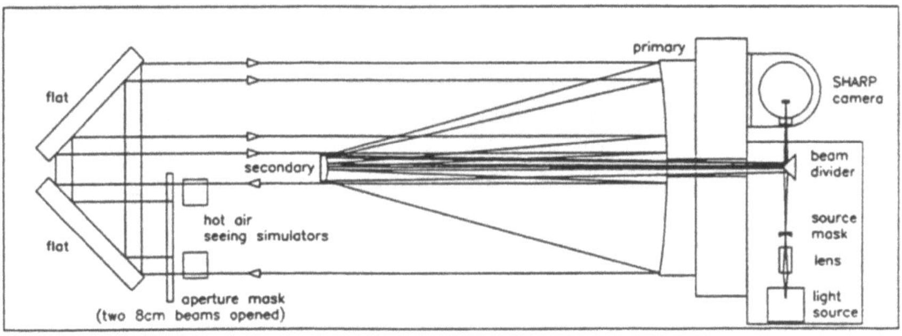

Fig. 4. The experimental setup of the beam combiner simulator COSI. The lower half of the experiment acts as a transmitter, the upper half as a receiver

4 The Beam Combiner Simulator COSI

The beam combiner simulator COSI is built around a 1 m telescope, which had been used as a far-infrared balloon-borne telescope in previous years (Drapatz et al. 1981). Since this telescope has about the same primary diameter as the beam combiner planned for the VLTI, we decided to put this telescope again into operation. The overall arrangement of the optical components is shown in Fig.4. Just in front of the focal plane, the beam is split into two halves. One half is fed by the light source. The light is then transferred through the corresponding section of the telescope to three plane mirrors. These mirrors reflect the light to the receiver side of the telescope onto three identical plane mirrors. From there, the radiation is transmitted through the telescope to the near-infrared camera. Between the telescope and the plane mirrors, aperture masks and a seeing simulator can be installed. The seeing simulator currently consists of a number of hot air cells mounted in front of each of the single sub-apertures in the aperture mask. Temperature and flow speed of the air can be regulated. A detailed description of the main parts of the instrument (telescope, light source, aperture masks, seeing simulator, and camera) is given in Katterloher et al. (1994). Further details are also given in Böker et al. (1994) and Hofmann et al. (1994). For additional information on hot air seeing simulators see also Cruzalèbes, Schumacher & Robbe (1993).

5 Results from COSI

The results of our first experiments have demonstrated that COSI is an excellent test bed to experimentally investigate methods of interferometric image reconstruction in the near-infrared. In addition to a phase retrieval procedure based on the bispectrum measurement (Lohmann, Weigelt, Wirnitzer 1983) our results show that in the case of a stable interferometer, in which the delays between rays from different (phased) telescopes are successfully compensated, images of bright sources (like in the Galactic Center region) can probably be obtained with relatively simple methods like: (1) direct deconvolution by division of complex Fourier spectra or by the nonlinear Lucy algorithm (Lucy 1974) or (2) a Fienup (1978) reconstruction from a more complete Fourier plane coverage (possibly after synthesizing the power spectra for observations taken with different interferometer configurations). A detailed description of the application of such methods to COSI data is given in Böker et al. (1994).

Future experiments will concentrate on extending the mapping method applied here to fainter sources and on an investigation of imaging methods for the multi-speckle case. COSI may also be used to test differential delay lines as they will be used to arrange the output pupil configuration of the different telescope beams over the beam combiner aperture. Furthermore we plan to investigate experimental aspects of beam combination using apertures of different sizes (2m and 8m telescopes) and non-homothetic observation modes.

6 Conclusions

We detected proper motions in the Galactic Center implying that it contains a dark central mass concentration with densities above 10^{12} $M_{\odot}pc^{-3}$. Given that the dark mass in the Galactic center is now constrained to be at least ten times more concentrated and therefore one thousand times denser than the visible stellar cluster we can now exclude with some confidence that it consists of a cluster of solar mass remnants (neutron stars or white dwarfs, see the detailed discussion in Genzel et al. 1996). Considering the stellar dynamics outside $R \approx 0.05$ pc, the remaining possibilities are a core collapsed cluster of 10-20 $M_\odot$ stellar black holes (Morris 1993, Lee 1995), or a single massive black hole, or a combination thereof. The proper motion data on the SgrA*(IR) cluster indicate that this must be a massive black hole. Future interferometric measurements of stellar proper motions using the VLTI will allow us to measure the orbits of stars in very fine detail. This will result in a very accurate determination of the stellar cluster mass distribution and concentration of the central dark mass.

References

Böker, T., Cruzalébes, P., Hofmann, R., Katterloher, R., Eckart, A., Genzel, R., Drapatz, S., "The MPE Imaging Beam Combiner Simulator COSI", 1994, *Astron. Astrophys.*, 288, 656.

Cruzalèbes, P., Schumacher, G., Robbe, S., 1993, *Astron. Astrophys.*, 272, 709.

Drapatz, S., Haser, L., Hofmann, R., Niekerke, J., Rothermel, H., Sämann, G., Stöcker, J., Wutz, H., 1981, Mitteilung der astron. Gesellschaft 54, 91.

ESO/VLT Interferometry Panel, 1989, The VLT Interferometer Implementation Plan, in: VLT Report No. 59b (European Southern Observatory, Garching bei München, Germany, 1989).

ESO/VLT Interferometry Panel, 1992, Coherent Combined Instrumentation for the VLT Interferometer, in: VLT Report No. 65 (European Southern Observatory, Garching bei München, Germany, 1992).

Eckart, A., Genzel, R., 1996, Nature Vol. 383, 415.

Eckart, A., Genzel, R., 1997, M.N.R.A.S 284, 576.

Eckart, A., Genzel, R., Krabbe, A., Hofmann, R., van der Werf, P.P. and Drapatz, S. 1992, Nature 355, 526

Eckart, A. Genzel, R., Hofmann, R., Sams, B.J. and Tacconi-Garman, L.E. 1993, Ap.J. 407, L77

Eckart, A. , Genzel, R., Hofmann, R., Sams, B.J., Tacconi-Garman, L.E. and Cruzalebes, P. 1994, in The Nuclei of Normal Galaxies, eds. R.Genzel and A.I. Harris, Kluwer (Dordrecht), 305

Eckart, A. Genzel, R., Hofmann, R., Sams, B.J. and Tacconi-Garman, L.E. 1995, Ap.J. 445, L26

Eckart,A., Böker, T., Hofmann, R., Katterloher, R., Quirrenbach, A., Löwe, M., Cruzalèbes, P., 1994, SPIE Proceedings Vol.2200, on "Amplitude and Intensity Spatial Interferometry II", 15-16 March 1994, Knoa, Hawaii, edt. J.B.Breckinridge, p.458.

Faucherre, M., et al., 1992, in: Proceedings of the ESA Coloquium on Targets for Space-Based Interferometry, ESA SP-354, p. 91.

Fienup, J.R., 1978, Opt.Letters 3, 27.

Genzel, R., Thatte, N., Krabbe, A., Kroker, H. and Tacconi-Garman, L.E. 1996, Ap.J. in press

Haller, J.W., Rieke, M.J., Rieke, G.H., Tamblyn, P., Close, L. and Melia, F. 1986, Ap.J. 456, 194

Hofmann, R., Blietz, M., Duhoux, P. , Eckart, A., Krabbe, A. and Rotaciuc, V. 1993, in Progress in Telescope and Instrumentation Technologies, ed. M.H. Ulrich, ESO Report 42, 617
Katterloher, R., Böker, T., Eckart, A., Hofmann, R., Jakob, G., Quirrenbach, A., Löwe, M., 1994, SPIE Proceedings Vol.2200 on "Amplitude and Intensity Spatial Interferometry II", 15-16 March 1994, Knoa, Hawaii, edt. J.B.Breckinridge, P.469.
Kormendy, J. and Richstone, D. 1995, Ann.Rev.Astr.Ap.33, 581
Klein, B.L., Ghez, A.M., Morris, M., Becklin, E.E., 1996, in: The Galactic Center ASP Conference series, Vol. 102, 1996, Roland Gredel, ed.
Lee, H.M. 1995, Mon.Not.Roy.Astr.Soc. 272, 605
Lindqvist, M., Habing, H. and Winnberg, A. 1992, Astr.Ap. 259, 118
Lohmann, A.W., Weigelt, G., Wirnitzer, B., 1983, Appl. Opt. 22, 4028.
Lucy, L.B., 1974, *Astron. Journ.*, 79, 745.
McGinn, M.T., Sellgren, K., Becklin, E.E. and Hall, D.N.B. 1989, Ap.J. 338, 824
Menten, K.M., Eckart, A., Reid, M.J. and Genzel, R. 1996, Ap.J. submitted
Morris, M. 1993, Ap.J. 408, 496
Reid, M. 1993, Ann.Rev.Astr.Ap. 31, 345
Rieke, G.H. and Rieke, M.J. 1988, Ap.J. 330, L33
Sellgren, K., McGinn, M.T., Becklin, E.E. and Hall, D.N.B. 1990, Ap.J. 359, 112

The VLTI and the Galactic Center He I Cluster

Francisco Najarro[1], Alfred Krabbe[2], Reinhard Genzel[2], Dieter Lutz[2], Rolf P. Kudritzki[1,3]

[1] Universitäts-Sternwarte München, Scheinerstraße 1, D-81679 München, Germany
[2] Max-Planck-Institut für extraterrestrische Physik, 85740 Garching bei München, Germany
[3] Max-Planck-Institut für Astrophysik, 85740 Garching bei München, Germany

Abstract. The infrared spectra of early-type stars are powerful diagnostic tools for studying the process of mass-loss via a stellar wind and for determining fundamental stellar parameters. First quantitative infrared spectroscopic studies of the brightest members of the cluster of HeI objects in the center of our Galaxy have not only confirmed the stellar nature of these objects, but have also yielded stellar parameters. These indicate that the He I emission line stars are evolved blue supergiants close to the evolutionary stage of Wolf-Rayet stars. The HeI objects belong to a a stellar cluster of massive stars formed some 5 million years ago. Our results suggest that this stellar cluster may dominate the energetics and radiation field of the central parsec of the Galaxy. It is, therefore crucial to perform quantitative infrared spectroscopy of the other, fainter early-type stars of the cluster to better understand their role in powering the central parsec. To achieve this task we need to obtain not only moderate high S/N spectra of the fainter stars but also accurate continuum measurements for each object. Because of the enormous crowding present in the Galactic Center, the VLTI presents a unique opportunity to perform these flux measurements.

1 Introduction

The He I emission line cluster in the Galactic center discovered by Krabbe et al. (1991) provides a unique opportunity to test star formation and evolution in connection with the ionization and energetics of the Galactic center. Results of a detailed spectroscopic investigation of the brightest source (AF star) of the cluster in the He I 2.058μm line by Najarro et al. (1994) revealed that the AF star is a helium-rich blue supergiant/Wolf-Rayet star. It is characterized by a strong stellar wind and constitutes a moderate source of Lyman continuum photons. The latter result clearly suggested that the cluster of He I stars can make a significant contribution to the total luminosity and the Lyman continuum flux of the inner parsec. Therefore, it was crucial to analyze the spectra of other He I objects of the cluster and to obtain their stellar parameters to better constrain their role in the energetics of the central parsec. We have carried out spectroscopic observations of the He I cluster using CSG4, FAST, SHARP and 3D (Krabbe et al. 1995) and derived stellar parameters for eight He I objects.

2 Quantitative Infrared Spectroscopy of the He I Cluster

With the observational contrains given by the K continuum flux and the Bγ, He I 2.058μm and He I 2.112μm line profiles, we proceeded as in Najarro et al. (1994) and used the iterative, non-LTE method presented by Hillier (1987, 1990) which solves the radiative transfer equation in spherical geometry, subject to the constraints of statistical and radiative equilibrium, for the expanding atmospheres of early-type stars. We considered a pure H-He atmosphere with 12 H, 49 He I (n$\leq$10) and 12 He II levels. The stellar parameters obtained for the eight brighter He I objects are presented in Table 1. Agreement between observed and computed values was always better than 10%. Below we discus our results for the analyzed objects.

Table 1. Derived stellar parameters for GC He I sources.

	IRS Object							
	16NE	16C	16SW	16NW	7W	13E1	15NE	15SW
R_* ($R_\odot$)	85	85	90	70	23	60	45	62
L_* ($10^5 L_\odot$)	22.0	19.0	25.9	10.3	4.10	22.6	9.10	10.5
T_{eff} (10^4K)	2.41	2.32	2.44	2.20	3.04	2.89	2.66	2.35
He/H	1.0	3.0	1.3	1.3	>500	>500	>100	>100
$\dot{M}$ ($10^{-5} M_\odot\,yr^{-1}$)	9.50	10.5	15.5	5.30	20.7	79.1	18.0	16.5
V_∞ (km s^{-1})	550	650	650	750	1000	1000	750	700
V_{rad} (km s^{-1})	-40	200	300	-100	-250	0	-150	-200
Log Q(H^+)	49.6	49.5	49.7	49.0	49.2	49.9	49.4	49.3
Log Q(He^+)	< 45	< 45	< 45	< 44	< 44	45.0	< 44	< 44
Log Q(He^{++})	< 32	< 32	< 32	< 31	< 33	< 33	< 33	< 32

IRS 7W & IRS 13E. IRS 7W and IRS 13E are the hottest ($T_{eff}\approx$30kK) objects of the sample. They show no stellar hydrogen features in their spectrum, which confirms their highly evolved status. The stellar parameters derived are similar to those of WN8 and WN9 stars Crowther et al. 1995a), while the observed near-IR spectrum clearly indicate a later spectral type than WN8 (Hillier 1985), and probably corresponds to a WN9 or WN10 type.

Figure 1 shows the general good agreement of our computed line profiles for IRS 7W and 13E with the observed He I 2.058μm, Bγ and He I 2.112μm profiles. Both He I 2.058μm lines (including the strong electron scattering wings observed in IRS 13E) are well reproduced. For IRS 13E the computed absorption dip is clearly deeper than the observed one. This discrepancy may be due to stellar rotation. The fits to the observed He I (7–4) lines confirm the absence of H in the winds of IRS 7W and IRS 13E. Our models reproduce as well the central part ($\pm V_\infty$) of the 2.11μm observed feature corresponding to the He I 2.112μm line

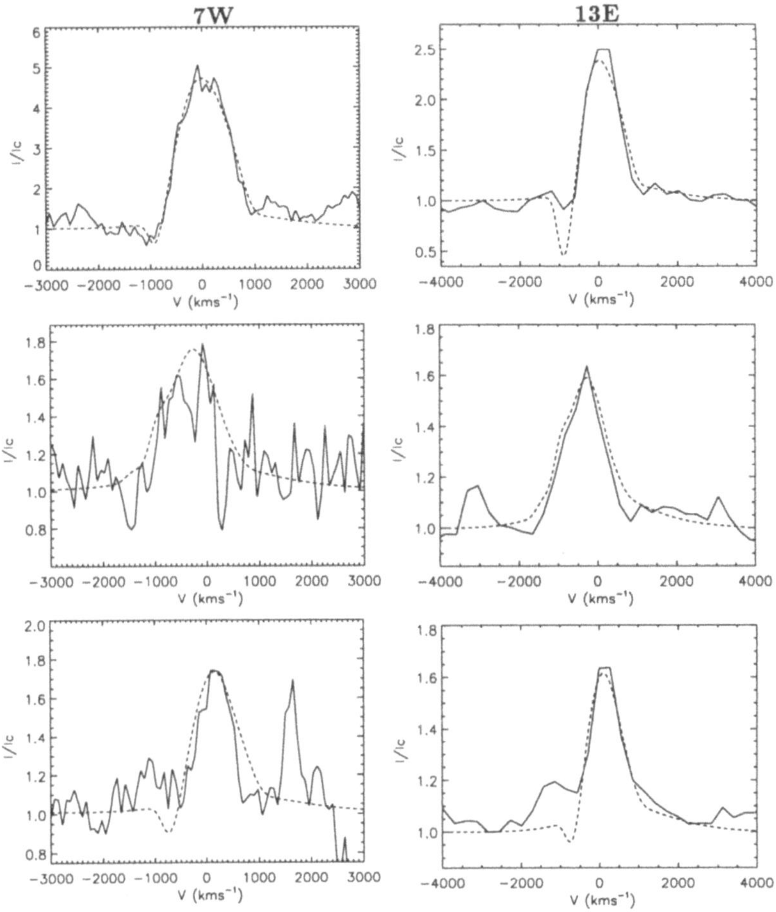

Fig. 1. left (from top to bottom): IRS 7W observed **CGS4** (solid) and computed (dashed) He I 2.058μm, Bγ and He I 2.112μm profiles. **right (from top to bottom)** IRS 13E observed **3D** (solid) and computed (dashed) He I 2.058μm, Bγ and He I 2.112μm profiles. Diffuse emission has been removed for both objects.

and reveal the important contribution of the N III/C III (8-7) lines (especially in IRS 13E).

IRS 16 sources. We obtain similar values for the He abundance, stellar luminosities and effective temperatures for all four IRS 16 sources, which indicates that they have similar ages. The derived stellar parameters suggest that these objects are probably at the end of the LBV evolutionary phase. Figure 2 shows the excellent agreement between the observed **3D** and computed He I 2.058μm, Bγ and He I 2.112μm profiles for the IRS 16 sources. The obvious P Cygni shape of the observed He I 2.058μm profiles, confirms the wind nature of the lines. Once more, our results are able to reproduce the He I 2.112μm component quite well. Despite the relatively low S/N ratio, Fig. 2 confirms the presence of the

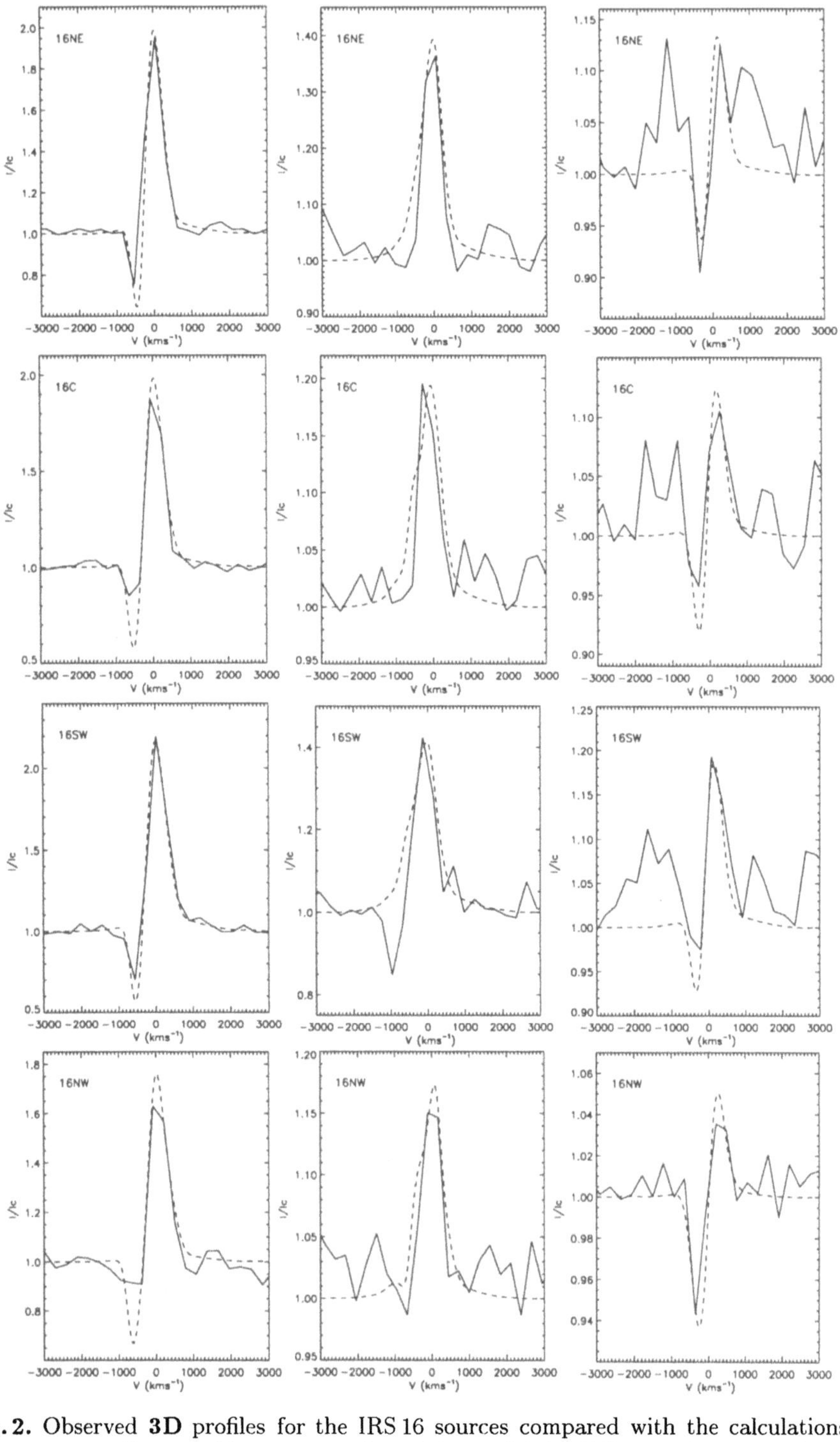

Fig. 2. Observed **3D** profiles for the IRS 16 sources compared with the calculations (He I 2.058μm, Bγ and He I 2.112μm from left to right). Diffuse emission has been removed.

N III/C III (8-7) lines in three of the IRS 16 sources (NE, C and SW), while they seem to be absent in the spectrum of IRS 16NW. This result is consistent with the lower temperature (ΔT_{eff} ~1500 K) and wind density obtained for the latter, as we expect the strength of the N III/C III (8-7) lines to decrease with decreasing T_* and wind density. This would also explain the absence of these features in the spectrum of the AF star (lower T_{eff}) and their relatively large strength in the hotter, denser IRS 13E and IRS 7W.

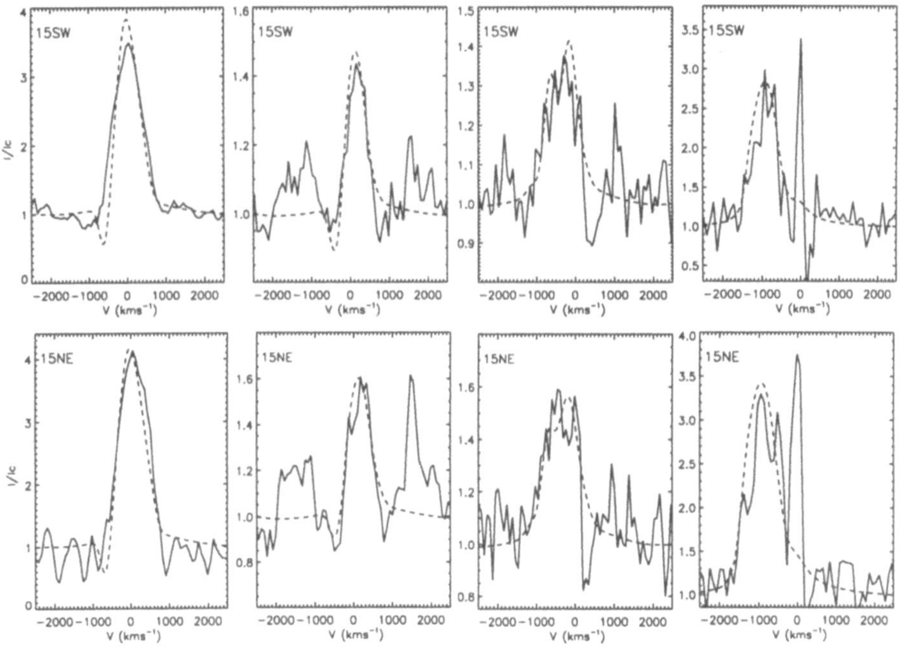

Fig. 3. IRS 15SW and IRS 15NE observed **CGS4** (solid) and computed (dashed) He I 2.058μm,He I 2.112μm,Bγ and Pα profiles. Most of the diffuse emission has been removed.

IRS 15SW & IRS 15NE. IRS15NE and IRS15SW may be considered as "transition" objects. These objects have effective temperatures only slightly higher (T_{eff}~25000 K) than those obtained for the IRS16 objects suggesting that these objects are finishing their LBV evolutionary phase. However, they appear to be pure He sources (as do IRS7W and IRS13E-1) indicating a Wolf-Rayet phase. Figure 3 shows the general good agreement of our model calculations with the observed IRS 15SW and IRS 15SW He I 2.058μm, He I 2.112μm, Bγ and Pα lines. Our computed profiles are able to reproduce the shapes of the observed profiles satisfactorily (see Fig. 3) and account for the broader blends due to the He I (4–3) and He I (7–4) lines. We note, once more, the clear presence of the N III/C III 2.103μm feature bluewards of the He I 2.112μm line.

Evolutionary Status, Ionizing Flux and GC Mass. Krabbe et al. (1995) have analyzed the nuclear star cluster using the star cluster models of Krabbe et al. (1994) and conclude that the observations fit a $7\pm1 \times 10^6$yrs old decaying burst in which ~30000 stars formed very well. Their results clearly favor a model in which the central parsec of the Galaxy is powered by a cluster of hot stars, as earlier proposed by Rieke & Lebofsky (1982) and Allen & Sanders (1986). The eight objects studied here alone, already supply $\approx 3 \times 10^{50}$ Lyman continuum photons per second (> 60% of the total Lyman flux as estimated from thermal radio continuum measurements by Genzel et al. 1994). Therefore, the hot star cluster can fully account for the bolometric and Lyman ionizing luminosities of the central parsec. None of these objects contribute to the He I-continuum. Even IRS13E1 does not provide more than 1% of the flux required for the He-ionization of the SgrA West HII region, $\log Q(\mathrm{He}^+) \approx 49$ (Krabbe et al. 1991). Hence, some hotter O and/or WR stars is required, the presence of which have been recently reported by Krabbe et al. (1995). Finally, we may use the derived radial velocities to obtain the velocity dispersion of the objects and therefore derive the enclosed mass in the central parsec. Krabbe et al. (1995) have obtained velocity measurements from Gaussian fits to the observed Pα, He I 2.058μm, He I 2.112μm and Bγ lines, and obtained a mass concentration of ~2 to $4\times10^6 M_\odot$, thus finding a high statistical significance for a dark mass concentration. The latter could be in the form of massive stellar remnants ($10 M_\odot$), i.e., the black hole cluster option suggested by Morris (1993), or more likely as a massive black hole (Genzel et al. 1994).

3 The VLTI and the Cluster of Early-Type Stars in the GC

Future, moderate S/N spectra and continuum measurements of the recently detected O and WR stars in the galactic center are crucial to check whether they can or can not account for the He I-continuum in the Galactic center. However, this is not an easy task. Because of the tremendous crowding in the central parsec of our galaxy, many of the bright sources that appear point like with an angular resolution of $0.5''$, are multiple. This, may lead to substantial errors when deriving the ionizing photons. Therefore, a high angular resolution is required to isolate the continuum sources. The VLTI constitutes the ideal instrument to obtain such measurements. At the distance of the Galactic center (8.5kpc) the angular resolution achieved by the VLTI in the K band will be about 20 AU, i.e., **only** 50 times the stellar radii of the He I stars. Further, since IRS 7 can be used as nearby bright fringe tracking source the "non-source-phased and - cophased" case applies, providing a high sensitivity. Thus, in principle we will be able to obtain accurate photometric measurements as well as moderate resolution spectra (R~3000) of O, B and WR stars in the central parsec of our Galaxy.

References

Allen, D.A., Sanders, R.H., 1986, NATURE, 319, 191

Crowther, P.A., Hillier, D.J., Smith, L.J., 1995, A&A, 293, 172

Genzel, R., Hollenbach, D.J., Townes, C.H., 1994, Rep.Progr.Phys., 57, 417

Hillier, D.J., 1985, AJ, 90, 1514

Hillier, D.J., 1987a, ApJ Suppl., 63, 947

Hillier, D.J., 1990, A&A, 231, 116

Krabbe, A., Genzel, R., Drapatz, S., Rotaciuc, V., 1991, ApJ, 382, L19 A cluster of HeI emission line stars in the Galactic Center

Krabbe, A., Stenberg, A., Genzel, R., 1994, ApJ, 424, 72

Krabbe, A., Genzel, R., Eckart, A., Najarro, F., Lutz, D., Cameron, M., Kroker, H., Tacconi-Garman, L.E., Thatte, N., Weitzel, L., Drapatz, S., Geballe, T., Sternberg, A., Kudritzki, R. , 1995, ApJ 447, L95

Morris, M., 1993, ApJ, 408, 496

Najarro, F., Hillier, D.J., Kudritzki, R.P., Krabbe, A., Genzel, R., Lutz, D., Drapatz, S., Geballe, T.R., 1994a, A&A, 285, 573

Rieke, G.H., Lebofsky, M.J., 1982, in AIP Conference Proceedings 83: The Galactic Center, eds. G. Riegler, R. Blandford, New York, 194

Globular Cluster Proper Motions

Georges Meylan

European Southern Observatory, Karl-Schwarzschild-Strasse 2,
D-85748 Garching bei München, Germany

Abstract. Although proper motion measurements provide two components of the space velocity, the third one being given by radial velocity measurements, the proper motions of stars in globular clusters have not yet been intensively used in the studies of the internal dynamics of these stellar systems. This is because the small size of space motions of stars in globular clusters has made their proper motions difficult to measure with the required precision. We briefly review hereafter (i) the different astrophysical problems related to globular clusters which would benefit from the knowledge of stellar proper motions, (ii) the main outputs from studies using stellar radial velocities, (iii) the quality of the existing proper motions, (iv) the few programs of measurement of proper motions in globular clusters with the WFPC2 camera on board the Hubble Space Telescope, and (v) the proper motion capabilities of the ESO-VLTI program.

1 Scientific Problems Related to Globular Cluster Proper Motions

In theory, proper motions provide more dynamical information than radial velocities, since they are two- instead of one-dimensional (see, e.g., Wybo & Dejonghe 1995, 1996). Proper motions of stars in globular clusters can potentially improve our knowledge of numerous fundamental astrophysical problems, such as, e.g.:

• The distances to galactic globular clusters, derived by comparison between proper motions and radial velocities of cluster stars, using dynamical models (Lupton et al. 1987). The resulting distances can be used to put rather stringent limits on the absolute magnitude of stars in the color-magnitude diagram (Rees 1996).

• The dynamics of the globular cluster system around our Galaxy, although already partly known, would benefit from high-quality proper motions, especially for globular clusters located deep in the bulge or way out in the halo (Dauphole & Colin 1995). Underlying is the problem of galaxy formation (see Larson 1996 for a review).

• The dynamical pruning of globular clusters during their successive crossings of the galactic plane, with creation of extra-tidal tails visible in deep surface density maps (Grillmair et al. 1995), and, more generally, the consequences of the cluster orbital motions on the internal dynamics of individual globular clusters (see Meylan & Heggie 1996 for a review).

• The internal dynamics of a few nearby and rich globular clusters, such as, e.g., ω Centauri and 47 Tucanae, which are used, with a few other globular clusters, as testing-bench for our understanding of these dynamical systems. Most of our present knowledge comes from studies constrained by stellar radial velocity measurements (e.g., see Meylan et al. 1995 and Merritt et al. 1997 in the case of ω Centauri). The measurements of numerous stellar proper motions in such globular clusters would provide essential constraints on the velocity dispersion profile, ideally as a function of stellar masses, with potential insights into mass segregation and anisotropy of the velocity dispersion.

• The precise determination of the luminosity and mass functions, generally difficult to obtain for low-mass stars close to the hydrogen-burning limit. Proper motions should provide one of the best ways to distinguish back- and fore-ground stars from the stars belonging to the clusters (King 1996 pers. comm.)

• Estimates of the proportion of binary stars as a function of the stellar mass, and long-term monitoring for orbit determination (see, e.g., the proceedings edited by Milone & Mermilliod 1996).

2 Output from Studies Using Stellar Radial Velocities

Since the pioneering work of Griffin (1967, 1974), cross-correlation techniques have proven their exceptional efficiency in radial velocity determination. During the same period of time, no similar improvement has characterized proper motion measurements.

During the last two decades, the purpose of building dynamical models has been to construct various simplified mathematical descriptions of a star cluster, each of them easily comparable with observations (density profile from star counts and radial velocity dispersion profile from radial velocities). Only very few of these models, however, have achieved any prominence in applications. Foremost amongst these are King's (1966) models and their variants.

2.1 King-Michie Models

King models approximately incorporate three essential dynamical processes: (i) dynamical equilibrium, (ii) the effect of gravitational encounters between pairs of stars, which, like collisions in a gas, tend to set up a maxwellian distribution of velocities, and (iii) a cutoff in energy, ε_t, above which stars are deemed to have escaped. King's models depend on three dimensional parameters, which can be taken to be the central density ρ_c, the central velocity dispersion σ_c, and the tidal radius r_t. There is one dimensionless parameter, which can be taken to be the logarithm of the ratio of the tidal to core radius, i.e., the concentration $c = \log\,(r_t/r_c)$, or the dimensionless central potential.

After the introduction of a mass spectrum (Da Costa & Freeman 1976), the next important variant of King's original models was the inclusion of anisotropy of the velocity distribution, by making the distribution function dependent on the specific angular momentum (Gunn & Griffin 1979). The most commonly chosen form, defining the so-called King-Michie model (Gunn & Griffin 1979), is

$$f \propto \exp(-j^2 l^2/r_a^2)(\exp(-2j^2\varepsilon) - \exp(-2j^2\varepsilon_t)) \ . \tag{1}$$

for $\varepsilon < \varepsilon_t$, where r_a is a constant. When anisotropy is introduced in this way along with unequal masses, it is usual to take r_a to be the same for all masses, as it corresponds to the radius beyond which the anisotropy becomes important. Incidentally, the increase of anisotropy with radius may not be particularly appropriate for tidally bound models, because of the isotropising effect of the tidal field, which is important at large radii.

In a recent application of King-Michie models, the mean radial velocities obtained with CORAVEL for 469 individual stars located in the galactic globular cluster ω Centauri $\equiv$ NGC 5139 (Mayor et al. 1997) have been used to derive the radial velocity dispersion profile. It increases significantly from the outer parts inwards: the 16 outermost stars, located between 19.2′ and 22.4′ from the center, have a radial velocity dispersion $\sigma_p = 5.1 \pm 1.6$ km s^{-1}, while the 16 innermost stars, located within 1′ from the center, have a radial velocity dispersion $\sigma_p = 21.9 \pm 3.9$ km s^{-1}, the largest velocity dispersion value obtained in the core of any galactic globular cluster.

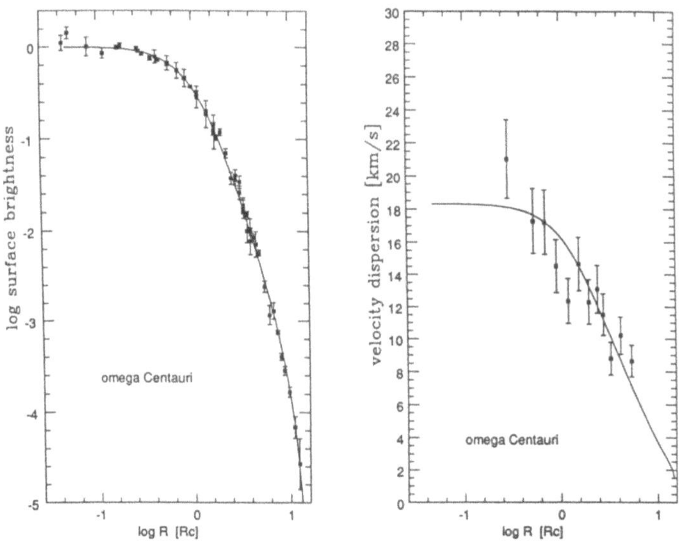

Fig. 1. ω Cen: logarithm of the surface brightness profile (left) and radial velocity dispersion profile (right) as a function of the logarithm of the radius. In both panels, the points represent the observations and the solid lines the King-Michie model with index # 1 in Table 2 of Meylan et al. (1995). The binning of the radial velocities into 12 boxes has only a display purpose and is not related to the actual way the velocity data are fitted to the model.

A simultaneous fit of these radial velocities and of the surface brightness profile to a multi-mass King-Michie dynamical model (see Fig. 1) provides mean estimates of the total mass equal to $M_{tot} = 5.1\ 10^6 M_\odot$, with a corresponding mean mass-to-light ratio $M/L_V = 4.1$. The present results emphasize the fact that ω Cen is the brightest and the most massive globular cluster in our Galaxy (Meylan et al. 1995).

2.2 Non-Parametric Techniques

The purpose of building models such as King's models and their derivatives is to construct a simplified mathematical description of a star cluster. They provide a number of parameters (mass spectrum, concentration, anisotropy radius, etc.) which can be adjusted to optimise the fit with observations. Nevertheless they are based on strict assumptions with regard to the form of the distribution function. These assumptions take account of dynamical theory, though in some respects they contradict or oversimplify it. The results can be strongly biased by the – arbitrary – choices of the integrals of motion and of the form of the functional dependence. Indeed it is true that, even when the profiles of mass density and velocity dispersion are known, the distribution function is still not uniquely determined (Dejonghe 1987, Dejonghe & Merritt 1992); for instance the anisotropy is still not completely constrained. For this and other reasons it is worth considering methods which attempt to construct the distribution function from the observations, with minimal assumptions.

There are quadratic programming techniques (Dejonghe 1989) which fall into this class. The distribution function is written in terms of basis functions, but the possible risks of such an approach have been discussed by Merritt & Tremblay (1994) in a related context. They recommend a modification which serves to smooth the resulting estimate of the distribution function. The general aim is to infer the gravitational potential $\Phi(r)$ and the phase-space distribution function $f(\varepsilon)$, given the observations of the surface density and velocity dispersion profiles of a "tracer" population. Briefly, in the case of a globular cluster, (i) the projected density $I(R)$ provides the space density $\nu(r)$, (ii) the projected velocity dispersion $\sigma^2(R)$ provides the space velocity dispersion $v^2(r)$, (iii) the Jeans equation provides the gravitational potential $\Phi(r)$, and (iv) the Eddington equation provides the phase-space distribution function $f(\varepsilon)$. Nevertheless, a disadvantage of such techniques arises from the delicate process of deprojection using Abel integrals (Merritt 1993a,b).

Non-parametric studies have been published for five globular clusters, using samples from a few hundred up to a few thousand stars. As a result, non-parametric mass density and M/L_V profiles can now be compared with more traditional theoretical models for core-collapse clusters. The two non-collapsed globular clusters, viz., NGC 362 and NGC 3201, seem to exhibit significant differences from the two possibly collapsed globular clusters, 47 Tuc and M15. The non-parametric estimates of the M/L_V for the four clusters show a remarkable difference between the M/L_V profiles of the two collapsed and the two other clusters (Gebhardt & Fischer 1995).

Côté et al. (1995), using King, King-Michie, and non-parametric models, present, for the globular cluster NGC 3201, an interesting comparison between the different results, and a discussion of how to disentangle the consequences of the assumptions and disadvantages of each approach.

In another non-parametric study, Merritt et al. (1997) use the same 469 mean radial velocities obtained in ω Cen with CORAVEL (Mayor et al. 1997) and already used in a recent application of King-Michie models (Meylan et al. 1995). The deduced surface brightness and space density profiles as a function of the logarithm of the radius are displayed in Fig. 2, from Merritt et al. (1997), who also derived the two-integral distribution function $f(\varepsilon, l_z)$. See §3 below.

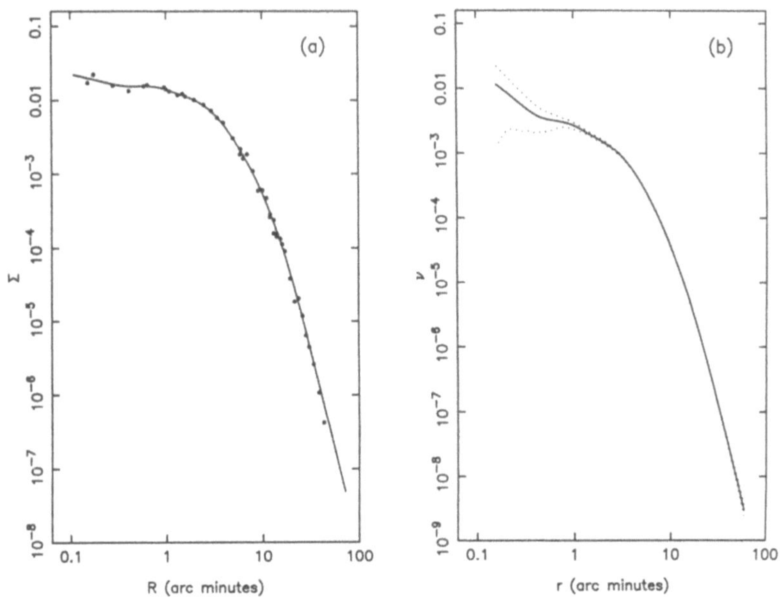

Fig. 2. ω Cen: (a) logarithm of the surface brightness profile and (b) space density profile as a function of the logarithm of the radius (Merritt et al. 1997). In the left panel, the points represent the observations and the solid line the solution to the optimization problem. In the right panel, the dashed lines are the 95% confidence bands on the estimate of the space density $\nu(r)$.

These models need large samples of stars, with ideally both radial velocities and proper motions measured. This has already started with fiber-fed, multi-object spectrographs like ARGUS at Cerro Tololo, HYDRA at Kitt Peak, and AUTOFIB at the Anglo Australian Observatory which can obtain radial velocities about 25 times faster than previously done. Similar gains result from using Fabry-Perot interferometers to measure radial velocities. With such a device, Gebhardt et al. (1997) have acquired radial velocities for 4,200 stars in the core of ω Centauri (see Merritt 1997). However, the same improvements in quality and sample size have still to be done for the proper motions. The announced data obtained in ω Cen (Reijns et al. 1993 and Seitzer, pers. comm.), viz., proper motions for about 7,000 stars and stellar radial velocities for about 3,500 stars, will permit investigation of the 3-D space velocity distribution and rotation. A quantum jump in the understanding of the internal dynamics of this globular cluster should result from the interpretation of such data samples.

3 Quality of the Existing Proper Motions

Proper motions in globular clusters have generally been measured from photographic plates obtained at various Schmidt telescopes, at the Palomar 200-inch and other 4-m class telescopes. First- and second-epoch data are generally separated by a few decades, although first-epoch data are sometimes obtained from significantly older Yerkes and Harvard plates. See, e.g., Cudworth & Hanson (1993) and Scholz et al. (1996). It is only recently that techniques for astrometry with CCDs have been developed (Monet et al. 1992, Tinney 1993, and §4 below), providing proper motions on rather small fields, when compared with photographic plates.

The small size of the internal stellar motions of stars in globular clusters – typically 5 to 10 $km\,s^{-1}$ – makes the proper motions difficult to measure, since a precision of about 1 $km\,s^{-1}$ or less is required. In the following, millisecond of arc is abbreviated by marcsec and microsecond of arc by μarcsec. For example, for a nearby cluster at a distance of 5 kpc, a velocity dispersion of 5 $km\,s^{-1}$ corresponds to a displacement of 20 $marcsec\,cent^{-1}$, which is the equivalent of 1.5 micron in 80 years on Yerkes plates. Consequently, the errors in the proper motions have been comparable in size to the motions themselves. This explains why the very few studies of proper motions in globular clusters have provided only little information on the internal dynamics of these stellar systems (see Lupton et al. 1987 and Leonard et al. 1992, both of which examined M13 using the data of Cudworth & Monet 1979). In the case of M5, new reduction and analysis of already existing material improved the quality of the proper motions (Rees 1993).

So far proper motions have essentially been used to deduce the space velocities of globular clusters. And again in this case, the uncertainties remain large. In the case of 47 Tuc, the three components of the space velocity are the following. The mean radial velocity of the cluster

- $V_r = -18.8 \pm 0.6$ $km\,s^{-1}$

is given by Meylan (1989), while the two components of the proper motions are given by Cudworth & Hanson (1993), namely

- $\mu_\alpha = +0.34 \pm 0.17$ $arcsec\,cent^{-1}$
- $\mu_\delta = -0.19 \pm 0.15$ $arcsec\,cent^{-1}$.

The three components of the space velocity, $(U,V,W) = (51 \pm 34, -42 \pm 29, 41 \pm 23)$ in $km\,s^{-1}$, relative to the local standard of rest, give $V_{spa} = 78 \pm 50$ $km\,s^{-1}$ (Cudworth & Hanson 1993). Most of the error on this measurement comes from the proper motion uncertainties.

It is worth emphasizing that the precise estimate of the space velocity of a globular cluster is not a useless academic exercise. The knowledge of this simple quantity may have some direct and deep consequences on the study of the internal kinematics and dynamics of this cluster.

For instance, if the cluster has a substantial proper motion, a systematic correction must be made to the measured radial velocities, as done by Merritt et al. (1997) in their non-parametric study of the rotational field in ω Cen. Feast et al. (1961) note that the translation of a solid object produces an apparent rotation, since the projection of the space velocity along the line-of-sight is different at different points of the image. Let X, Z be the coordinates on the plane of the sky, with X and Z parallel to the isophotal major and minor axes, respectively. The "perspective rotation" discovered by Feast et al. (1961) results in a radial velocity increment which depends on X, Z, V_X, V_Z, and D, where V_X and V_Z are the components of the cluster space velocity parallel to the X and Z axes and D is the distance to the cluster.

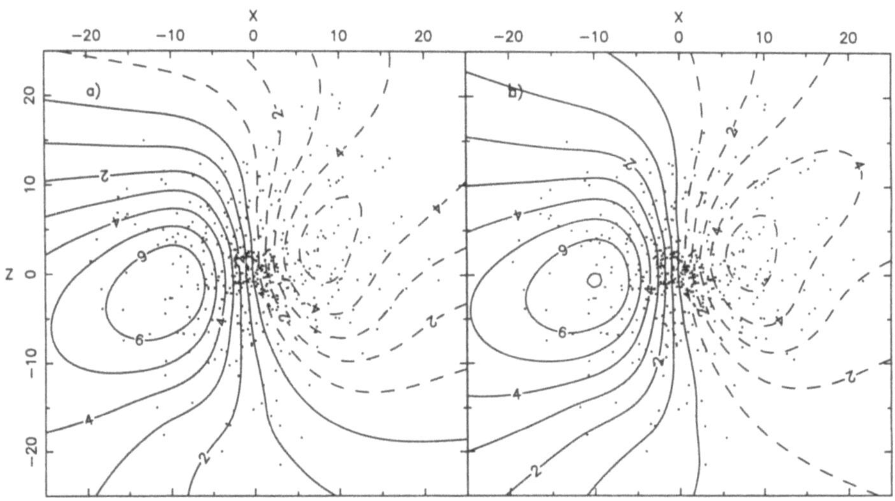

Fig. 3. ω Cen: a) Line-of-sight rotational field obtained in the non-parametric study by Merritt et al. (1997). The Z-axis is parallel to the isophotal minor axis and East is toward the left; distances are in arc minutes. Dots indicate positions of the stars in the kinematic sample (Mayor et al. 1997). Contours are labelled in km s^{-1}, measured with respect to the cluster mean radial velocity; dashed contours indicate negative velocities. b) Line-of-sight rotational field, after the perspective rotation due to the proper motion of ω Cen has been removed from the measured velocities (Merritt et al. 1997).

The proper motion of ω Cen has been determined by Murray et al. (1965). Corrected for differential galactic rotation (Cudworth 1994 pers. comm.), the components of the proper motion parallel to the X and Z axes are

- $\mu_X = +0.26 \pm 0.06$ arcsec cent^{-1}
- $\mu_Z = -0.74 \pm 0.05$ arcsec cent^{-1}.

Using a distance of 5.2 kpc and assuming an error in this number of ± 10%, Merritt et al. (1997) find

- $V_X = +64 \pm 16$ km s^{-1}
- $V_Z = -183 \pm 22$ km s^{-1}.

The resulting change in the radial velocity at point (X, Z) has an amplitude which is small but not negligible, amounting to about 1 km s^{-1} near the edge of the kinematic sample.

Fig. 3a shows the uncorrected and Fig. 3b the corrected rotational velocity field after removal of the perspective rotation. The contours in Fig. 3b are substantially more symmetric at large radii.

4 Proper Motions with the HST WFPC2

Two current programs aim at using the WFPC2 camera on board the Hubble Space Telescope (HST) to measure proper motions in globular clusters in order to study (i) high-velocity stars in the core of 47 Tuc (PI = G. Meylan) and (ii) the amount of very-low-mass stars in NGC 6397 (PI = I.R. King).

4.1 A Census of High-Velocity Stars in 47 Tucanae

The first program using HST (first-epoch data already acquired) is related to high-velocity stars in globular clusters. Such stars were observed, in ground-based studies, in two different globulars: viz. in M3 by Gunn & Griffin (1979) and in 47 Tuc by Meylan et al. (1991). The origin of the high velocities of these stars is very likely related to encounters with hard binary stars.

It is now well established that binary stars control the late stages of core collapse (see Hut et al. 1992 and Meylan & Heggie 1996 for reviews). Binaries are concentrated in the core due to mass segregation and they can delay, halt, and even reverse the collapse. Primordial binaries provide a finite "fossil fuel" because the binaries are eventually destroyed by being driven to coalescence, by being disrupted by an encounter with other binaries, or by being ejected from the cluster by the recoil from a strong interaction. At very high core densities, new binaries can be formed by encounters between three single stars (Heggie 1975) or by "tidal-capture", in which two stars approach so closely that they dissipate enough of their relative kinetic energy through tidal oscillations to remain bound (Fabian et al. 1975).

Unfortunately, the binary population in dense cluster cores is difficult to study observationally. Because of crowding in the core of 47 Tucanae, radial velocity searches are confined to the few bright giants (e.g., Gebhardt et al. 1995). Numerical simulations predict that many binaries in dense cores will contain at least one heavy, dark, white dwarf or neutron star. The numbers and kinds of exotic stellar objects in the core are a useful probe of the binary population because they are produced by stellar collisions that occur during gravitational encounters involving binaries, but these exotic objects are rare, making small number statistics a problem. Another possible probe of the binary population is suggested by the presence of high-velocity stars.

For decades, 47 Tuc has been intensively studied, both photometrically and kinematically. It is one of the closest ($D \simeq 4.6$ kpc; $(m-M)_V = 13.3$) and most massive ($M \simeq 10^6 M_\odot$) globular clusters in our Galaxy. Its high concentration $c = \log\,(r_t/r_c) \simeq 2.1$ suggests that it is dynamically evolved and, although well fitted by multi-mass anisotropic King-Michie models, that it is evolving through a sequence of quasi-equilibrium states towards its ultimate fate of core collapse (Meylan 1989). There is an increasing amount of direct and indirect observational evidence for stellar encounters and collisions in the core of 47 Tuc. The previously-known single X-ray source has been resolved by ROSAT into at least five components (Hasinger et al. 1994). Besides the high-velocity stars, other evidence for binaries in 47 Tuc comes from the discovery of 8 eclipsing binaries (Edmonds et al. 1996), of eleven millisecond pulsars (Robinson et al. 1995), and the very high density of centrally-clustered BSs observed by HST (Paresce et al. 1991). With one of the richest populations of collision products, 47 Tuc is clearly an ideal laboratory for confronting theoretical predictions with observations.

High-Velocity Stars from Radial Velocities. Radial velocities measured for 550 stars within two core radii of the center of 47 Tuc (Gebhardt et al. 1995) have recently increased the sample of high-velocity stars from 2 to 8. With velocities more than 32 km s^{-1} from the cluster mean (about three times the core radial velocity dispersion of $\simeq$ 10 km s^{-1}), these stars are moving at close to the cluster escape velocity. The original high frequency of rapidly moving stars (2 out of 50) stimulated theoretical studies of their expected numbers (Phinney & Sigurdsson 1991, Sigurdsson & Phinney 1995). While the larger sample yields a lower frequency, these studies have demonstrated that the number of high-velocity stars is a powerful tool for probing conditions in the core and testing the dynamical models. However, only proper motions can provide the larger samples (N $\geq$ 5,000) needed to exploit this tool.

High-Velocity Stars from CCD Astrometry with HST. Experience obtained with the techniques of CCD astrometry on the ground has shown that CCDs make almost ideal devices for faint, relative astrometry. Astrometric positions with a precision of 5 marcsec per epoch are now being routinely measured from the ground (Monet et al. 1992; Tinney 1993). In the absence of atmospheric seeing (the fundamental limitation for ground-based CCD astrometry), there is no reason why a suitably-designed astrometric program with WFPC2 cannot achieve precisions limited only by photon statistics. This means that measuring proper motions with 1-σ uncertainties of 0.2 marcsec year^{-1} should be achievable. This is a conservative approach, aiming first for the high-velocity stars, but estimates show that it should be possible to measure proper motions for most of the stars in the frames.

Consequently, the Planetary Camera is used to obtain a complete census of high-velocity stars inside half a core radius of 47 Tuc, down to a limit of $U \simeq 22$. The F300W filter has been chosen in order to reduce the range of brightness presented by the stars in the field, thus preventing the saturation of pixels by red giants and allowing more or less all stages of stellar evolution in the cluster to be seen. This census includes stars on the red giant branch (RGB), subgiant branch, red horizontal branch (RHB), upper and lower main sequence (down to $\sim$ K3V), and white dwarfs.

CCD astrometry from the ground has shown that three fundamental issues must be addressed in designing the observing program for high-precision relative astrometry. Once all three issues are solved, obtaining positions accurate to better than 1% of a pixel can easily be achieved, provided that the required photon counting limits ($\approx 10,000$ photons) can be reached.

• The first requirement is that the astrometry must be carried out in a *differential* mode. Traditional photographic astrometry has always been limited by the need for a detailed knowledge of the field distortions (if the field is not flat). If it cannot be guaranteed *a priori* where the astrometric stars will be placed in that field, it is essential to be able to flatten the field to high precision. On the ground, CCDs obviate this problem by allowing one to always place the target objects back on the same place on the CCD to within a few pixels, at which point the field distortions cancel out and high precision can be reached. This strategy is adopted: all HST frames are made with an identical roll angle and by placing the stars at the same position on the PC1.

• The second requirement is that the PSF must be well-sampled by the CCD. It has been realized in recent years that CCDs do not have uniform sensitivity across a pixel (e.g., Jorden et al. 1995). In the worst case, the sensitivity can vary by as much as 5% across a single pixel. This obviously has serious implications for a program which aims to measure positions to an accuracy of better than 1% of a pixel. Because HST images are undersampled even by the PC1, the program of 15 exposures per epoch incorporates a dithering pattern of sub-pixel shifts with different values in order to improve the spatial resolution. This enables to regain the oversampling of the PC1 images required for high-precision astrometry.

• The last requirement is to have a distant reference frame. Fortunately, this is a simple problem to solve in the case of 47 Tuc. A background of stars in the SMC provides a rich reference frame of stars which can be easily identified on color-magnitude diagrams (see Hesser et al. 1987). These numerous F-type stars should appear at $U = 21.0$-22.0 mag. The photometry published by Guhathakurta et al. (1992) is not deep enough to allow current identification of these SMC stars or any WDs in 47 Tuc.

Proper Motions with HST. Based on tests with WFPC1 frames, a realistic goal for the precision of the final proper motions is a 1-σ limit of 0.23 marcsec year^{-1}, which would allow a 5-σ detection of all stars with tangential velocities $\geq 22\ \mathrm{km\,s^{-1}}$ in the WFPC2 frames (PC field).

The fifteen WFPC2 frames constituting the first epoch observations for this project were obtained by HST in October 1995 (Cycle 5), with the PC field centered on the core of 47 Tuc. Each exposure lasted 350 seconds and was taken through the F300W filter, which has a passband similar to U. The final combined image has total exposure time of 1.46 hour and reaches F300W = 25 mag for 3-σ detections.

Scientific Goals. Given the strong limits imposed by seeing on the ground-based acquisition of radial velocities in the core of 47 Tuc, HST is the only telescope able to provide, in the very core of the cluster, a kinematic sample consisting of thousands of stars. The present census of proper motions will provide unique constraints on:

- relaxation processes (possible anisotropy of the velocity dispersion),
- collision and ejection rates (meaningful fraction of high-velocity stars from a large sample),
- the form of the velocity distribution in the core as a function of the stellar mass, and
- mass segregation.

The first-epoch observations of the project already provide a number of very interesting astrophysical by-products, namely:

- deep luminosity functions in the UV;
- deep color–magnitude diagrams ;
- variable stars (pulsating stars and binaries); and
- the identification of possible UV counterparts for known X-ray sources (these LMXRBs or CVs being accreting binary systems with neutron star or white dwarf primaries, respectively).

In addition, the fifteen images have revealed one star that increased in brightness by more than two magnitudes in less than an hour. By the end of the observations, this star was the brightest object in the core of the cluster in the bandpass used. Possible causes for such a brightening are (i) a very large flare on a magnetically active star (RS CVn), (ii) an increase in the brightness due to an accretion disk instability in a cataclysmic variable (CV) or (iii) in a soft X-ray transient (SXT), or (iv) a novae. There are arguments against every one of these possibilities (Minniti et al. 1996).

A preliminary report of the first-epoch HST observations, obtained in Cycle 5, has already been published (Meylan et al. 1996).

4.2 NGC 6397: Very-Low-Mass Stars Content

Fig. 4, from Cool et al. (1996), displays a $V_{555} - I_{814}$ vs. V_{555} color-magnitude diagram of the galactic globular cluster NGC 6397. The most prominent feature of this impressive diagram is the cluster main sequence which winds in a very narrow band from the upper left, at $V_{555} \sim 16$ mag, just below the turnoff, to the lower right, at $V_{555} \sim 26$ mag, spanning nearly 10 magnitudes in the V_{555} filter.

The majority of points to the left and right of the main sequence represent background halo and foreground disk stars, respectively. Their numbers are in reasonable agreement with those predicted in this cluster's direction by Ratnatunga & Bahcall (1985) based on the Bahcall & Soneira (1980) model of our Galaxy. The fainter end of the main sequence, at $V_{555} \sim 26$ mag, i.e., $M_V \sim 14$ mag, corresponds to stellar masses of about 0.10 $M_\odot$, which is close to the hydrogen-burning limit.

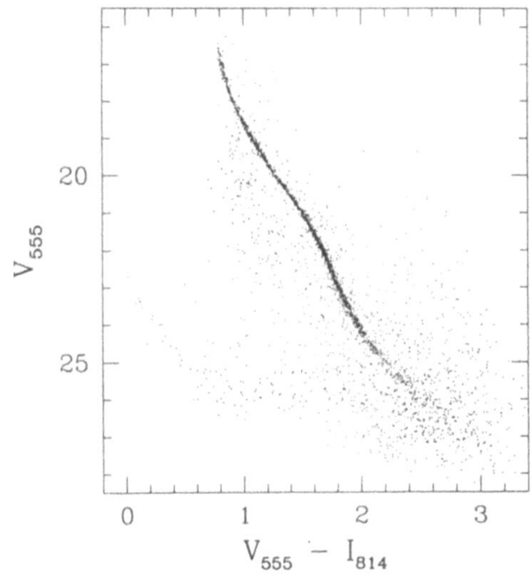

Fig. 4. HST colour-magnitude diagram of NGC 6397, in the filters I_{814} and V_{555}, from Cool et al. (1996).

Very Few Low-Mass Stars. The luminosity functions of two other clusters, M30 and M15, are very similar, over a range of more than 6 magnitudes, while NGC 6397 is markedly deficient in faint stars (King 1996, Cool et al. 1996). This implies that the mass functions of M30 and M15 are very similar in the range 0.12 $M_\odot \leq M \leq 0.8\ M_\odot$. This may be the consequence of both very similar initial conditions and very similar evolution, a scenario which is less contrived than assuming that the present similarity has been created by evolution from different initial conditions. But, since all three clusters are very similar in metallicity — [Fe/H] $\sim$ -2.0 — and morphology or dynamical status (all are collapsed), what can be the reason why there are so many fewer low-mass stars in NGC 6397 ? The reason is that these clusters have very different orbits around the galactic centre. Although the traditional theories of tidal shocks have never been well enough quantified and are known to be unreliable (Weinberg 1994a,b,c), the effect of tidal shocks is certainly present in the parameters governing the dynamical evolution of globular clusters. NGC 6397, much closer to the galactic plane and galactic centre than M15, is clearly hit much harder and much more often by passages through the galactic plane (King 1996 and Dauphole et al. 1996).

NGC 6397, being the globular cluster with the smallest apparent distance modulus, viz. $(m - M)_V = 12.3$, offers the best opportunity of studying the termination of the main sequence, i.e., of observing the mass function close to the hydrogen-burning limit, commonly believed to be at about 0.08 $M_\odot$ for stars of solar abundance. At lower metallicity, this mass limit is predicted to be slightly higher (D'Antona & Mazzitelli 1996), a good point to add to the nearby character of NGC 6397, since its [Fe/H] = −1.9.

Unfortunately, it is impossible to determine, from photometric data only, whether the stars in the lower part of Fig. 4 are cluster members.

Proper Motions with HST. King et al. (1996 pers. comm.) propose to use proper motions deduced from HST images to separate faint cluster stars from faint field stars. The general proper motion of the cluster, 15.5 marcsec year^{-1} (Cudworth & Hanson 1993), is large enough to allow such a sorting process over a reasonably short time base. From their observations of NGC 6397 taken in August and October 1994 till the middle of Cycle 6, the cluster will have moved, relative to distant background stars, by nearly 0.4 pixel of the Wide Field Camera (WFPC2), i.e., about 40 marcsec.

Tests were performed on two different Wide Field Camera images (WFPC2) taken in August and October 1994 at different orientations. By using a method establishing for each individual star a network of 10 reference stars, they compare two samples, one made of stars on the main sequence and the other made of stars whose colors suggest they are not cluster members. The median displacement of the non-member stars is noticeably different from the cluster stars. The shift is consistent with that expected from the cluster motion and strongly suggest that King et al. (1996 pers. comm.) have detected the proper motion of the cluster from a pair of WFPC2 images separated by just two months.

The expected time baseline of nearly 3 years, between their 1994 observations and the ones in Cycle 6, will allow a clear detection of proper motions. This will offer the capability to discriminate, in the lower parts of the color-magnitude diagram in Fig. 4, the field stars from the cluster stars, and to get closer than anyone ever has to the hydrogen-burning limit at the bottom of the main sequence.

5 Proper Motions with the VLTI

The interferometric mode of the ESO-VLT is conceived as a pure imaging array (about VLTI, see Beckers 1991, von der Lühe et al. 1994). The precision required on the separation between the points of intersection of azimuth and altitude increases with the separation between the sources whose relative positions are to be measured. Since the 4 large 8-m units of the VLT cannot be optimized for astrometric requirements, wide-angle astrometry will never be possible with the VLTI.

5.1 Narrow Angle Astrometry

However, the above requirements on precision are relaxed if the astrometry is performed only over narrow angles ($\lesssim 15''$). Shao & Colavita (1992) described the scientific potential of narrow angle astrometry (NAA). Their considerations led von der Lühe et al. 1995) into analyzing the present VLT project and finding that the already existing VLTI capabilities offer many advantages with respect to NAA.

Beckers (1993) estimated $K = 20$ as limiting magnitude for observing conditions which would apply to NAA, viz., presence of a nearby bright reference star which is used for image tracking and fringe co-phasing. Even if one allows for the reduced sensitivity of a dedicated astrometric beam-combining instrument, which will be required for NAA with the VLTI, sensitivity is not going to be a severe limitation.

Shao & Colavita (1992) estimate an ultimate accuracy of about 10 μarcsec in one hour of integration time, based on the use of a 17.5 K-mag (20.5 V-mag) reference star within $15''$ of a target star that is 13 K-mag or brighter, using a 200-m baseline interferometer with 1.5-m apertures. The astrometric error, based on measurements of turbulence and wind speed profiles at Paranal, expected for a 30-minute integration with the VLTI, is about 10 μarcsec for a separation of $10''$. Quirrenbach (1995) estimated that, with NAA at the VLTI, planets could be detected through the reflex motion of the central star at considerable distances, viz., Jupiter masses out to about 1 kpc.

This would allow the observation of proper motions of stars in nearby globular clusters like, e.g., NGC 6397, 47 Tucanae, ω Centauri, and M15. It is worth mentioning that such proper motions are not absolute but relative ones.

Table 1. Velocity dispersions - Proper motions

Cluster	Distance [kpc]	Core Vel. Disp. [$\mathrm{km\,s^{-1}}$]	Proper Motion [μarcsec year^{-1}]
NGC 6397	2.2	4.5	430
ω Cen	4.9	16	690
47 Tuc	4.6	11.5	530
M15	10.5	12	240

5.2 NAA in Globular Clusters

When observing, in NAA mode, two stars in the field, their (α,δ) motions are due essentially to three effects: (i) their relative trigonometric parallaxes (consisting of two homologous ellipses), (ii) the possible binary nature of each of them, and (iii) their relative proper motions. In the framework of globular clusters, the effects due to the trigonometric parallax cancel out, and only orbital and proper motion effects remain. Consequently, NAA would allow, e.g., the precise investigation of binary orbits, of the fraction of binaries among giants and main-sequence stars as well as the study of the variation of the velocity dispersion, with possible anisotropy, as a function of the distance to the center of the cluster.

Table 1 gives for the four already mentioned galactic clusters their distances in kpc, their core velocity dispersions in $\mathrm{km\,s^{-1}}$, and the corresponding proper motions in μarcsec year^{-1}. The uncertainties on these proper motion measurements have to be about an order of magnitude smaller than the quantity observed. This is of the same order of magnitude, or larger than, the estimated NAA astrometric errors of about 10 μarcsec.

The velocity dispersion in globular clusters could be investigated in regions well outside of the cluster core, i.e., where the anisotropy may be expected. The velocity dispersion could be measured in rather remote globular clusters, e.g., in the Large Magellanic Cloud, where $\sigma_p = 10\ \mathrm{km\,s^{-1}}$ would translate into a proper motion of 40 μarcsec year^{-1}. It is also conceivable to study the possible velocity dispersion cusps in core-collapsed globular clusters like M15 by measuring the proper motions of numerous stars in the cores of such clusters.

6 Conclusion

Most of the information related to proper motions of stars in globular clusters is still locked in the sky. It is up to us, with the next generations of instruments, to bring it down to the earth. Classical astrometry with CCDs and narrow angle astrometry with instruments such as the VLTI should be among the most promising techniques.

7 Acknowledgements

The author thanks his collaborators, David Merritt and Michel Mayor, for allowing figures to be quoted before publication.

References

Bahcall J.N., Soneira R.M., 1980, ApJS, 47, 357
Beckers J.M., 1991, J. Optics, 22, 73
Beckers J.M., 1993, in VLT Interferometer: Science Objectives and Requirements, VLT Program Review 1993
Cool A.M., Piotto G., King I.R., 1996, ApJ, 468, 655
Côté P., Welch D.L., Fischer P., Gebhardt K., 1995, ApJ, 454, 788
Cudworth K.M., Hanson R.B., 1993, AJ, 105, 168
Cudworth K.M., Monet D.G., 1979, AJ, 84, 774
Da Costa G.S., Freeman K.C., 1976, ApJ, 206, 128
D'Antona F., Mazzitelli I., 1996, ApJ, 456, 329

Dauphole B., Colin J., 1995, A&A, 313, 119

Dauphole B., Geffert M., Colin J., et al., 1996, A&A, 313, 119

Dejonghe H., 1987, MNRAS, 224, 13

Dejonghe H., 1989, ApJ, 343, 113

Dejonghe H., Merritt D., 1992, ApJ, 391, 531

Edmonds P.D., Gilliland R.L., Guhathakurta P., et al., 1996, ApJ, 468, 241

Fabian A.C., Pringle J.E., Rees M.J., 1975, MNRAS, 172, 15P

Feast M.W., Thackeray A.D., Wesselink A.J., 1961, MNRAS, 122, 433

Gebhardt K., Fischer P., 1995, AJ, 109, 209

Gebhardt K., Pryor C., Williams T.B., Hesser J.E., 1995, AJ, 110, 1699

Gebhardt K., Pryor C., Williams T.B., Hesser J.E., 1997, in preparation

Griffin R.F., 1967, ApJ, 148, 465

Griffin R.F., 1974, ApJ, 191, 545

Grillmair C.J., Freeman K.C., Irwin M., Quinn P.J., 1995, AJ, 109, 2553

Guhathakurta P., Yanny B., Schneider D.P., Bahcall J.N. 1992, AJ, 104, 1790

Gunn J.E., Griffin R.F., 1979, AJ, 84, 752

Hasinger G.R., Johnston H.M., Verbunt F., 1994, A&A, 288, 466

Heggie D.C., 1975, MNRAS, 173, 729

Hesser J.E., Harris W.E., VandenBerg D.A., et al., 1987, PASP, 99, 739

Hut, P., McMillan, S.L.W., Goodman, J., et al., 1992, PASP, 104, 981

Jorden P.R., Deltorn J.-M., Oates A.P., 1995, in Instrumentation in Astronomy, eds., Crawford D.L. & Craine E.R., SPIE proc., 2198, 836

King I.R., 1966, AJ, 71, 64

King I.R., 1996, in Dynamical Evolution of Star Clusters: Confrontation of Theory and Observations, IAU Symp. 174, eds. Hut P. & Makino J. (Dordrecht: Kluwer), p. 29

Larson R.B., 1996, in Formation of the Galactic Halo...Inside and Out, ASP Conference Series Vol. 92, eds. Morrison H. & Sarajedini A., (San Francisco: ASP), p. 241

Leonard P.J.T., Richer H.B., Fahlman G.G., 1992, AJ, 104, 2104

Lupton R.H., Gunn J.E., Griffin R.F., 1987, AJ, 93, 1114

Mayor M., Meylan G., Udry S., et al., 1997, AJ, submitted

Merritt D., 1993a, ApJ, 413, 79

Merritt D., 1993b, ApJ, 409, 75

Merritt D., 1997, in press

Merritt D., Meylan G., Mayor M., 1997, AJ, submitted

Merritt D., Tremblay B., 1994, AJ, 108, 514

Meylan G., 1989, A&A, 214, 106

Meylan G., Dubath P., Mayor M., 1991, ApJ, 383, 587

Meylan G., Heggie D.C., 1996, A&AR, 8, 1

Meylan G., Mayor M., Duquennoy A., Dubath P., 1995, A&A, 303, 761

Meylan G., Minniti D., Pryor C., Tinney C., Phinney E.S., Sams B., 1996, in the ESO/STScI workshop on Science with the Hubble Space Telescope - II, eds. Benvenuti P., Macchetto F.D., & Schreier E.J., (Baltimore: STScI), p. 316

Milone E.F., Mermilliod J.-C., eds., 1996, The Origins, Evolution, and Destinies of Binary Stars in Clusters, ASP Conference Series Vol. 90, (San Francisco: ASP)

Minniti D., Meylan G., Pryor C., Phinney E.S., Sams B., Tinney C., 1996, ApJ, in press

Monet D.G., Dahn C.C., Vrba F.J., et al., 1992, AJ, 103, 638

Murray C.A., Jones D.H.P., Candy M.P., 1965, Royal Obs. Bull., 100, E81

Paresce F., Shara M.M., Meylan G., et al., 1991, Nature, 352, 297

Quirrenbach A., 1995, in Science with the VLT, eds., Walsh J.R. & Danziger I.J., (Berlin: Springer), p. 33
Ratnatunga K.U., Bahcall J.N., 1985, ApJS, 59, 63
Rees R.F., 1993, AJ, 106, 1524
Rees R.F., 1996, in Formation of the Galactic Halo...Inside and Out, ASP Conference Series Vol. 92, eds. Morrison H. & Sarajedini A., (San Francisco: ASP), p. 289
Reijns R., Le Poole R., de Zeeuw T., Seitzer P., Freeman K.C., 1993, in Structure and Dynamics of Globular Clusters, ASP Conference Series Vol. 50, eds. Djorgovski S.G. & Meylan G., (San Francisco: ASP), p. 79
Phinney E.S., Sigurdsson S., 1991, Nature, 349, 220
Robinson C., Lyne A.G., Manchester R.N., et al., 1995, MNRAS, 274, 547
Scholz R.D., Odenkirchen M., Hirte S., et al., 1996, MNRAS, 278, 251
Shao M., Colavita M.M., 1992, A&A, 262, 353
Sigurdsson S., Phinney E.S., 1995, ApJS, 99, 609
Tinney C.G., 1993, AJ, 105, 1169
von der Lühe O., Ferrand D., Koehler B., Zhu N., Reinheimer Th., 1994, in Amplitude and Intensity Spatial Interferometry II, ed., Breckinridge J.B., SPIE proc., 2200, 168
von der Lühe O., Quirrenbach A., Koehler B., 1995, in Science with the VLT, eds., Walsh J.R. & Danziger I.J., (Berlin: Springer), p. 445
Weinberg M.D., 1994a, AJ, 108, 1398
Weinberg M.D., 1994b, AJ, 108, 1403
Weinberg M.D., 1994c, AJ, 108, 1414
Wybo M., Dejonghe H., 1995, A&A, 295, 347
Wybo M., Dejonghe H., 1996, A&A, 312, 649

Part 9

STELLAR DIAMETERS AND SURFACE STRUCTURE

Indirect Imaging of Stellar Surface Structure

A.P. Hatzes

McDonald Observatory, University of Texas, Austin, TX.

Abstract. The VLT interferometer may one day directly image surface structure on a number of stars. On the other hand, indirect imaging techniques (Doppler tomography) have been studying the surface of wide variety of rapidly rotating stars for a number of years. In this review the basic principles of Doppler tomography are discussed and recent results for active late-type stars are presented. Contributions that the VLTI can make to the study of stellar surface structure are also discussed and it is argued that single K giant stars may be ideal candidates for direct imaging with the VLTI.

1 Introduction

The Sun, a typical star, shows a wide variety of stellar structures: sunspots, faculae, plage, prominences, coronal holes, convection cells, etc. The study of such structures on the Sun is relatively easy due to its resolved disk. Optical interferometry may enable astronomers to resolve the stellar disk of a large number of stars and consequently to image *directly* stellar surface structure. There is considerable evidence for surface structure on a variety of stars and much of the details of this structure have come through *indirect* imaging techniques such as Doppler tomography. In this review the basic principles behind Doppler tomography are briefly described and typical results are presented for active late-type stars. The final part of this contribution addresses areas in which the VLT interferometer (VLTI) can make significant contributions to investigations of stellar surface structure.

2 Doppler Tomography

For rapidly rotating stars there is a mapping between wavelength location in the stellar absorption line profile and a corresponding location on the stellar disk – all regions of the star lying along a constant radial velocity chord are mapped into the same location in the spectral line profile. If a cool starspot is present on the surface then the reduction of continuum flux due to the spot results in a pseudo-emission distortion appearing in the line profile (see Vogt & Penrod 1983a). At a given phase the distortion pattern represents a one dimensional projection of the stellar surface structure and as the star rotates a different projection of the spot feature is presented. Given a sufficient number of these projections the two dimensional distribution of the surface structure can be derived.

Figure 1 demonstrates how Doppler imaging derives latitude, longitude, and shape information of surface structure from a series of spectral line profiles. It shows a star with 3 different cool spot distributions: a circular spot near the pole (top), a circular spot near the equator (middle), and an 'L'-shaped spot at low latitude. The dotted line represents the spectral line profile from an unspotted star rotating at 80 $\mathrm{km\,s^{-1}}$, while the line represents the spectral line profile from the spotted star.

Longitude information about the spot feature is trivially determined by the rotation phase when the distortion is at line center. Latitude information is gleaned from a sequence of profiles by several effects: 1) The maximum velocity displacement of the distortion feature in the line profile is largest for low-latitude spots. These appear and disappear in the far wings of the spectral line whereas high-latitude features always stay near line center. 2) The fraction of a rotation phase that a spot remains in view is an indication of its latitude; circumpolar features are always in view, low-latitude features can spend half a rotation phase out-of-view. 3) The speed with which a distortion propagates through a line profile also yields latitude information with low-latitude features traveling through the line profile faster than high-latitude features. Finally, shape information of the spots comes from differences in the shape of the distortions. Note that the 'L'-shaped spot produces a similar distortion pattern in the wings of the line as does a circular spot, but it is a more sharply peaked distortion at line center indicating that a feature lies primarily along a longitude line.

What star can be Doppler imaged? First it must have appreciable rotation. To first order the number of Doppler image "resolution" elements is given by $v\,sin\,i/W$ where $v \sin i$ is the projected rotational velocity and W is the full width of the intrinsic spectral line profile from a star with no rotation. Although very rapidly rotating stars have more Doppler resolution elements these stars are not practical candidates due to line blending and the fact that the spectral lines become more shallow at higher rotational velocities which causes the spectral distortions to be lost in the noise. Ideally, one would like a star to have a projected rotational velocity of 30–80 $\mathrm{km\,s^{-1}}$. For cool stars the intrinsic line width W is about 10 $\mathrm{km\,s^{-1}}$ which yields about 2–8 resolution elements for an image.

A star must also have an intermediate inclination. If the inclination is too high there is complete ambiguity as to which hemisphere a spot feature lies (the Doppler image of the surface structure on such of star will be a mirror image about the equator). If the inclination is too low the star's projected rotational velocity will be small. An ideal Doppler imaging candidate should have an inclination of 20-70°.

In recent years Doppler imaging has developed into a mature field and there are a number of mathematical techniques that have been developed for inverting a time sequence of spectral line profiles. These include the maximum entropy method (Vogt, Penrod, & Hatzes 1987), Tikhonov regularization (Goncharsky et al. 1982), and a CLEAN-like algorithm (Kürster 1993). In instances where the different techniques were applied to the same data set consistent results were obtained (Strassmeier et al. 1991; Kürster et al. 1992). Doppler tomography has been used to map cool starspots on active late-type stars, abundance features on Ap stars (see Hatzes 1996 for a recent review), and to resolve high-order

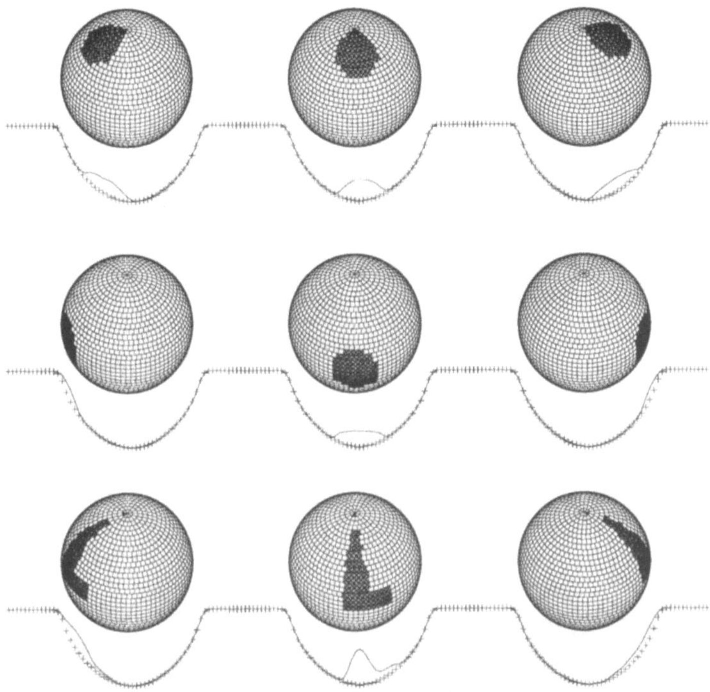

Fig. 1. Spectral distortions due to a cool circular spot at high latitude (top), low-latitude (center), and an L-shaped spot (bottom). Each feature is shown at the same rotation phase. The crosses represent the spectral line profile from an unspotted star and the line for a star with the spot.

pulsation modes in non-radially pulsating stars (e.g. Vogt & Penrod 1983b). Space only permits a brief description of results for active late-type stars.

3 Doppler Imaging Results for Active Late-Type Stars

The key ingredients of stellar magnetic activity are rotation and convection along with a dynamo mechanism that converts the mechanical energy of these two quantities into magnetic energy. There are several classes of stars which possess high rates of rotation and deep convection zones. Consequently, these stars also show high levels of stellar activity including large starspots which are believed to trace the surface magnetic flux. Their large rotation rates also makes them amenable to Doppler imaging studies.

RS CVn- and FK Comae-Type Stars. The RS CVn-type and FK Comae-type stars are a class of rapidly rotating late-type giant stars. The high rotation in the RS CVn stars are due to tidal spin-up by a binary companion. FK Comae stars are single stars whose high rotation rates result from the fact that they are coalesced binary stars. These stars exhibit stellar activity in the form of spectacular flares, chromospheric emission, enormous cool starspots, and plage-like regions, but on a level several orders of magnitudes greater than is found on the Sun. For instance, sunspots rarely cover more than 3% of the solar surface, but on RS CVn and FK Comae stars they can cover 10–40% of the stellar surface.

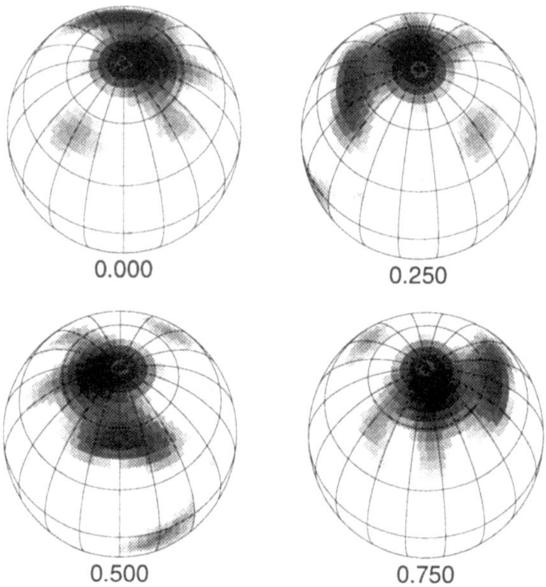

Fig. 2. The cool spot distribution on the RS CVn-type star HR 1099 in 1986 derived via Doppler tomography

Many RS CVn and FK Comae stars are relatively bright, so they have been the primary targets of many Doppler imaging studies. Images have been derived for numerous RS CVn stars (e.g. Vogt & Penrod 1983a; Vogt & Hatzes 1991; Strassmeier et al. 1991; Strassmeier 1994; Hatzes 1995a) and a few FK Comae stars (Kürster et al. 1992). The spot morphology of many of these stars was characterized by a large polar spot with several appendages extending to lower latitudes and several isolated low-latitude spots. Figure 2 shows the spot distribution of HR 1099 in 1986 and is a typical spot morphology for these types of stars.

T Tauri and Weak T Tauri Stars. Magnetic activity in the form of starspots and chromospheric emission is also seen in young stars such as classic T Tauri stars (CTTS) and weak T Tauri (WTTS) stars. (CTTS have strong Hα emission and an infrared excess associated with the presence of an accretion disk;

WTTS have weaker emission and no accretion disk.) In the case of these pre-main sequence stars the rapid rotation stems not from tidal spin-up, but rather their young age (residual angular momentum from the star's collapse from the proto-cloud).

Fewer Doppler images of CTTS and WTTS stars have been derived primarily due to the relative faintness of the stars. Recently, images of these stars have begun to appear in the literature and these also show evidence for high-latitude spot activity. The first Doppler image of a WTTS (V410 Tau) was presented by Strassmeier et al. (1994) and it showed a large polar spot whose center was displaced from the rotation pole of the star. This was in contrast to the RS CVn stars which showed centered polar spots. This spot distribution was also confirmed by Joncour et al. (1994a) and Hatzes (1995b). Figure 3 shows the Doppler image for the spot distribution on the WTTS HDE 283572 in 1994 derived from spectral data taken at McDonald Observatory. It is similar to the image presented by Joncour et al. (1994b) which also showed a large, slightly decentered polar spot. Polar spots have also been found on the TTS star Sz 68 (Johns-Krull & Hatzes 1996) and on two zero-age main sequence stars in the Pleiades (Stout-Batalha & Vogt 1996). Such high-latitude activity seems to be a common feature among rapidly rotating stars in spite of their evolutionary status.

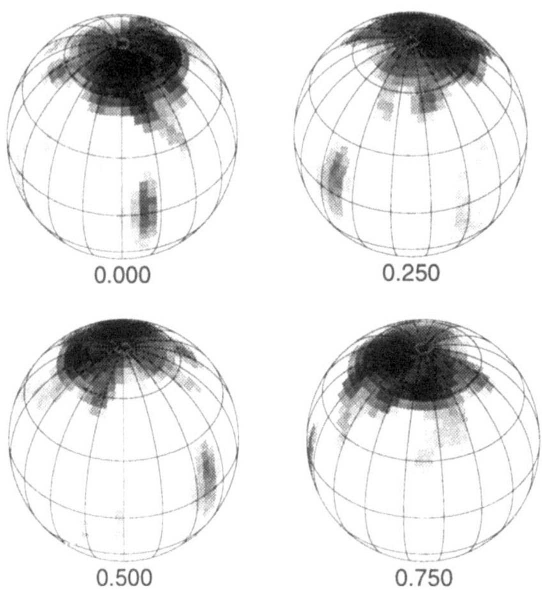

Fig. 3. The spot distribution on the WTTS HDE 283572 in 1994

Why Doppler Image a Star? A single Doppler image is often not of great use, scientifically. What then, are the scientific objectives of Doppler imaging? First, it is to study spot morphology and evolution as it occurs in a single star over a long time base by studying a time series of Doppler images. After all, the ability of solar physicists to resolve the stellar disk and to study in detail sunspot behavior led to some of the most fundamental discoveries of solar activity: differential rotation, the butterfly diagram of spot migration patterns, the magnetic nature of sunspots, just to name a few. These discoveries have led to the use of the solar paradigm for interpreting virtually all of stellar activity. Doppler imaging offers the opportunity to study spot behavior in other stars in almost the same exquisite detail as on the Sun provided one is given a sufficient number of images with good time resolution.

Vogt et al. (1996; see Vogt & Hatzes 1996 for a conference version of the paper) in a recent study presented Doppler images of the RS CVn star HR 1099 that spanned nearly 12 years. By tracking the motion of spot features they were able to establish 1) that the polar spot is long-lived with an area that cyclicly varies with a period of 3 years, 2) that the differential rotation rate of HR 1099 is a factor of 50 less than on the Sun and in the opposite sense, i.e. the poles rotate faster than the equator, and 3) that spots emerge near the magnetic equator and often migrate poleward in a clockwise spiraling pattern (slower than the rotation period) finally merging with the polar spot flux. All these point to a spot morphology vastly different than the Sun. Clearly, the solar paradigm may be inappropriate for other, more active stars.

Another goal of Doppler imaging is to compare the spot morphology as it appears in different classes of active stars. How does spot (and thus magnetic) activity differ between various stars and what stellar parameters influence these differences? Even though polar spots may seem to be an ubiquitous phenomenon among rapidly rotating stars, there do seem to be some morphological differences among different stars even in the same class. Figure 4 shows a gallery of Doppler images for various stars in a flattened polar projection as viewed from the rotation pole of the star. Circles are drawn every 30° down to a latitude of −30° and with the heavy circle representing the stellar equator. The top panel of stars shows 4 young stars: The TTS Sz 68, the WTTS star V410 Tau, the pre-main sequence (but post-T Tauri) stars V824 Ara A and AB Dor (see Kürster 1994). The lower panel shows four evolved stars, two RS CVn stars and two FK Comae stars.

There appear to be distinct differences in spot morphology between the various stars. The TTS Sz 68 shows a centered polar spot, whereas the V410 Tau and V824 Ara have a decentered polar spot. AB Dor on the other hand, has a spot distribution dominated by an active latitude band, similar to the solar case. The evolved stars also show differences. One RS CVn star (HR 1099) is dominated by a polar spot, whereas σ Gem has high latitude spot activity but in a band as opposed to a polar cap. The FK Comae star HD 199178 has a strong polar cap, yet YY Men has a polar cap *and* an active latitude band. What causes these differences? Age? Rotation period? Depth of convection zone? Such understanding

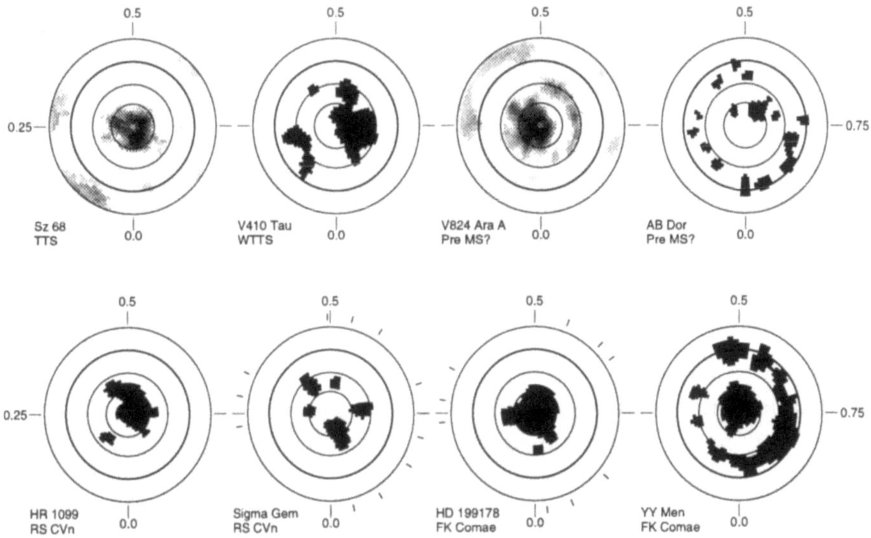

Fig. 4. Doppler images of various stars. (Top row) Young stars: Sz 68 (TTS), V410 Tau (WTTS), V824 Ara A, and AB Dor (both pre-main sequence). (Bottom) Evolved stars: HR 1099, σ Gem (both RS CVn stars), HD 199178, and YY Men (both FK Comae stars).

can only come by imaging a wide variety of active late-type stars.

4 Direct Imaging via Interferometry

4.1 Interferometry vs. Doppler Imaging

What contributions can the direct imaging capabilities of the VLTI make to our understanding of stellar surface structure on stars typically used for Doppler imaging studies? Although many active late-type giants have large radii, their distances are so large as to render them marginally resolvable with the VLTI; many of these stars have angular diameters much less than 0.5 mas. There are a few active stars, however, that can be resolved with the VLTI. Table 4.1 lists the active RS CVn-type stars with the largest angular diameters (1–8 resolution elements) that are accessible with the VLTI.

Specific areas for which the VLTI can make significant contributions:

- Confirm the existence of polar spots.
  Although polar spots have appeared in many Doppler images derived with a variety of techniques, these features are still controversial. The final confirmation of such features will undoubtedly have to await direct imaging.

Table 1. Candidate Late-type stars for the VLTI

Star	Diameter (mas)
ζ And	4.0
5 Cet	2.8
IM Peg	$\geq$2.3
33 Psc	1.7
BD Cet	$\geq$ 1.3
AS Cap	$\geq$1.3
AT Cap	$\geq$1.0

- Relationship of chromospheric structure to photospheric structure.
  Few attempts have been made to relate the photospheric tracers of activity (spots) to the chromospheric tracers (plage) due to the complicated shape of chromospheric lines (Ca II H and K and Hα), but also to the poorly understood line formation of chromospheric lines. Broad Hα emission profiles have been seen in many active stars (e.g. Hatzes 1995a) and one interpretation is that they arise from a thick ($\sim 2R_*$) co-rotating shell. Direct imaging of chromospheric structure can better probe the relationship between chromospheric and photospheric structure. Also by comparing a photospheric image to an Hα image one could confirm the "thick shell" nature of the Hα emission.
- Are hot spots really present?
  Donati et al. (1992) as well as Vogt et al. (1996) found evidence for hot spots on the HR 1099. Without accurate simultaneous photometry to constrain the image, hot spots may be an artifact of the reconstruction. Direct imaging can confirm the presence of such features.

4.2 K Giants – The Ideal VLTI Imaging Candidates

It is clear that the VLTI cannot compete with Doppler imaging in studying surface structure on *rapidly rotating* stars; Doppler imaging can sample a larger number of these stars (since the technique is independent of the stellar distance) and with considerably better imaging "resolution." The greatest scientific return for the VLTI, as far as imaging of surface structure goes, is to image a complementary sample of active stars, in particular slowly rotating stars that are inaccessible to the Doppler imaging technique.

Single K giants (i.e. non-RS CVn systems) may be such candidates. They have large angular diameters and are the largest sample of stars that can be directly imaged with the VLTI. There is also evidence that these stars have surface structure. Choi et al. (1995) found Ca II variability with periods of 60–160 days in 10 of 12 K giant stars. This was interpreted as rotational modulation by stellar active regions.

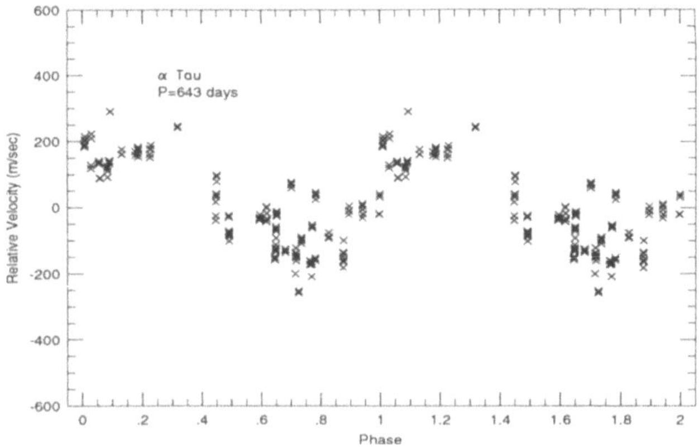

Fig. 5. The RV variations of α Tau phased to a period of 643 days

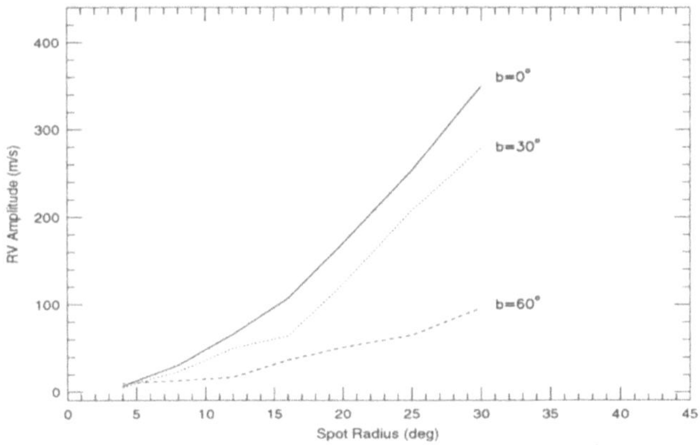

Fig. 6. The expected amplitude of RV variations due to a cool spot ($\Delta T = 1200$ K) on the stellar surface as a function of spot radius. The 3 curves are for a spot at latitude, $b = 0°$, $30°$, and $60°$.

Many K giants are also radial velocity (RV) variables. Hatzes & Cochran (1993) found RV variations in three K giants with periods of 233–643 days and amplitudes of $\sim$100 m s^{-1} . Figure 5 shows the RV measurements of α Tau phased to a period of 643 days. There are three interpretations as to the cause of this variability: nonradial pulsations, rotational modulation by surface structure, or planetary companions (mass $\sim 6\ M_{Jupiter}$).

If K giants have surface features, can they be imaged with the VLTI? Figure 6 shows the expected RV amplitude as a function of spot radius due to a cool spot (1200 K below photosphere). The three curves are for spots located at 3 different latitudes. If the RV variations on α Tau were due to a cool starspots, then the spot radius would have to be about 20° or 1/3 the fraction of the projected stellar disk. The angular diameter of α Tau is 21 mas or about 40 VLTI optical resolution elements. So the VLTI should easily be able to resolve a spot feature on α Tau, *if* it were responsible for the RV variability. (Of course a spot of this size can also be detected via photometry, but no such photometric observations have been performed on α Tau. These would be difficult due to finding suitable comparison stars and the long period of the variations.)

The VLTI should also be able to confirm whether the RV variations are due to planetary companions. The astrometric perturbation on these stars due to a planetary companion is about ~ 0.3 mas, at least a factor of 10 greater than the astrometric precision of measurements that can be made with the VLTI.

The VLTI should be able to image surface features on slowly rotating K giants which would be complementary to Doppler imaging studies. Together they can provide a more fundamental understanding of the dynamo mechanism as a function of rotation period by making available a wide range of stellar "laboratories" other than the Sun for testing theories of stellar activity.

The author wishes to thank M. Kürster for providing the spectral data used in deriving the Doppler images of V824 Ara A and AB Dor.

References

Choi, H.-J., et al. 1996, PASP, 107, 744.

Donati, J.-F. et al. 1992, A&A, 265, 682.

Goncharsky, A.V., Stepanov, V.V., Khokhlova, V.L., & Yagola, A.G. 1982, Astr. Zh., 59, 1146.

Hatzes, A.P. 1995a, AJ, 109, 350.

Hatzes, A.P. 1995b, ApJ, 451, 784.

Hatzes, A.P. & Cochran, W.D. 1993, 413, 339.

Hatzes, A.P. 1996, in *IAU Symposium 176, Stellar Surface Structure*, eds. K.G. Strassmeier and J.L Linsky, p. 305.

Joncour, I., Bertout, C., & Menard, F. 1994, A&A, 285, L25.

Joncour, I., Bertout, C., & Bouvier, J. 1994, A&A, 291, L19.

Johns-Krull, C.M. & Hatzes, A.P., 1996, in preparation.

Kürster, M., Hatzes, A.P., Pallavicini, R. & Randich, S. 1992, in *Proc. Seventh Cambridge Workshop on Cool Stars, Stellar Systems, and the Sun*, M. Giampapa & J. Bookbinder (eds.), Astron. Soc. Pacific Conf. Ser., Vol. 26, p249.

Kürster, M. 1993, A&A, 274, 851.

Kürster, M. 1994, A&A, 289, 899.

Stout-Batalha, N.M. & Vogt, S.S. 1996, in *IAU Symposium 176, Stellar Surface Structure*, eds. K.G. Strassmeier and J.L. Linsky, p. 337.

Strassmeier, K.G. 1994, A&A, 381, 395.

Strassmeier, K.G., et al. 1991, A&A, 247, 130.

Strassmeier, K.G., Welty, A.D., & Rice, J.B. 1994, A&A, 285, 17.

Strassmeier, K.G. 1994, A&A, 281, 395.

Vogt, S.S & Penrod, G.D. 1983a, PASP, 95, 565.

Vogt, S.S & Penrod, G.D. 1983b, ApJ, 275, 661.

Vogt, S.S., Pendrod, G.D., & Hatzes, A.P. 1987, ApJ, 321, 496.

Vogt, S.S. & Hatzes, A.P. 1991, in *The Sun and Cool Stars: Activity, Magnetism, and Dynamos*, ed. I. Tuominen, D. Moss, G, Rüdiger, (Springer-Verlag: Berlin) p. 297.

Vogt, S.S. & Hatzes, A.P. 1996, in *IAU Symposium 176, Stellar Surface Structure*, eds. K.G. Strassmeier and J.L. Linsky, p. 245.

Vogt, S.S., Hatzes, A.P., Misch, A., & Kürster, M. 1996, New Astronomy, submitted.

Stellar Surface Observations with the VLTI

O. von der Lühe

European Southern Observatory
Karl-Schwarzschild-Str. 2, D - 85748 Garching

Abstract. The resolution of the VLTI approaches 0.5 milli-arcsec at visible wavelengths and 1 milli-arcsec in the near infra-red. This is sufficient to resolve about 2000 near-by bright stars. The surface of about 400 stars, mainly K and M giants, can be mapped with a resolution of a dozen pixels or better across the stellar surface. This resolution permits detailed studies of the structure of stellar atmospheres, their hydrodynamics and magneto-hydrodynamics. VLTI can easily resolve starspots and very likely also convection cells on an active K-type giant at 10pc distance. A marginal detection of the largest starspots on a very active solar-type star at the same distance may be possible, but it appears very unlikely that the surface structure on an inactive solar-type star can be imaged.

Computer simulations of stellar surfaces were used to study in detail the response of the VLT interferometer to an extended, complex source. First results which simulate the end-to-end process in a simplified manner indicate that useful maps can be produced provided that a sufficient number of baselines are combined in Earth-rotation synthesis mode. We investigate the signal to noise ratio which which one could expect to observe stellar surface structure and conclude that the VLTI will provide sufficient sensitivity.

1 Introduction

Stars will be one of the major science targets of the VLTI. The surface of many near–by stars can be mapped with a resolution of better than a dozen pixels across the stellar surface. High angular resolution permits detailed studies of the structure of stellar atmospheres, their hydrodynamics and magneto-hydrodynamics. Although potential target stars are usually bright, the large collecting area of VLTI is essential for measuring the faint fringes resulting from the resolved stellar surface at spectral resolution high enough to resolve spectral line features. In order to better understand the requirements, we are in the process of developing detailed models of stellar surfaces to simulate observations with VLTI. We have already presented some science goals and results in two earlier papers (von der Lühe et al. 1995, von der Lühe et al. 1996). Some of these results are reproduced here. In addition, we present detailed analyses of the expected sensitivity of the VLTI. It is shown that stars with apparent diameters which can be resolved by the VLTI are also bright enough for on-source fringe tracking at shorter baselines, allowing to integrate the faint fringe signal at longer baselines with high sensitivity at high spectral resolution.

2 Statistics

The apparent diameter of a star must at least be of the order of the narrowest fringe spacing to be resolved by an interferometer. The VLTI with its baselines of about 100 m will be capable to resolve 0.5 milli-arcsec at visible wavelengths and about 1 milli-arcsec at 1 μm. This will suffice to derive a few fundamental parameters of stellar structure, such as diameters and center-to-limb variation as a function of the wavelength, for stars with diameters of the order of the milli-arcsec. Detailed surface structure will be observable for stars with diameters of about 5 milli-arcsec and bigger. The *Catalogue of Apparent Diameters and Absolute Radii of Stars* (CADARS; Fracassini et al. Fracassini, et al., 1988) was used to investigate the potential of VLTI for resolving stellar surfaces. The catalogue data should be within a factor of two or three from the truth, sufficient to assess the order of magnitude of potential targets for VLTI.

Fig. 1 presents a histogram of the diameter distribution of all sources in the catalogue with an apparent diameter larger than a milli-arcsec and a declination of less than 40 degrees, which is within reach from Cerro Paranal with VLTI baselines. There is a total of 1309 such sources. The figure shows cumulative histograms for different spectral types, as well as the overall cumulative distribution. Virtually all stars which are potentially resolved by VLTI are late types, in particular K Giants. About 50% of all stars shown have apparent diameters of less than 2.5 milli-arcsec; little more than just estimating diameters and CLVs will be possible for those. About 20% or some 260 are larger than 5 milli-arcsec. It should be possible to produce fairly detailed maps, with a resolution of at least 5 by 5 pixels. Only 7% or some 90 stars, and all of them K and M stars, are larger than 10 milli-arcsec, and are already resolved by the VLT Unit telescopes at visible wavelengths. Detailed surface maps will be possible for those stars provided that good baseline coverage is achieved. VLTI has a particular edge here, because there are four big telescopes which can be combined with relocatable auxiliary telescopes.

Fig. 2 shows a scatter diagram of the angular diameters vs. visual magnitude of the selected CADARS sources. The catalogue is evidently not complete since stars with $m_V > 6.5$ and apparent diameters of less than 5 milli-arcsec are underrepresented. The main conclusion here is that resolved stars are bright. There are no stars with visual magnitudes fainter than 12 in the sample, the bulk has magnitudes between 2 and 7. These sources can therefore serve as their own reference sources for adaptive optics which will perform near the diffraction limit at near infrared – and at later stages of the observatory's development also visible – wavelengths. This circumstance makes VLTI very sensitive for measuring low visibilities. Because of the bias towards brighter stars one can expect many more stars, about a factor 0.5 to 1 of the present sample, at magnitudes above 6.5 and with apparent diameters less than 10 milli-arcsec. These will be predominantly K and M stars. In total, one can expect about 2000 stars to be resolved with VLTI, out of which some 400 will be large enough to obtain maps of surface detail.

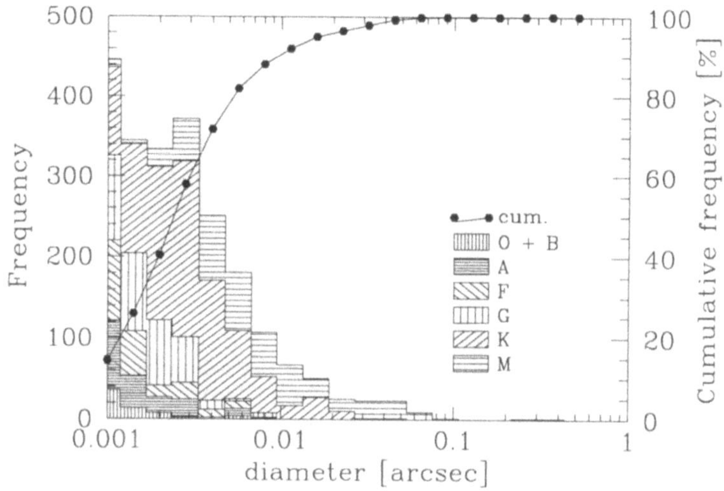

Fig. 1. Histogram of CADARS stars with apparent diameters larger than 0.001 arcsec and declinations below 40 degrees.

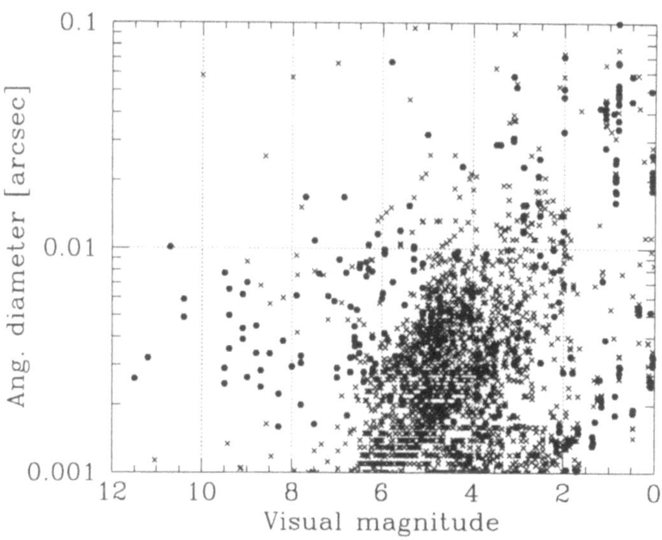

Fig. 2. Apparent angular diameters vs. visual magnitude of CADARS stars with declinations below 40 degrees. Solid dots: direct techniques, crosses: indirect techniques.

3 Stellar Surface Phenomena and Visibility Functions

The range of surface phenomena on stars has been discussed in depth in von der Lühe et al., (1996). We considered convection, starspots and active regions, and distinguished between solar-type and evolved stars. Estimates of the visibility function, i.e., the normalized Fourier transform of the expected intensity distribution, are presented in fig. 3 for a solar-type star and in fig. 4 for a giant. The figures show the various contributions to visibility magnitude. Due to the vast range in both angular frequencies and visibility magnitude, the figures show log-log plots. In order to be independent from assumptions about the distance to the star, angular frequencies are shown in "inverse stellar radii" (this unit corresponds to one interferometer fringe across the stellar diameter). The scale in arcsec shown on the top of the plots illustrate the scales which result from placing the Sun (fig. 3) and a K giant with 25 solar radii (fig. 4) at a 10 pc distance. The contributions to the visibiltiy result from the following phenomena.

Stellar disc. Its sharp edge produces the bulk of the interferoemtric signal at low angular frequencies. Fig. 3 shows the Airy function-like visibility of a disc with a limb darkening coefficient of 0.32. The visibility at higher frequencies is indicated as the dot-dashed line with a high frequency turn-off related to the pressure scale height in the photosphere. The stellar disc component in Fig. 4 is shown for a limb darkening coefficient of 1.0.

Star spots and active regions. Both spots and plages would contribute signals to the low to medium frequency range. The visibility contribution shown in Fig. 3 represents five starspot groups, each with a diameter of 2% of the stellar diameter and a rest intensity of 0.05 times the disc center intensity. The dotted lines represent a variation of the signal by a factor of $\pm\sqrt{5}$. Spots with a diameter of 10% of the stellar diameter and a rest intensity of 40% were assumed for the evolved star. Lower spot contrasts on giants are expected due to the larger pressure scale heights and the lower magnetic field strengths.

Small-scale magnetic fields. The solar surface is dotted with magnetic field concentrated in small elements which manifest themselves as network bright points (NBPs). The contribution shown in Fig. 3 assumes a solar density of NBPs with a contrast of 1.6 and a 50 km diameter. There is no such contribution for the evolved star within the range covered by the plot in Fig. 4.

Convection. The contribution shown in Fig. 3 represents solar granulation with an *rms* contrast of 0.15 times disc central intensity and an upper cutoff frequency of 1400 km. The lower cutoff was arbitrarily set to 10 times the upper cutoff. Little is known about the properties of convection for stars of types much later than the sun. We show in Fig. 4 the range of visibility signal one would expect for an intensity contrast varying from 1% to 25% for convective cells ten times the size of solar granules, to account for the increase in size with decreasing surface gravity.

The observation of surface phenomena on late dwarfs with VLTI is next to hopeless. The disc of the Sun at 10 pc would be just barely resolved by VLTI baselines. Spots and active region on these low activity stars just produce too little variation in visibility to be measurable. Convection is only accessible to

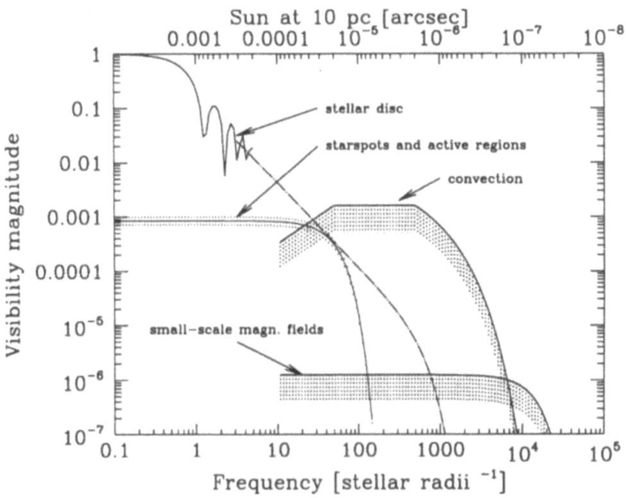

Fig. 3. Visibility magnitude for a solar type star.

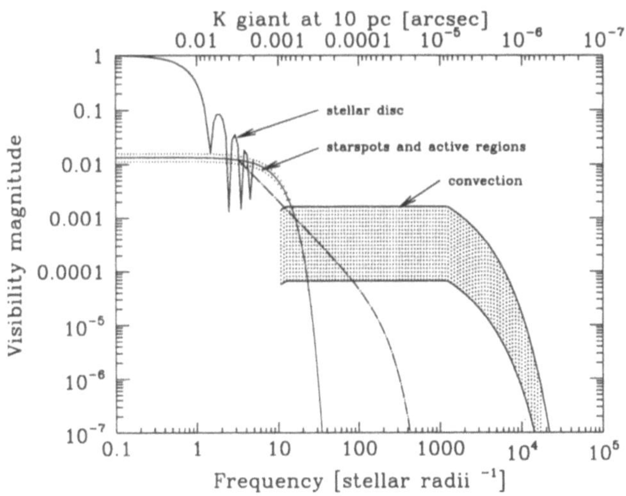

Fig. 4. Visibility magnitude for a giant star.

baselines of order of several to several tens of kilometers. The situation looks much better for active stars. Starspots and active regions of the sizes implied by Doppler imaging should be easily observable with VLTI on a number of stars. However, the stars one would observe with the VLTI are not necessarily good targets for Doppler imaging and *vice versa*.

Figs. 5 and 6 illustrates the effect of a starspot on the visibility function. Fig. 5 shows a contour plot of the surface intensity distribution, with some center-to-

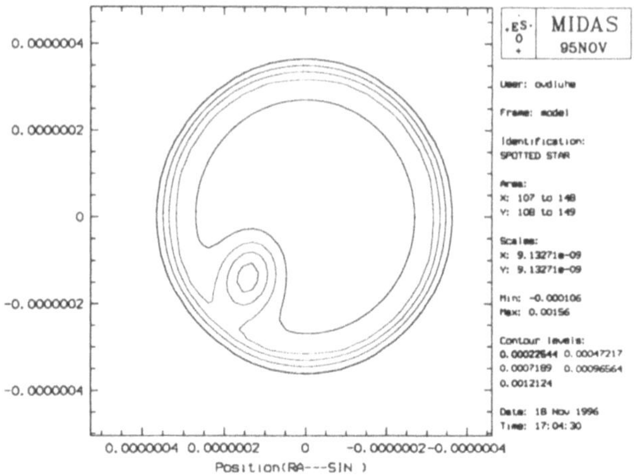

Fig. 5. Intensity distribution of a star with a spot.

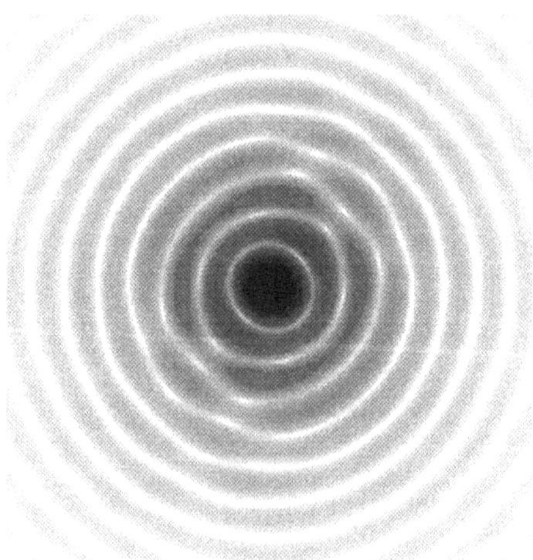

Fig. 6. Visibility magnitude of the spotted star.

limb variation and a dark spot off the center near the lower left limb. Fig. 6 shows the resulting visibiltiy magnitude in a logarithmic scale over three decades. The most prominent feature is the maximum near the origin at the center and the rings, those represent the sharp stellar limb. The spot does not manifest itself by adding a signal to the visibility function, but rather by distorting the shape of the rings. This is the signature which must be detected by the interferometer in order to reconstruct images with starspots.

4 Simulations of Observations with VLTI

We have performed simulations of observations of stellar surfaces using the Cal-Tech VLBI data reduction package (Pearson, 1991). An example is shown in figures 7 and 8. The simulated configuration of the VLTI included the four large 8m telescopes and three 1.8m Auxiliary telescopes at locations where they optimally match the configuration of the large telescopes in order to achieve a compact, well covered UV plane. The resulting synthesis coverage using the Earth's rotation for a source at $-65°$ declination is shown in fig. 7. The same configuration was used to simulate observations of the star model shown in fig. 5 at declinations of $+15°$, $0°$, $-15°$, $-35°$, and $-65°$, thus covering the range of declinations which are relevant to the VLT. The zenith distance was always limited to $60°$.

The simulated star has an apparent diameter of 5 mas and a flux of 0.9 Jy (m_R = 8.89). The observing wavelength is Hα (6563 Å). Simulated visibility and closure phase data were subjected to Fourier inversion which yields a dirty map. Several thousand iterations of CLEAN were used to obtain clean maps of the stellar surface. A rendition of the model source at infinitely good resolution (top) and the CLEANed maps are shown in fig. 8. This simulation is best suited to assess the aperture synthesis capability of various configurations of the VLTI when using standard radio interferometry data analysis procedures, rather than to assess sensitivity issues.

The CLEANed maps always show the spot at its correct position near the lower left limb. The maps for declinations near the celestial equator appear mottled, while the maps for $-35°$ and $-65°$ seem to represent more faithfully the stellar surface at the resolution which can be obtained with the array. The reason is that the UV coverage near the equator is always worse than the one for southern declinations. The inversion procedure used here is easy to implement in the given context, but is not the best suited for an extended source like a stellar surface. CLEAN attempts to fit point sources to a dirty map, and a stellar surface is not very well represented by that process, which explains the somewhat lumpy reconstructed maps. A model fitting procedure, with initial model parameters taken from the CLEANed maps, may have produced better results, but this was not done here. All in all, one can conclude that the VLTI will be able to produce maps of stellar surface in Hα with a quality which would allow one to study large starspots on the sample of stars whose apparent diameters exceed some 3 mas.

5 Sensitivity

There are two factors which limit the sensitivity of the VLTI to map surfaces of stars. First, in order to obtain necessary spectral diagnostics, the spectral resolution must be rather high, of order 50000. The flux per spectral channel is therefore rather low, which requires integration times that exceed the atmospheric correlation time scales. Active fringe tracking guarantees long coherent

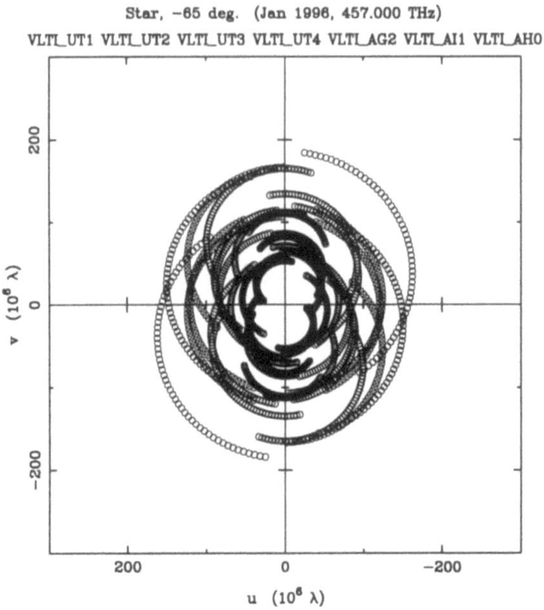

Fig. 7. Fourier domain (UV) coverage of the simulated observations for a source declination of -65 degrees.

integration times of fringes in narrow spectral bands. The fringe tracker sensor can be used with broad spectral bands, but requires integration times shorter than the atmospheric correlation time. The sensitivity of the fringe tracker therefore determines the limiting magnitude and must be used on shorter baselines for which the intrinsic visibility of the source is high. For an array with N elements, fringe tracking must be performed on at least $N-1$ baselines, but the fringe signal will be stabilized on all $N/2(N-1)$ baselines. A proper selection of the baseines used for finge tracking is therefore necessary in order to maximise the sensitivity.

The second factor is the low visibility of the source. Figs. 3, 4 and 6 demonstrate that the intrinsic visibility of the phenomena of interest is of order $10^{-2} \ldots 10^{-3}$ which is rather low. The visibility determines directly the signal to noise ratio. The detection of intrinsic visibilities of 10^{-2} is nowadays demonstrated, some further development for detecting 10^{-3} will be necessary.

The sensitivity of an optical interferometer depends on several factors. If the beam combiner rejects all photons which are scattered by atmospheric and instrumental aberrations by, e. g., using an Airy-disc sized mode stop in an image conjugate just before beam combination, the signal to noise ratio SNR of a photon and detector noise limited visibility measurement is given with

$$\mathrm{SNR} = \frac{SN_p}{\sqrt{1 + SN_p + N_d^2}} \frac{V}{N_{Tel}} \sqrt{K} \ . \qquad (1)$$

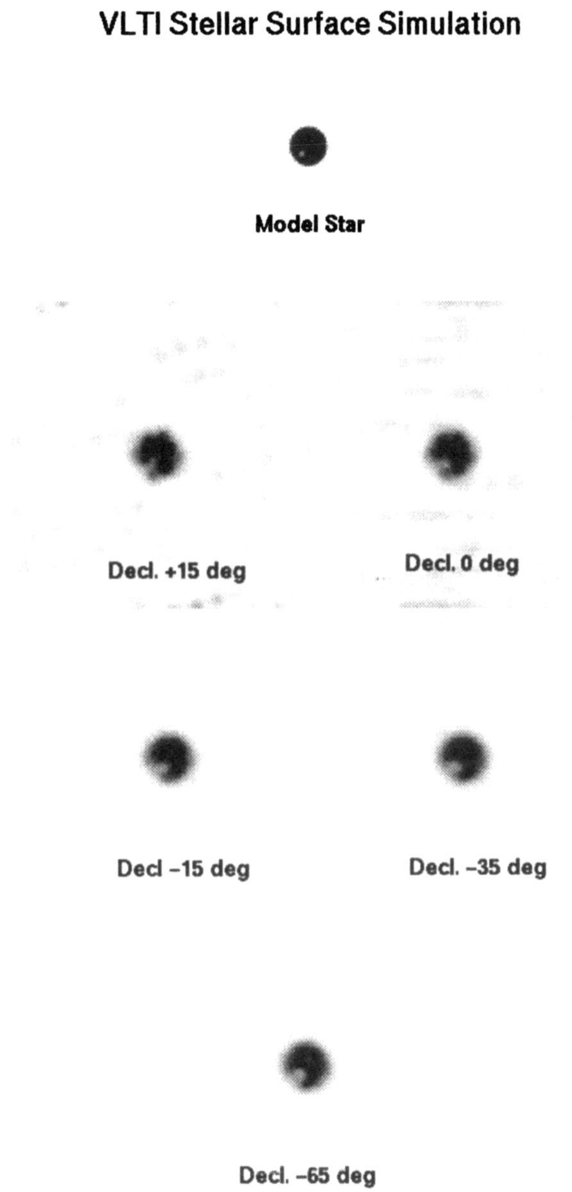

Fig. 8. Sample CLEAN maps of a stellar surface at various declinations.

N_p is the number of photons collected by all telescopes, including losses due to atmosphere, transmission of optics, and detector response. S represents the Strehl ratio of the optical train, including atmosphere effects and any compensation by adaptive optics, at the point of beam combination. The product SN_p is the number of detected *coherent* photons – these are the photons which contribute to the visibility signal. Because of the mode stop the number of photons which contribute to the noise is also SN_p. N_d is the number of photons which are equivalent to the *rms* detector readout noise. V is the intrinsic visibility of the source at the observed baseline. N_{Tel} is the number of elements in the array. K is the number of *incoherently* averaged visibility measurements. The latter comes about when the interferometer operates in open loop geometric delay tracking mode, which makes it necessary to average the visibility modulus of many short exposures.

Figs. 9 and 10 show the visibility amplitude signal–to–noise ratio for a single baseline VLTI with two 8m Unit telescopes and with two 1.8m Auxiliary telescopes in the Johnson H band. Each curve within a plot represents a different intrinsic visibilitiy of the source, ranging from 1 to 10^{-3}. Each plot within a set represents a different fringe integration mode. The top left curves represent the SNR for short integration times needed for fringe tracking using the entire H band. The top right curve applies to long on-detector integration times (900 s) but with a narrow spectral band of $0.2\AA$ and represents the SNR in a single high resolution spectroscopic channel. The bottom two sets of curves apply to incoherent averaging of short exposures at medium passbands ($100\AA$), the left with adaptive optics, the right without.

An adaptive optics system which improves the Strehl ratio at each telescope to about $0.4 \cdots 0.5$ in the H band was considered. This applies to all except to the bottom right plots, where the Strehl ratio is determined by atmospheric conditions (Fried parameter $r_{o,H} = 75$cm which results in $S = 0.009$ for 8m telescopes and $S = 0.17$ for 1.8m telescopes). Also, the term SN_p in the square root in the denominator of eqn. 1 was replaced by N_p in the bottom right plots because a mode stop would make little sense in a seeing limited configuration. The short exposures are determined by the atmospheric correlation timescale. With a turbulence weighted mean wind speed of $5ms^{-1}$, the resulting short exposures amount to 0.023s for the plots shown here. The overall throughput including detector sensitivity was set to 0.1 for the two top plots, and to 0.05 for the plots at the bottom. This was to account for solid state detectors in the first two cases, which include detector readout noise equivalent to a NEP of $10^{-16}W/\sqrt{Hz}$. For the bottom two curves, photon counting detectors with zero readout noise were assumed.

Most curves exhibit a bend which separates two regimes. For fainter sources, the detector noise dominates and the SNR is proportional to the signal, whereas for bright sources, photon noise dominates and the SNR grows with the square root of the signal. Fringe tracking is possible with the 8m telescopes for sources with $m_H \leq 15$ provided the intrinsic visibiity is high. Stars with $m_H \leq 10$ can be fringe tracked on baselines with intrinsic visibilities as low as 0.02. The

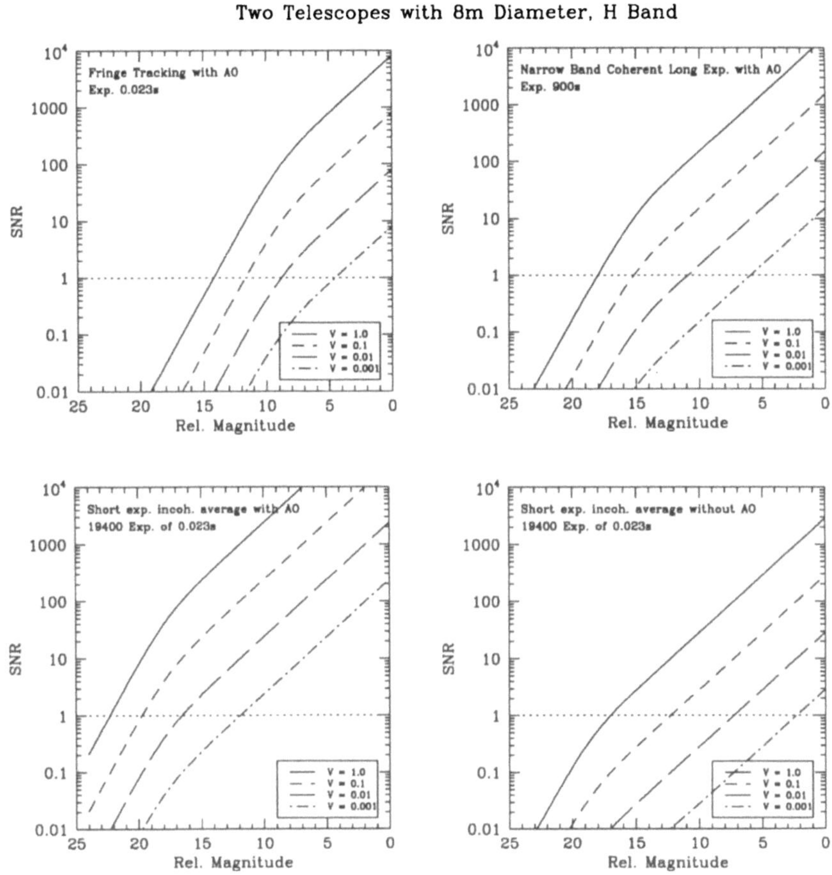

Fig. 9. Signal to noise ratio of two 8m telescopes in the H band. See text.

long exposure integrations can go considerably fainter, it should be possible to detect visibilities between 210^{-2} to 10^{-3} for the magnitude range from 12 to 6. The 1.8m Auxiliary telescopes are considerably less sensitive for the fringe tracking and long integration modes with adaptive optics, which is what one would expect. The magnitude range does not go much fainter than $m_H = 8$. The incoherent modes which require less real time control can go faint only when adaptive optics are used with the 8m telescopes. A totally passive mode (no adaptive optics and no fringe tracking) would be best done using only the 1.8m Auxiliaty telescopes.

The conclusion of this analysis is that the sensitivity of the VLTI is well matched to the problems one can expect. Faint sources are likely to be less resolved than bright ones implying higher intrinsic visibilities and thus more signal. Stars with m_H fainter than 10 have apparent diameters less than 5 mas,

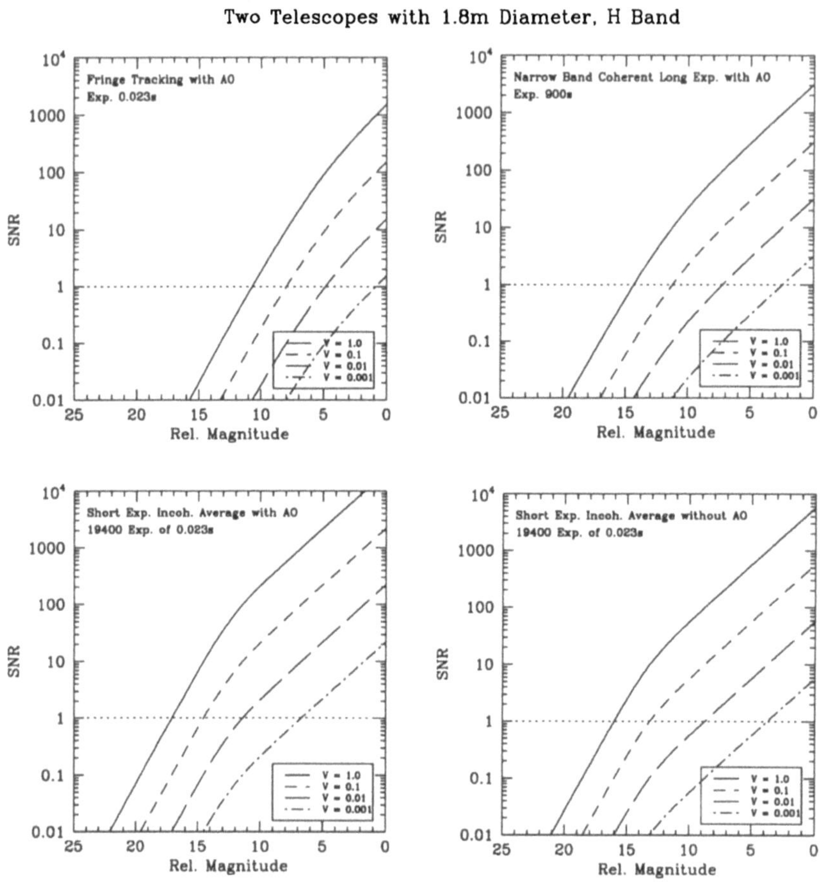

Fig. 10. Signal to noise ratio of two 1.8m telescopes in the H band. See text.

and will require measurements of visibilities of order 10^{-2}. This can be relatively easily achieved with the 8m telescopes. Obviously, well resolved stars tend to be rather bright, and an intrinsic visibility of order 10^{-3} should be easier to detect. The fringe tracking mode appears to be more demanding in terms of sensitivity compared to observations with narrow spectral bands in conjunction with long exposures.

6 Conclusions

Interferometry with the VLT will produce crucial new insight in the understanding of stellar atmospheres by providing direct images of stellar surfaces. By assessing the potential of VLTI we conclude that about 2000 stars will be resolved to VLTI baselines and that it should be possible to obtain maps with useful detail for some 400 stars. A large fraction of the latter are late type giants, which indicates that at least for this class of stars, a fairly complete sample of maps should be doable. We expect that the available baseline length will limit the resolution to structures of the order of large spots and active regions on magnetically active stars.

We use computer simulations to study in detail the performance of VLTI, to investigate operational requirements of a given science program, and to study data analysis methods. Standard VLBI data reduction techniques will be sufficient for broadband imaging, but more sophisticated analysis and model fitting may be necessary for high resolution spectroscopy. It will also be necessary to include many baselines into the array if detailed maps of the stellar surface are to be made.

A sensitivity analysis shows that Auxiliary telescopes suffice for observing the brighter stars and / or with baselines at which the visibiity of the source is high. This is mainly due to sensitivity limitations of the fringe tracking mode. Unit telescopes will become necessary to measure fainter stars with magnitudes between 10 to 15 and with high spectral resolution. The operation of Unit telescopes at near infra-red wavelengths will require adaptive optics.

References

Fracassini, M., Pasinetti-Fracassini L. E., Pastori, L., Prioni, R. 1988, *Bull. CDS* 35, 121.

von der Lühe, O., Schüssler, M., Solanki, S., Caligari, P. 1995, *Science with the VLT*, J. R. Walsh and I. J. Danziger (eds), ESO Astrophysics Symposia, Springer, p. 94.

von der Lühe, O., Solanki, S., Reinheimer, Th. 1996, *Stellar Surface Structure*, K. Strassmeier and J. Linsky, (eds), IAU Coll. 176, pp. 147–163.

Pearson, T. J. 1991, "Introduction to the CalTech VLBI Programs," California Institute of Technology VLB Manual.

On the Scale of Photospheric Convection

Bernd Freytag[1], Hartmut Holweger[1], Matthias Steffen[2,1], Hans-Günter Ludwig[3]

[1] Institut für Astronomie und Astrophysik der Universität Kiel, D-24098 Kiel, F.R.G.
[2] Astrophysikalisches Institut Potsdam, D-14473 Potsdam, F.R.G.
[3] Max-Planck-Institut für Astrophysik, D-85740 Garching, F.R.G.

1 Stellar Surface Structure: Size of Granules

Improvements in observational methods have opened the possibility to resolve structures on the surface of stars other than the Sun. With interferometric techniques on a ground based telescope (Buscher et al. 1990) and with the HST (Gilliland & Dupree 1996) a bright spot has been detected on the surface of Betelgeuse. Such surface inhomogeneities may result from magnetic activity, global pulsations, or perhaps from convection.

Schwarzschild (1975) estimated the scale of photospheric convection in red giants and supergiants by extrapolating the size of granules on the Sun to those stars: First he compared the typical horizontal size of a solar granule x_{gran} (his adopted value was 2000 km) with characteristic depth scales of a mean solar stratification. Prime candidates for these scales were the subphotospheric pressure scale height H_p and the extent of the strongly superadiabatic layers $H(\Delta\nabla)$. In a second step he analyzed a red giant model with $M=0.65\,M_{\odot}$, $L=8000\,L_{\odot}$, $T_{\text{eff}}=3600\,\text{K}$. Assuming that the ratio between horizontal extent of a granule and characteristic depth scale is a constant he concluded for $x_{\text{gran}}/H(\Delta\nabla)=\text{const}$ that the surface of a red giant is covered by a few hundred granules, which would cause hardly any easily observable effect. In the extreme case $x_{\text{gran}}/H_p=\text{const}$ only a small number of granules would occupy the surface of the star at any time causing large scale surface inhomogeneities and total brightness fluctuations.

2 Numerical Simulations

Based on 2D numerical radiation hydrodynamics calculations of time-dependent compressible convection, we studied a variety of stellar surface convection zones. The non-local equation of radiative energy transport is solved along a number of representative rays, assuming LTE, with realistic opacities (see Ludwig et al. 1994). The equation of state accounts for ionization processes.

The list of models of the surface layers of *deep convection zones* covers stars with $4\,300\,\text{K}\leq T_{\text{eff}}\leq 7\,300\,\text{K}$, $2.54\leq\log g\leq 4.74$. Additionally there are models of *shallow convection zones* which are completely included in the computational volume: main-sequence A-type stars: $7\,300\,\text{K}\leq T_{\text{eff}}\leq 9\,000\,\text{K}$; cool DA white dwarfs: $11\,400\,\text{K}\leq T_{\text{eff}}\leq 14\,200\,\text{K}$, $\log g\approx 8.00$. Despite the absence of models of red giants or supergiants, the range in effective temperature, gravity, and composition allows the analysis of the size dependence of convective elements and a cautious extrapolation towards giants and supergiants.

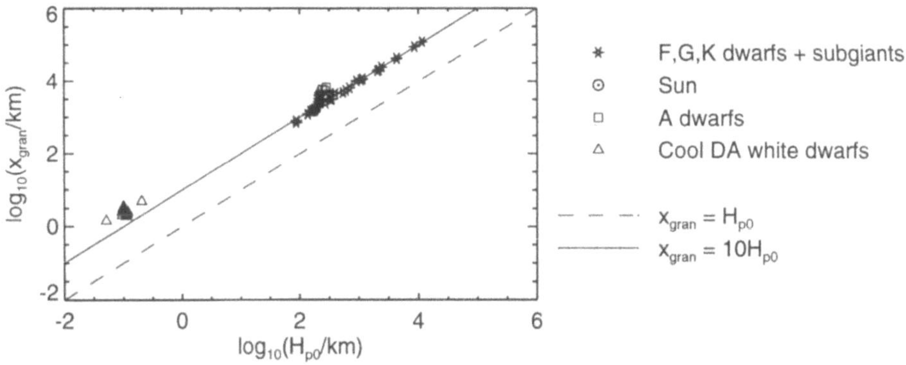

Fig. 1. Horizontal extent of granules versus pressure scale height

The most promising candidate as characteristic length for granules is the *pressure scale height* H_p (see Antia et al. 1984). In Fig. 1 the mean horizontal size of a granule $x_{\rm gran}$ is plotted as a function of the characteristic pressure scale height $H_{p0} = \mathcal{R}T_{\rm eff}/g$. All dwarfs with deep adiabatic convection zones closely follow the relation $x_{\rm gran} \approx 10\,H_{p0}$ with only a small scatter although the models cover more than 2 orders of magnitude in gravity. The models of A stars and white dwarfs with radiatively dominated convection zones have larger granules because efficient radiative exchange tends to favour larger cells.

Other depth scales are less useful as characteristic sizes of granules even if they are important for the detailed behaviour of granulation: the density inversion in many convection zones makes the *density scale height* a problematic measure for the size of granules, especially in very thin convection zones like those of A stars. The run of the *opacity* with height differs strongly between the DA white dwarfs and the A-type stars with no noticeable effect on the granular scale. In the Sun the *ionization zones* of hydrogen and helium start well below the surface layers with their granular convection, whereas in A-type stars these ionization zones are located just within the thin convection zones, which does not seem to affect the scale of granulation strongly. In our sample of models the thickness of the layers with strongly *superadiabatic temperature gradient* $H(\Delta\nabla)$ shows the same trend as the H_{p0}. However, there exist small but systematic deviations from a linear relationship between $x_{\rm gran}$ and $H(\Delta\nabla)$, indicating an increasing ratio of them towards the red giants.

For this reason we favor the extrapolation based on $x_{\rm gran} \approx 10\,H_{p0}$. Application to Schwarzschild's sample star results in the extreme picture of at most a handful of granules on the entire stellar surface with typical diameter of $x_{\rm gran} \approx 10^8\,{\rm km} \approx 0.6 R_*$ which could be resolvable with the VLTI.

References

Antia H.M., Chitre S.M., Narasimha D. (1984): ApJ **282**, 574–583
Buscher D.F., Haniff C.A., Baldwin J.E., Warner P.J. (1990): MNRAS 245, 7p–11p
Gilliland R.L., Dupree A.K. (1996): ApJ **463**, L29–L32
Ludwig H.-G., Jordan S., Steffen M. (1994): A&A **284**, 105–117
Schwarzschild M. (1975): ApJ **195**, 137–144

Observing Stellar Surfaces with a High Precision Infrared Interferometer

Guy Perrin[1], Vincent Coudé du Foresto[2], Stephen T. Ridgway[3], Jean-Marie Mariotti[1], Nathaniel P. Carleton[4], and Wesley T. Traub[4]

[1] Observatoire de Paris, DESPA, F-92195 Meudon, France
[2] Max-Planck Institut für Astronomie, Königstuhl, D-69117, Heidelberg, F.R.G.
[3] National Optical Astronomy Observatories, Tucson, Arizona 85726-6732, U.S.A.
[4] Harvard-Smithsonian Center for Astrophysics, Cambridge, Mass. 02138, U.S.A.

Abstract. In this paper we report the results of a fiber optics beam-combiner at the focus of the IOTA interferometer. The spatial filtering of the atmospheric turbulent optical modes by single-mode fiber optics enabled us to measure fringe contrasts with precisions superior to that of classical all-optics beam-combiners. Diameter measurements of late-type stars at 2.2 μm with accuracies better than 1% are presented. Preliminary effective temperatures are derived for a sample of giant stars. These results demonstrate the potential of single-mode fibers for light recombination in imaging interferometers with high-dynamic range. A similar instrument set on the VLTI would have a much higher sensitivity. Among all possible applications, study of stellar surfaces would benefit of high accuracy measurements provided by that instrument.

1 Introduction

In actual optical interferometers, the accuracy of visibility calibration is limited by the non-stationarity of the atmospheric turbulence. The average reported values are more than one percent, i.e. consistent with the accuracies obtained in speckle-interferometry, a technique indeed limited by the same phenomenon.
For several years (Coudé du Foresto & Ridgway 1991), we have developed the technique of single-mode interferometry which provides visibility moduli measurements that are independent of atmospheric turbulence, thus opening the way to interferometric image reconstruction with a very high dynamic range in the optical domain. We report some illustrative results of the advantage of this technique obtained at 2.2 μm on the IOTA interferometer with a fiber optics beam combiner. No bias due to atmospheric turbulence is present in the data at a 0.5% accuracy level allowing future programs of stellar surface imaging.

2 Experimental Set-Up

Recombination is achieved at 2.2 μm with single-mode fiber optics (Perrin et al. 1995), the Fiber Linked Unit for Optical Recombination (FLUOR), set at the focus of the Infared-Optical Telescope Array (IOTA). IOTA is located at the Smithsonian Institution's Fred Lawrence Whipple Observatory on Mount Hopkins in Arizona, and is a collaborative project of the Smithsonian with Harvard

University, the University of Massachusetts, the MIT (Lincoln Lab) and the University of Wyoming. IOTA is currently a two-element interferometer with 45 cm entrance pupils (Carleton et al. 1994). The available baselines range from 5 to 38 m. It is operated both in the visible and in the infrared. First fringes were obtained at 2.2 μm with an all-optics beam-combiner (Dyck et al. 1995).
The beams picked up by the two collectors are injected into fibers after they have been delayed. Fringes are detected by InSb hybrid detectors cooled with liquid nitrogen at 77 K. The fringe packet is scanned in optical path difference by an air-bearing continuous delay line at a 300 Hz frequency. A fraction of the input energy is sampled in each arm before recombination. The two detected fluxes (the photometric outputs) are proportionnal to the coupling efficiencies in the two arms and allow full calibration of the loss of coherence due to atmospheric turbulence (Coudé du Foresto et al. 1997).

3 Performance

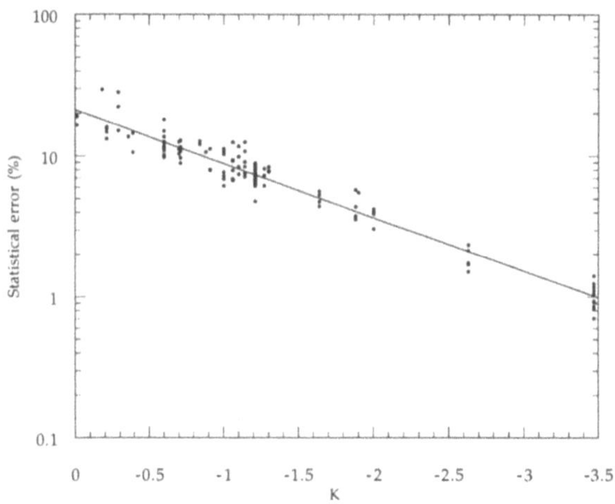

Fig. 1. Statisical error on contrast vs. K magnitude. The error is the dispersion of the contrast distribution or the statistical error for one acquisition. It is fitted by a power law.

The performance of the interferometer is characterized by its sensitivity, its precision and its capacity to produce high quality measurements.
The current sensitivity of FLUOR is limited to K=0. The main is reason is that only 10% of injected photons are detected in the interferometric outputs because we do not use appropriate optics. We had to use non-optimized, off-the shelf optics in our current set-up. New custom-made focusing optics will be tested

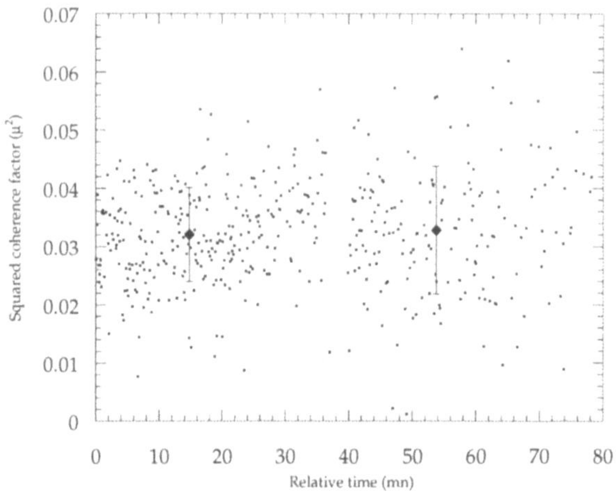

Fig. 2. Effect of seeing variations on contrast. Two contrast etimates with corresponding dispersions are plotted before and after 40 minutes. The seeing is degrading during the whole sequence. The dispersion thus inreases as less flux is coupled to the fibers but the contrast estimates remain steady and free of atmospheric bias.

in the next months. We also plan to replace InSb detectors by a more sensitive NICMOS III detector. A realistic estimate of the future K limit magnitude should be of the order of 3.

In this paper, the contrast is the amplitude of the fringe packet and visibility is the contrast devided by the estimated transfer function (point source response of the instrument). The constrast estimate computed by the data reduction process has fluctuations whose origins are diverse. Contrast measurements are sensitive to detector, photon and optical-path-fluctuations noise. In the infrared, detector noise is predominant over photon noise which can be neglected for all astronomical sources. Atmospheric optical path fluctuations (or piston) are the first order of turbulent modes. This is the only mode left unfiltered by the single-mode fibers. It is a fluctuating delay between the two beams. It could be reduced down to a negligeable level by scanning fringes more rapidly than the 300 Hz fringe frequency used for the acquisitions reported here. From standard models of turbulence it is shown that for average conditions the piston is attenuated by more than a factor of ten for scanning frequencies of the order of 3 kHz (Perrin 1997). Piston noise is the main noise for very bright sources. For most sources, contrast accuracies are limited by detector noise. The statistical error on contrast measurements normalized to one acquisition is plotted versus K magnitude on Fig. 1. It can be fitted by a power law, hence the normalized rms error as a function of K or source brightness (Φ):

$$\sigma(\%) = 21.31 \times 10^{0.38K} = 21,31 \times \Phi^{-0.95} \tag{1}$$

The dominant term is thus detector noise for which the error is proportional to the reverse of the source flux. Piston noise accounts for most of the error for bright sources hence the -0.95 exponent in the power law instead of the -1 expected exponent. For sources brighter than K=-0.5, acquisition of 100 interferograms (which takes 7 minutes of observing time at the current acquisition rate) is enough to get an average statistical error smaller than 1%.

The statistical error on constrast is part of the final error on visibility. The

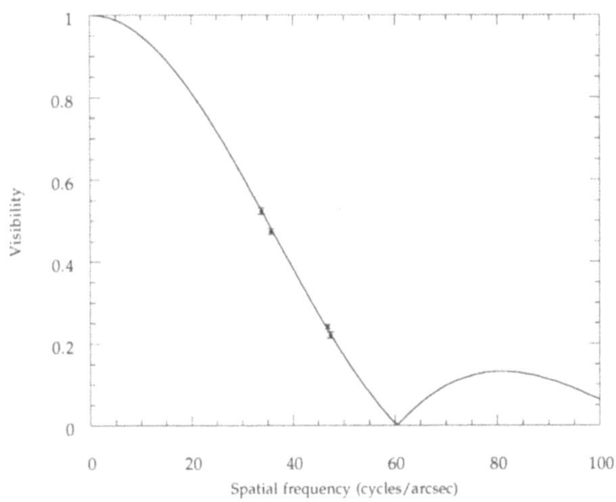

Fig. 3. Visibilities measured on Arcturus (α Boo). The data are fitted by a uniform disk model with diameter $\oslash_{UD} = 20.20 \pm 0.08\,\mathrm{mas}$.

advantage of a FLUOR-type recombination unit is that the transfer function is independent of seeing. Fig. 2 is an example of an observing sequence with a fluctuating seeing. The contrast and the statistical dispersion have been computed for two 40 minutes long subsequences. The statistical precision gets poorer with the degrading seeing since less flux is coupled in the fibers, but the contrast remains the same. This example is illustrative of the stability of the transfer function. Contrasts are calibrated by observing sources of known visibility. The final error on the calibrated visibility is deduced from the statistical error on the source fringe contrast and on the transfer function. Calibrated visibilities measured on Arcturus are displayed on Fig. 3. They are fitted by a uniform disk model. The residual normalized error of the least square fit (or χ^2) is 0.75, indicating that error bars have not been underestimated. After calibration, differences between models and data points are typically of the order of 0.5% for bright sources and no bias has been detected at this level of precision.

Table 1. Uniform disk model fit to FLUOR data. The data points are those used for disk model fitting.

Source	# data points	$\oslash_{UD}$ (mas)
α Boo	4	20.20 ± 0.08
τ^4 Ser	2	11.21 ± 0.18
BK Vir	4	10.73 ± 0.23
RT Vir	4	13.06 ± 0.15
SW Vir	3	16.81 ± 0.12
EU Del	2	10.65 ± 0.25
LY Ser	2	8.97 ± 0.31
RS Cnc	4	14.27 ± 0.09
RX Boo	2	18.23 ± 0.12
δ Oph	1	10.08 ± 0.48
α Tau	2	19.75 ± 0.11
α Her	7	30.90 ± 0.02
α Ori	4	42.71 ± 0.02

Table 2. Limb-darkened disk diameters and effective temperatures for giants.

Source	Sp	[V-K]	$\oslash_{LD}$ (mas)	Flux (10^{-13} W cm^{-2})	T_{eff} (K)
α Boo	K1.5 III	2.90	20.82 ± 0.08	50.4 ± 2.0	4321 ± 44
τ^4 Ser	M7.4 III	7.56	11.60 ± 0.19	3.76 ± 0.07	3026 ± 29
BK Vir	M6.9 III	8.52	11.11 ± 0.24	3.09 ± 0.03	2944 ± 33
RT Vir	M8 III	8.38	13.52 ± 0.16	3.41 ± 0.07	2740 ± 21
SW Vir	M7.3 III	9.66	17.50 ± 0.13	7.27 ± 0.05	2905 ± 12
EU Del	M6 III	7.19	11.02 ± 0.26	4.00 ± 0.04	3153 ± 38
LY Ser	M8 III	7.19	9.28 ± 0.32	1.92 ± 0.07	2860 ± 57
RS Cnc	M6.4 III	7.49	14.77 ± 0.09	6.79 ± 0.07	3108 ± 12
RX Boo	M8 III	9.74	18.87 ± 0.12	7.12 ± 0.05	2783 ± 10
δ Oph	M0.5 III	3.94	10.43 ± 0.50	7.10 ± 0.19	3740 ± 95
α Tau	K5 III	3.67	20.44 ± 0.11	33.8 ± 1.4	3947 ± 41

4 Diameters and Effective Temperatures for Late-Type Stars

Among the topics that can be adressed by interferometry, measuring stellar diameters is the most common and allows effective temperature calibration, an important parameter for stellar physics since it measures the energy radiated at all wavelengths.

Table 1 is a list of diameters measured with FLUOR on IOTA for non-Miras giants and supergiants so far. Most stars have a diameter determined with a relative precision better than 1% with very few data points. In the case of α Boo,

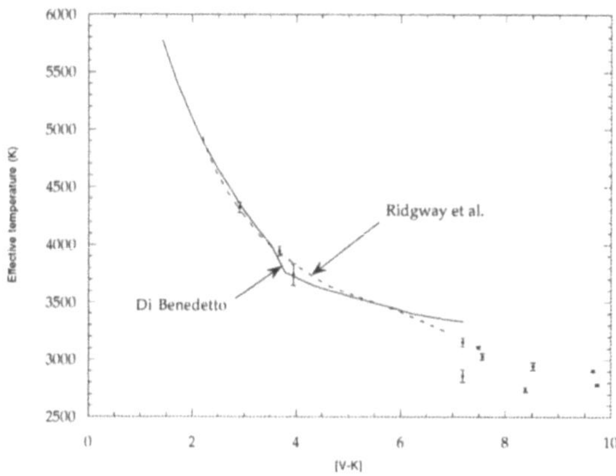

Fig. 4. Effective temperature vs. [V-K]. Solid line: calibration by Di Benedetto (1993). Dotted line: calibration by Ridgway et al. (1980). Dots: FLUOR calibration.

α Her and SW Vir whose visibilities have been obtained with high accuracy around and before the first zero with two different baselines, we have been able to fit data points by limb-darkened disk models found in the litterature (Manduca 1979; Scholz and Takeda 1987). In these three cases, the residuals of the fit by a limb-darkened disk are smaller than those with a uniform disk model. We thus derive an average scaling factor between the uniform disk diameter and the limb-darkened disk diameter of 1.035 that is used for the other giants and supergiants of our sample. In Table 2 are listed the derived limb-darkened diameters. The effective temperature is computed for giants from the limb-darkened diameter and the bolometric flux according to the relation derived from the Stefan-Boltzman's law:

$$T_{eff} = 1.316\,10^7 \left(\frac{F_{\rm bol}}{1\,{\rm W\,cm^{-2}}}\right)^{1/4} \left(\frac{1\,{\rm mas}}{\O_{LD}}\right)^{1/2} \tag{2}$$

Some bolometric fluxes have been found in the literature (Blackwell et al. 1986; Di Benedetto & Rabbia 1987; Dyck et al. 1996) and others have been provided by one of us. The case of supergiants is more complicated and effective temperatures have not been derived. Their radiation is intrinsically reddened by their circumstellar dust and makes it difficult to compute a bolometric flux. The FLUOR temperature measurements are compared to previous determinations on Fig. 4 (Ridgway et al. 1980; Di Benedetto 1993). Effective temperature is plotted versus the [V-K] color index. Our results are consistent with lunar occultation and I2T results for stars hotter than 3500°. They extend the previous scales to giant stars later than M6. Most of these stars are variable and we do not have a precise measurement of their flux and of their color index at the time when they were observed. We have not investigated this point yet and error

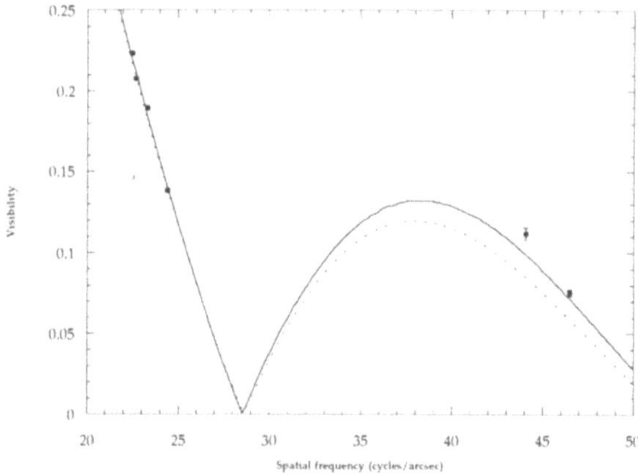

Fig. 5. Visibilities measured on Betelgeuse and best uniform disk model fit of first lobe data points (solid line). A limb-darkened disk model is also plotted (dashed line). Visibilities at the shortest and at the longest baselines have been measured at different epochs.

bars on temperatures may be optimistic. Hence an incertainty in their effective temperature. This preliminary work requires to be accompanied by photometric measurements, our main source of errors indeed.

5 Study of Stellar Surfaces

Other areas of interest are binary stars, circumstellar dust, astrometry, etc... Another paper in this conference adresses the issue of detecting exoplanets with a single-mode $2\,\mu$m instrument on the VLTI (Coudé du Foresto et al. 1996). Diameters are the easiest information to get from a stellar interferometer. Once the diameter has been determined it is possible to detect and calibrate the limb-darkening effect. The next step towards the knowledge of stellar objects is the study of their surface. This information is of prime importance since surface activity can be connected to internal stellar activity and structure. We have initiated the study of the surface of the supergiant Betelgeuse with FLUOR. A bright spot has been recently detected at its surface in the UV by the Hubble Space Telescope (Gilliland and Dupree 1996). The star has been observed with FLUOR/IOTA at two different baselines (11.2 and 21.2 m). The data points at the shortest baseline allowed us to derive the diameter of Table 1. It is then remarkable that the points measured at the longer baseline have a visibility excess compared to the uniform and limb-darkened disk models (see Fig. 5). It is difficult to connect this excess to that produced by a spot as imaged by the HST in the UV since we only have two data points. Some more visibility points are required to go any further.
Nevertheless, this example shows that a FLUOR-like instrument is adequate to

study stellar surfaces. Betelgeuse may not be representative of all supergiant stars. A study by von der Lühe et al. (1996) predicts that the amplitude of star spots in the visibility function is of the order of 0.1% for solar-type stars and 1% for K giants. Accuracies on visibilities such as those provided by FLUOR are thus mandatory to detect and monitor these spots. A FLUOR-like instrument set on the VLTI would benefit of a much higher sensitivity. This would open the way to a statistical study of the surface of stars.

6 Conclusion

The fiber optic recombination technique in optical interferometry has proved to be efficient in measuring spatial frequency components of bright stars with a high accuracy. A similar experiment set on the VLTI would benefit of a large sensitivity which would give access to interesting astrophysics studies. Among them, the study of stellar surfaces is definitely within the possibilities.

References

Blackwell, D.E., Booth, A.J., Petford, A.D., et al., 1986, MNRAS **221**, 427

Carleton N. P., Traub W.A., Lacasse M.G., 1994, in SPIE Conf. 2200 "Adaptive optics in Astronomy", Kona (Hawaii), 17-18 March 1994, p. 152

Coudé du Foresto, V., Ridgway S. T., 1991, " High Resolution Imaging by Interferometry II", Beckers J. M. & Merkle F. (eds.), Garching (Germany), p. 731

Coudé du Foresto, V., Mariotti, J.-M. and Perrin, G., 1996, this conference

Coudé du Foresto, V., Ridgway, S.T. and Mariotti, J.-M., 1997, A&AS, in press

Di Benedetto, G.P. and Rabbia, Y., 1987, A&A **188**, 114

Di Benedetto, G.P., 1993, A&A **270**, 315

Dyck H. M., Benson J.A., Carleton N.P., et al. 1995, AJ **109** (1), 378

Dyck H. M., Benson J.A., Van Belle, G.T., et al. 1996, AJ **111** (4), 1705

Gilliland, R.L., Dupree, A.K., 1996, ApJ **463**, L29

Manduca, 1979, A&AS **36**, 411

Perrin G., Coudé du Foresto V., Ridgway S.T., et al, 1995, in SPIE Conf. 2476 "Fiber optics in astronomical applications", Orlando (Florida), 17-21 April 1995, p. 120

Perrin, G., 1997, A&AS, in press

Ridgway, S.T., Joyce, R.R., White, N.M. and Wing, R.F., 1980, ApJ **235**, 126

Scholz, M. and Takeda, Y., 1987, A&A **186**, 200

von der Lühe O. et al., 1996, IAU Symposium 176 "Stellar surface structure", K.G. Strassmeier Ed., (in press)

Studying the Cooolest Giant Stars at Milliarcsecond Resolution

Andrea Richichi

Osservatorio Astrofisico di Arcetri, Largo E. Fermi 5, I-50125 Firenze, Italy

Abstract. Measurements of the angular diameters, surface structure and near–stellar environment of late-type stars have been until now one of the major applications for a wide variety of high angular resolution techniques, and they will probably remain one of the priorities for the VLTI. The proposed angular resolution and sensitivity of the VLTI, as well as the wavelength coverage of its first generation instruments, match quite well those presently achieved by the technique of lunar occultations. Given the premium on VLTI time, it is useful to explore the possibilities opened by coordinated observations in which the VLTI is used to follow up interesting sources previously discovered by lunar occultations. We will review the current status of high angular resolution investigations of cool giant stars, and the impact that the VLTI could have in this field. In particular, attention is given to the coolest spectral types, below M5.

1 The Experience of Lunar Occultations as a Background for VLTI Studies

The technique of lunar occultations (LO) has been used for some decades to obtain high angular resolution (HAR) measurements. Although it has been applied from the X-ray to the radio wavelengths, it has been used most frequently in the visual and IR ranges, typically on stellar sources. In the most recent years, it has been used mostly in the near–IR, where its efficiency is maximum due to several factors which include the lunar background, the characteristics of the diffraction phenomenon at the Moon's limb, and the typical spectral energy disatribution of the target sources. By this method, it is possible to achieve an angular resolution which is close to the level of 1 milliarcsecond (mas), and to measure sources having $K > 10$ on a 4 m-class telescope. Fig. 1 illustrates the actual performance of the techinque as determined on the basis of several years of experience at the TIRGO 1.5 m telescope (Richichi et al. 1996). The use of array detectors allows to push the sensitivity limits by at least 2 mag (Richichi 1994).

The angular resolution, sensitivity and range of operating wavelengths of LO thus match very closely those currently specified for the VLTI (Paresce et al. 1996). No other method or technique available at present can offer all these characteristics at the same time. Therefore, it appears that prospective users of the VLTI could find it useful to take advantage of the experience developed with LO, in doing HAR investigations with similar resolution and sensitivity. Expecially so, since time on the VLTI will be necessarily precious, due to the high demand for a facility which will be unique in the southern emisphere, and because of the intrinsic observational methods of long baseline interferometry.

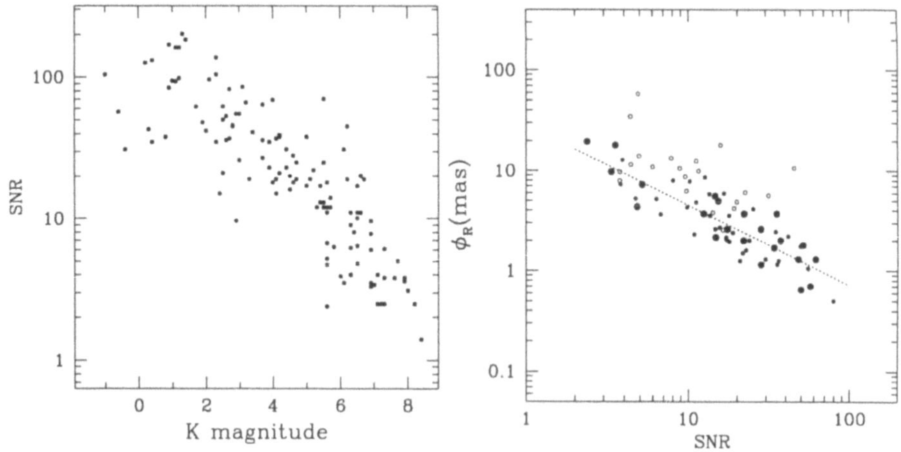

Fig. 1. *Left:* SNR values for a sample of LO lightcurves obtained at a 1.5 m telescope (Richichi et al. 1996), as a function of the K magnitude of the occulted star. The large spread of magnitudes at any given value of the SNR is indicative of the strong influence of the background. *Right:* Limiting resolution ϕ_R as determined on a sample of unresolved sources (Richichi et al. 1996). Different dot types are used to illustrate the effect of the integration time used to sample the lightcurves, from short ($\tau \leq 2.0$ ms, large solid dots) to long ($\tau > 4.0$ ms, open dots). The line is a fit to the solid dots only, and is consistent with $\phi_R \propto \mathrm{SNR}^{-1}$.

Furthermore, an initial phase will exist when only a limited number of baselines will be available, and only one or two instruments which presumably will offer only some of the observational capabilities planned for the later fully operational phase. One should also consider that, again under the present development plans, it is likely that the VLTI will operate for a significant fraction of its time either in pure VISA mode with two (or later three) ATs, or in hybrid mode with one UT. In short, it can be foreseen that during its initial phase, which could last a few observing periods, the VLTI will be more suited to investigate relatively simple objects such as binaries and stellar disks and envelopes, rather than doing actual imaging of complex sources. This offers one more parallel with the situation encountered in LO work, where in general it is possible to obtain only 1D scans across the sources, and more rarely obtain quasi-2D information by combining observations from different sites (Richichi et al. 1995).

2 Current Status and Impact of VLTI Observations

We shall restrict the present discussion to the topic of late-type giant stars. Other fields of investigations which have benefitted from LO research and are likely to be a priority for the VLTI, expecially in its initial phase, include those of binaries among the main sequence stars, and of T Tauri stars. These topics are covered, at least partially, in other contributions to these proceedings (notably those by Davis, Aristidi et al., Natta, Menard among others), and we will not deal with them here. Also, the connection between LO and the VLTI for what concerns the thermal IR has been the subject of another contribution (Stecklum et al. in these proceedings) so we can restrict the following discussion to the near-IR, typically the K band. Finally, we mention that HAR investigations of main sequence stars have been the subject of another contribution to these proceedings, by Booth and Davis.

For some decades now, the measurement of the angular diameters of late-type giants has been a traditional application of all HAR techniques. In this field, LO have played a major role, thanks expecially to their higher angular resolution. A review by White and Feierman (1987) listed 124 stars with an angular diameter measured by LO (admittedly not all of late type), and Ridgway et al. (1980b) based their pioneering calibration of the effective temperatures of late–type giants on LO results. Since the review mentioned above, it can be estimated that an additional 30-40 measurements have been collected, partly available in the literature and partly yet unpublished. Also, an increasingly larger number of angular diameters are being obtained also by long–baseline interferometry (LBI). Two effective temperature calibrations have appeared based at least partially on IR LBI measurements (Di Benedetto 1993, Dyck et al. 1996), basically confirming the early results of Ridgway et al. (1980b), and more groups are providing contributions in this area (see Perrin, these proceedings).

The question naturally arises, whether it is still important to collect these kind of measurements, and if so what is the role that the VLTI could play. It can be shown by semi-quantitative arguments (Richichi et al. 1996) that, using the Two Micron Sky Survey data, about 50% of the stars with an angular diameter presumably $\gtrsim$3 mas lying on the Zodiacal belt still remain to be measured. Interestingly, these candidates are mostly in the southern emisphere, and a rough estimate of those who could be accessible to the VLTI - -including an extrapolation to those outside the Zodiacal belt- is probably of order 500 sources. This could represent a significant improvement, although admittedly not a dramatic one when compared to the roughly 200 sources measured to date. Rather, what makes the use of the VLTI very appealing is its potential to increase the accuracy on the individual measurements, and to detect sources which are very cool.

At present, one of the limitations in the empirical calibration of the effective temperatures, is the scatter within a given spectral type. If one inspects the $T_{\rm eff}$ calibration by Ridgway et al. (1980b), it is apparent that there is a wide scatter, due partly to the fact that the data came from several different groups and were obtained at a time when data analysis methods were less accurate than it was claimed. Reversely, if one inspects a $T_{\rm eff}$ calibration obtained by LBI, such as that of Di Benedetto and Rabbia (1987), it is evident that the sample is limited to one or two stars at the most per spectral (sub)type, thus giving an impression of better accuracy which could be artificial. This is due to the fact that with baselines limited to only a few tens of meters, there are not very many giant stars that can be fully resolved. Gaining a factor of 5-10 in the baseline, the VLTI could effectively resolve hundreds of giant stars, thus combining the

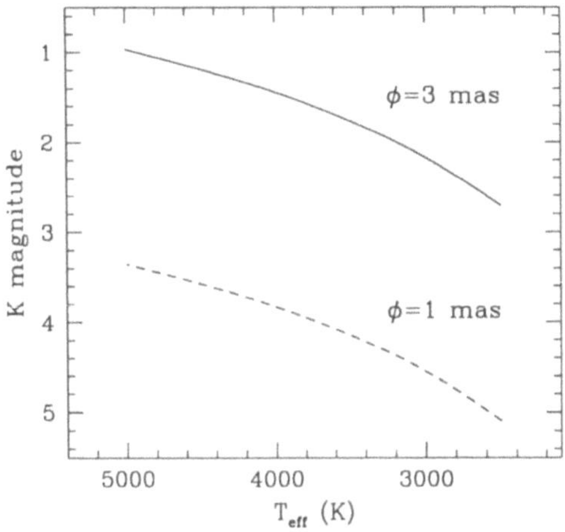

Fig. 2. The relation among the angular diameter of a star characterized as an ideal black–body, its effective temperature and its K magnitude (from Richichi et al. 1996).

sample richness of LO with the accuracy of LBI methods.

The second aspect concerns the cooler spectral types. So far, the available calibrations are reliable only for spectral types warmer than M6. Giant stars cooler than this class, equivalent to $T_{\rm eff}{\approx}3300\,{\rm K}$, have been measured very seldom, the reason being their relative scarcity when one is subject to the contrains of bright magnitudes (LBI) or position on the sky (LO). From Fig. 2 it can be seen that the relationship between angular diameter, $T_{\rm eff}$ and K magnitude is rather steep (in the approximation of an ideal black–body). Therefore, the number of cooler stars which can be resolved with a fixed angular resolution increases significantly if one can gain appreciably in limiting magnitude. With its sensitivity limit well below that of current LBI facilities, and without the pointing limitations of LO, the VLTI is in a position to measure a very large number of giant stars in the spectral classes cooler than M7.

The combination of the two aspects described above opens up a new possibility, namely that of studying in detail the effects of time variability and of chemical composition. In fact, it is clear that some -if not most- of the scatter in the $T_{\rm eff}$ calibration as we know it today could be due to changes in temperature for a same star observed at different phases of its variability cycle. Mass loss and pulsation are very common phenomena among late–type stars, with a frequency which increases for the cooler types to the point that they probably affect all giants later than M7. Investigations by LO (Ridgway et al. 1980a, Richichi et al. 1988, 1995) have evidentiated this behaviour in a few cases, but have lacked the necessary time coverage for extensive investigations. It is hoped that the gain offered by the VLTI could give us a new insight in the role played by these effects.

In the same way, the VLTI could prove to be instrumental in studies of mass loss and circumstellar matter around late–type giants. The experience

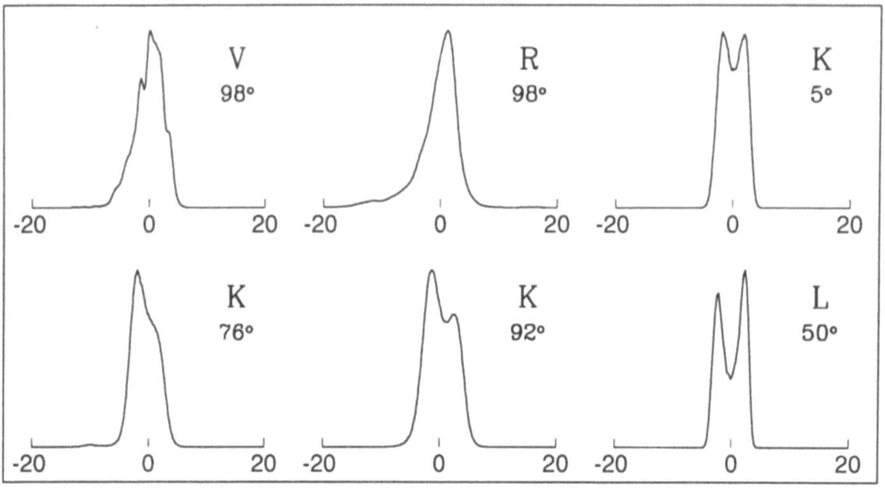

Fig. 3. Brightness profiles for TX Psc derived by LO, renormalized to the same arbitrary value. Horizontal axes are in milliarcseconds (1 mas $\approx$0.3 AU at the distance to TX Psc). Adapted from Richichi et al. (1995).

of LO shows that, for spectral types later than M5, circumstellar matter is very common on scales of few stellar radii. Emission with typical brightness ratios of 10-20% to the photospheric brightness are encountered at K, and often with non symmetric shapes. An illuminating example is that of the carbon star TX Psc, measured by LO in several wavelengths and along different position angles (Richichi et al. 1995), revealing circumstellar emission with a rich structure on scales of $\lesssim 0''.02$. Examples like this are at the limit of current LBI capabilities, while LO cannot provide the desired coverage. The VLTI in turn would be the ideal instrument to follow up individual interesting cases discovered by LO.

3 Concluding Remarks

From the experience gained by LO, we conclude that an instrument like the VLTI, capable of obtaining angular resolutions close to the milliarcsecond level in the near–IR and of surpassing the limit of K=10, could provide useful results in the field of cool giant stars, by significantly extending the number of measured angular diameters, by solving the problem of the effective temperature calibration for spectral types later than M6-M7 (still largely unknown at present), by providing the means for accurate investigations of the effects of stellar pulsation and mass loss, and by extensively probing the immediate circumstellar emission in those cases previously discovered by LO and which cannot be followed due to the limitations of the technique.

It is noteworthy to stress that this type of applications are ideally suited to the initial phase of VLTI operation, because they can be satisfied even with relatively simple instrumentation and with a limited number of baselines (i.e. actual closure-phase imaging is helpful but not strictly necessary).

The use of the auxiliary telescopes only would be sufficient. In this respect, we mention that all the target sources would be effectively unresolved for the individual ATs, and bright enough to serve as intrinsic references for a low order wavefront correction. This is a considerable advantage, since the implementation of laser guide star adaptive optics is likely to be postponed to a later phase, and at K the individual ATs can be made diffraction-limited with a relatively simple wavefront correction. The problem of fringe tracking is slightly more severe, since the stars are resolved by definition. In general, it is hoped that it will be possible to find fringe-tracking references within the 1-arcminute field of view, since the target stars are typically in the galatic plane and therefore in relatively crowded fields. Even if this condition should not be satisfied, it is conceivable that due to the flexibility of the VLTI (VISA) scheme, several sets of baselines could be found, which would allow to measure several points in the visibility curves, while maintaining enough fringe contrast for tracking.

References

Di Benedetto, G.P., Rabbia, Y. (1987): A&A, **188**, 114

Di Benedetto, G.P. (1993): A&A, **270**, 315

Dyck, H.M., Benson, J.A., van Belle, G.T., Ridgway, S.T. (1996): AJ, **111**, 1705

Paresce, F., Mourard, D., Bedding, T., Beletic, J., Haniff, C., Leinert, C., Malbet, F., Mariotti, J.M., Mozurkewich, D., Mundt, R., Petitjean, P., Quirrenbach, A., Reinheimer, T., Richichi, A., Röttgering, H., von der Lühe, O., Waters, R. (1996): The Messenger, **83**, 14

Richichi, A. (1994): IAU Symposium 158, W.J. Tango, J.G. Robertson (eds.), 71

Richichi, A., Salinari, P., Lisi, F., ApJ, **326**, 791

Richichi, A., Chandrasekhar, T., Lisi, F., Howell, R.R., Meyer, C., Rabbia, Y., Ragland, S., Ashok, N.M. (1995): A&A, **301**, 439

Richichi, A., Baffa, C., Calamai, G., Lisi, F. (1996): *submitted*

Ridgway, S.T., Jacoby, G.H., Joyce, R.R., Wells, D.C. (1980a): AJ, **85**, 1496

Ridgway, S.T., Joyce, R.R., White, N.M., Wing, R.F. (1980b): ApJ, **235**, 126

White, N.M., Feierman, B.H. (1987): AJ, **94**, 751

The Temperature Scale of the M Dwarfs

Andrew J. Booth[1], John Davis[1] and Mike Bessell[2]

[1] Astronomy Dept., School of Physics, University of Sydney, NSW 2006, Australia
[2] Mount Stromlo Observatory, Australian National University, Canberra, Australia

Abstract. The effective temperature scale for M dwarfs is largely undetermined. We show that the VLTI can make a significant contribution to the data for these stars.

1 Discussion

At the lower end of the main sequence only two eclipsing binary systems, YY Gem (M0.5, 3770 K), and CM Dra (M4.5, 3150 K) (Habets & Heintze, 1981) have accurate measured temperatures (Fig. 1). Thus, it is difficult to plot an accurate observational HR diagram for this region. However, these stars are intrinsically important and interesting (see eg. Bessell & Stringfellow, 1993).

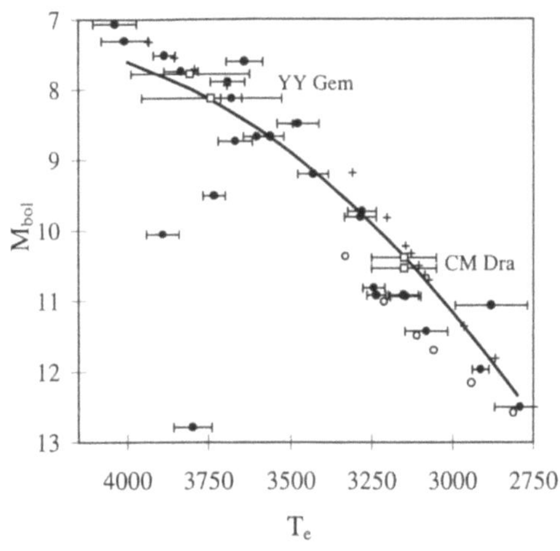

Fig. 1. Calculated HR diagram; • this work, □ existing data, — calibration of Bessell, 1994, o models of Burrows et al., 1993, + models of Baraffe & Chabrier, 1996

Currently the best approach to obtaining a temperature calibration is by model atmosphere fits to flux distributions and broad band colours. Figure 1 shows the empirical calibration of Bessell (1994). Whilst the model atmosphere fits are generally good, there are still problems with water and grain opacities (eg. Allard & Hauschildt, 1994). Evolutionary models tend to be ~200 K hotter or cooler than the empirical calibration (eg. Burrows et al., 1993, Baraffe & Chabrier, 1996 in Fig. 1). Clearly, more empirical temperatures are needed with accuracies at least as good as CM Dra and YY Gem (±100 K - ±200 K).

From the integrated flux at the earth, f_e, and angular diameter, θ, the effective temperature, T_e, can be obtained using $4f_e = \theta^2 \sigma T_e^4$. The VLTI with the Auxiliary Telescopes in the near IR is well suited to obtaining θ for these stars which are bright in the IR. This, with the large VLTI(AT) apertures, gives high photon rates and good signal to noise in visibility measurements. Unfortunately, in the near IR a ~ 200 m baseline means that no M dwarfs can be completely resolved. Thus, the accuracy of the measurements of θ and hence T_e depend on how well small changes in visibility can be measured. We use the calibration of Bessell (1994) and data from Leggett (1992) to predict θ for nearby M dwarfs. The most resolved of these has a visibility of $\sim 50\%$ of peak in J band at a 200 m baseline. In Fig. 1 we plot all stars whose visibility is $\lesssim 90\%$ of peak at 200 m.

The accuracy of the visibility measurements is crucial. At high photon rates the signal to noise is dominated by residual atmospheric variations. Experience with SUSI (Davis et al., 1994) suggests an accuracy in visibility of 0.5% (though in the near IR this could be better), and we have used this value to compute errors in T_e values for these stars. It must be noted that the accuracy of θ values depends on the accurate *calibration* for these fringe visibility measurements which requires unresolved reference sources (assumed accurate to 0.5%).

Figure 1 plots these results as likely errors for T_e in an observational HR diagram. $M_{\rm bol}$ values are from Bessell (1994) and the actual T_e values are not significant. Observation are assumed to be made in J band at a 200 m baseline.

2 Conclusion

Figure 1 shows that the VLTI can make a major contribution to this part of the HR diagram. Atmospheric and evolutionary models will be better constrained and the observational temperature scale will be extended to at least M6. There will be ~ 160 stars with T_e errors of $\lesssim 5\%$.

Some caveats should be placed on these T_e errors. There is no allowance for limb darking corrections, though these should be small. We have also made no allowance for errors in the bolometric flux measurements. This is unlikely to be realistic and good IR spectrophotometry will be needed for these stars.

References

Allard, F. & Hauschildt, P. H. (1994): E.S.O. Astrophysical Symposia, The Bottom End of the Main Sequence – and Beyond, Ed. Tinney, C. G., 32

Baraffe, I. & Chabrier, G. (1996): ApJ. **461**, L51–L54

Bessell, M. S. (1994): E.S.O. Astrophysical Symposia, The Bottom End of the Main Sequence – and Beyond, Ed. Tinney, C. G., 123

Bessell, M. S. & Stringfellow, G. S. (1993): ARA&A. **31**, 433–471

Burrows, A., Hubbard, W.B., Saumon, D., & Lunine, J. I. (1993): ApJ. **406**, 158–171

Davis, J., Tango, W. J., Booth, A. J., et al. (1994): Proc. of the SPIE Symp. Amplitude and Intensity Spatial Interferometry II, Vol. 2200, 231

Habets, G. M. H. J. & Heintze, J. R. W. (1981): A&AS **46**, 193–237

Leggett, S. K. (1992): ApJS. **82**, 351–394

Observations of Cepheids with the VLTI

Denis Mourard

Observatoire de la Côte d'Azur, Dépt. Fresnel, URA1361 du CNRS,
2130 Route de l'Observatoire
06460 Saint Vallier de Thiey, France

Abstract. The potential of VLTI for the observations of galactic Cepheids is investigated.

1 General Context

The main goal of this program concerns the improvement in the accuracy for the point 0 of the Period-Luminosity relationship for cepheids, a fundamental parameter for the determination of the distance scale in the universe. The program is based on simultaneous measurements of accurate radial velocities curve and cyclic variations of the angular diameter. Many other exciting points could be adressed by this program, such as detecting surface structure variations along the pulsation cycle or improving the models of pulsating atmospheres.

2 Preparation of the Sample

The selection of stars is made using the Simbad database at CDS with the following criteria: visual brightness $V < 10$, declination $-80° < \delta < +30°$ for a latitude of the Paranal site of $lat = -25°$, and objects of type "Cepheid". The second step is a cross-check of the result with the catalog of Caldwell et al. (1987) in order to obtain the period. In order to estimate the angular diameter of each stars, we use the F(V,V-K) relationship established by Di Benedetto (1993).

$$log\, \Phi = 0.5334 \; - \; 0.2V \; + \; 0.266(V - K), \tag{1}$$

where we estimate the value of (V-K) from the Johnson table using the mean spectral type for each star.

The interferometric observation is define by the length of the projected baseline onto the sky and the wavelength of observation. For the Cepheids program, our goal is to optimize for each star the ratio $\frac{\Delta V}{\sigma_V}$ between the variation of visibility induced by the cyclic variation of the angular diameter and the standart deviation of the visibility measurement. If σ_V is related only to the photon noise, then σ_V is proportional to V and the optimal baseline is the one corresponding to the first zero of the visibility function. However, I prefer to choose a more realistic criterion where the asymptotic value of σ_V is of the order of 1%, limited by the random visibility attenuation factors. This strategy has been adopted for the observations of δ Cephei with the GI2T interferometer (Mourard et al. (1996)). Under these conditions, the optimal baseline corresponds to the maximum of the derivative of the visibility function. With the assumption of a uniform disk of angular diameter Φ in mas, a baseline B in meters and a wavelength λ in nanometers, it corresponds to the following relation:

$$15.23\frac{B\Phi}{\lambda} = 2.3 \tag{2}$$

The maximum ground baseline of the VLTI is 200 meters for the ATs but after the projection onto the sky the apparent baseline at transit is limited to $200.cos(\delta - lat)$. For each star of the sample we calculate the optimum baseline and the corresponding mean visibility and amplitude of variation of visibility, assuming a variation of angular diameter $\frac{\Delta\Phi}{\Phi} = 15\%$. At $\lambda = 600nm$ we obtain 44 stars with a ΔV larger than 3% and around 10 stars with a ΔV larger than 10%, whereas at $\lambda = 2.2\mu m$ the sample is reduced to only 4 stars having a ΔV larger than 3%.

3 Discussion

The sample of Cepheids selected has a mean m_V magnitude of 6.6 and a typical period of 11 days. Using the VLTI with two ATs on the longest baseline and observing at the shortest wavelength corresponds to the best strategy for this program. The observations will be done every nights at the same baseline during at least two cycles and with a particular attention paid to the absolute calibration of visibilities. With a good coverage of the sample, one can expect to reach an absolute calibration of the zero point of the period-luminosity relationship at the level of less than 0.1 magnitude.

The four brightest Cepheids of the program (magnitude m_V smaller or equal to 4) have a predicted angular diameter larger than 1 mas. Their disks could be studied by the interferometer and we could measure the variation of the limb-darkening as a function of the pulsating phase. With a rather more extensive (u,v) plane coverage it will be possible to test the fundamental hypothesis of pulsation for these stars and bring strong constraints on pulsating models of atmosphere (Sasselov and Karovska (1994)). The influence of metallicity in the Period-Luminosity relationship will be investigated in order to increase the accuracy of the absolute calibration.

Acknowledgements: *I wish to thank Daniel Bonneau and Farrokh Vakili for the numerous and fruitful discussions that we have had on this program*

References

Caldwell, J.A.R. and Coulson, I.M. (1987): AJ **93**, 5, 1090
Di Benedetto, G.P. (1993): A&A **270**, 315
Mourard, D., Bonneau, D., Koechlin, L., et al. (1996): A&A in press
Sasselov, D.D. and Karovska, M. (1994): ApJ **432**, 367

CONFERENCE SUMMARY

The VLTI and the Universe: Conference Summary

Andreas Quirrenbach[1] and Jean-Marie Mariotti[2]

[1] Max-Planck-Institut für Extraterrestrische Physik, Giessenbachstraße, Postfach 1603, D-85740 Garching, Germany
[2] Observatoire de Paris, Section de Meudon, DESPA, 5, Place Jules Janssen, F-92195 Meudon Cedex, France

1 Interferometry: From Fringe Science to Mainstream Astronomy

When the coherent combination of the four telescopes of the VLT was first discussed during the design phase of the observatory, interferometry seemed to be a small field with very specialized applications. However, the reports issued by ESO's successive advisory groups on interferometry (ESO 1986, 1989, 1992) already showed the large potential of the VLTI for a broad range of astronomical problems. This became even more apparent during the ESO Workshop on "Science with the VLT", which took place in June 1994 (see Walsh and Danziger 1995). Several contributions to this workshop were dedicated to the interferometric mode of the VLT, and many others discussed the use of the VLTI in the context of larger multi-instrument programs. The strong interest in interferometry expressed by astronomers working in a wide variety of fields and the necessity to build the VLTI within tight financial constraints and optimized for the most promising scientific programs prompted ESO to form a new Interferometry Science Advisory Committee in April 1995. The ISAC was charged with updating the science case for the VLTI and prioritizing the technical options in order to maximize the astronomical return from the interferometer. A summary of this work has been published in a recent issue of the Messenger (Paresce et al. 1996).

At the same time ISAC served as the Scientific Organizing Committee for this workshop. In the minds of its members, the aim was clearly to present in more detail the ideas and views of ISAC, to gather new developments in astronomy since 1994 and their implications for the VLTI, and to solicit input and comments from potential users on what they expect from the VLTI. All speakers were asked to address explicitly the question of how their project would be carried out with the VLTI, and which capabilities would be required to accomplish the scientific goals. Throughout the workshop, it has been a real pleasure to notice that this challenge has been met: almost all participants succeeded in making the connection between their aspirations and the harsh reality of instrumental limitations. In our view, this has been one of the great successes of the workshop: mainstream astronomers have started to feel comfortable talking about fringe tracking and visibility calibration, just as if they were talking about photometric accuracy or signal-to-noise in a spectrum. Indeed, "Science with the VLTI" was perhaps the most comprehensive overview ever of the scientific questions that can be addressed with ground-based interferometry.

2 Opportunities for the VLTI, Old and New

As expressed by the choice of title, the workshop's primary goal was a discussion of the scientific potential of the VLTI. It has been clear since the time of Michelson's and Pease's pioneering work that interferometry can be applied to a broad range of problems in stellar and circumstellar astronomy. The capabilities of the VLTI, in particular its sensitivity and wavelength coverage, will substantially expand the range of problems that can be addressed with the instruments that are currently operational. The high expectations of stellar astronomers were reflected in the large number of sessions dealing with stellar diameters and fundamental parameters of stars, imaging of stellar surfaces, binary star research, and the analysis of stellar populations and determination of proper motion in extremely crowded regions like globular clusters and the Center of the Galaxy. Observations of circumstellar matter, related to accretion or mass loss phenomena, was another topic that received a lot of attention. Obtaining images of disks around pre-main-sequence objects is certainly one of the key scientific goals of the VLTI. Of equal importance are observations of dust disks around main-sequence stars, which range from β Pictoris-like objects down to the level of the zodiacal light in the Solar system, a topic that is still good for more surprises.

On the whole, however, we have a fairly clear picture of what we can expect from the VLTI for stellar astronomy, since there are well-developed models and theories that make clear predictions, and since some programs can be direct follow-ups to current efforts at interferometers with smaller aperture. In contrast, observing extragalactic objects with optical and near-infrared interferometry means entering completely uncharted territory. This implies that predictions of the scientific return from the VLTI are much more uncertain; on the other hand, the potential for breakthroughs in long-standing important questions is very high. Measuring the sizes of the dusty tori that are at the core of unified theories of Active Galactic Nuclei, resolving the broad line region in nearby Seyfert galaxies, determining the importance of star formation for the energy balance of Seyfert nuclei, and studying the structure of galaxies at high redshift are exciting prospects for the VLTI.

Stellar and extragalactic astronomy are the two pillars on which planning for the VLTI has always rested, but there is more. There are quite a few questions concerning small bodies in our own Solar System which could potentially be addressed with the VLTI. The main issue here is in which cases fringe tracking on the object itself is possible, or how suitable reference objects can be found. Among the recent developments in astronomy, the detection of brown dwarfs and extrasolar planets stands out as an event that opens a completely new area where observations with the VLTI will probably have a large scientific impact. Interferometric astrometry is a powerful method to search for Jupiter-like planets out to nearly 1 kpc, and for planets with 10 Earth masses orbiting the nearest stars. The surprising detection of a giant planet in a very close orbit around 51 Peg has established a new class of objects, which may be sufficiently bright to be detected directly with the VLTI. The VLTI could thus become the first instrument to perform spectroscopy and photometry of some extrasolar planets, and therefore to determine their temperatures, composition, and rotation periods. Some flexibility is required, of course, to incorporate the technical requirements for these challenging observations into the development plan for the VLTI.

3 Interferometric Life Before (and Beyond) Imaging

During this workshop, we have seen an expansion of the scientific agenda of the VLTI into new areas, and refined ideas about realistic observing programs. As a consequence, it has been realized that much, perhaps most, of the science with the VLTI can be done without "imaging". For example, there is a lot of interest in detecting faint companions (stars at the bottom of the main sequence, brown dwarfs, giant planets) near much brighter stars. Several talks dealt with methods to do this, involving relative phase information, either the phase difference between two wavelength bands, or the phase difference between the primary star and a nearby reference (i.e., narrow-angle astrometry). These approaches are orders of magnitude more sensitive than synthesis imaging of the field around the primary using closure phase, and they require fewer telescopes (three are adequate, four are good). As a bonus, information on scales smaller than the resolution limit of the interferometer can be obtained. This compensates partly for the fact that the baselines of the VLTI are somewhat smaller than what one would ideally like to have.

Another important point is the realization that even in situations where synthesis imaging is the "best" method, much of the scientifically interesting information can be extracted from much more limited data, if adequate models are available. Recall, for example, the discovery of apparent superluminal motion in extragalactic radio sources, which involved fitting simple few-parameter models to data obtained with a VLBI "array" consisting of only two antennas (Whitney et al. 1971; it is very instructive to read this classical paper with 25 years of hindsight)! In many cases, a simple measurement of the size or overall geometry of an object contains much useful information. This has also been demonstrated by results from single-baseline optical interferometers, such as the I2T, GI2T and Mk III instruments, which show that the Hα emission region of Be stars has a disk-shaped geometry. Data of similar quality on T Tauri and β Pictoris disks, and on dust tori in Seyfert galaxies is exactly what we hope to get from the VLTI soon after "first fringes".

4 The VLTI – a Community Instrument

At this point, it might be useful to recall that the VLTI is unique among the world's interferometer projects insofar as it has been conceived within the framework of a large observatory, and with a broad community of users in mind. In contrast to the situation at all operational interferometers, the typical observer using the VLTI will not be a member of the highly specialized team that built the instrument, but rather an astronomer who needs data at milliarcsecond resolution to complement his or her work with other methods. This concept of the VLTI as a community instrument will certainly enhance its scientific output, since it can capture the imagination of users from all fields of astronomy, observers and theorists alike. The broad range of topics covered at the workshop is an encouraging indication that this process is already well on its way.

However, additional ideas for applications of the VLTI are very welcome, especially in the areas of extragalactic astronomy and cosmology. Here, as in other fields, it must always be kept in mind that interferometry on its own will rarely lead to substantial progress; its real power lies in the unique capability to test predictions that have been formulated on the basis of observational work at lower angular resolution and theoretical modelling. Many projects therefore require the combination of data from the VLTI and more traditional techniques (spectroscopy, photometry, adaptive optics imaging etc.); "one-stop-shopping" at ESO will facilitate the planning of such observations tremendously. Open access to the VLTI will also ensure that programs of all sizes can be carried out, from a snapshot of a single target (e.g., measurement of the diameter of a particular star as a function of wavelength) to long-term monitoring of large samples (e.g., astrometric surveys for low-mass companions), with detailed studies of individual key objects (such as 51 Peg, α Ori, β Pic, the Galactic Center, NGC 1068) ranging somewhere in between.

While setting the scientific agenda for the VLTI is one important task for ESO's user community, involvement in technical development is equally important. In spite of many advances in the key technologies of beam combiner design, detectors, laser metrology, vibration suppression, and instrument control, astronomical interferometry is still difficult. The workshop saw several debates among the specialists about the merits of different instrumental concepts, and about the accuracy of detailed predictions of their performance. It appears unavoidable that the VLTI will have to support both astronomical research and development of interferometric techniques with a fairly large share of technical aspects for quite a while to come. Fortunately, ESO can rely on a number of institutes and observatories with experience in interferometry and related areas for building the "focal plane" instrumentation of the VLTI, beam combiners and fringe detectors. In short: to realize the full potential of the VLTI, ESO must draw on the astronomical *and* technical expertise available within the European astronomical community.

5 The Next Steps for the VLTI

Now that we have discussed our dreams about the science we wish to do with the VLTI, we have to turn back to the question of how we can make these dreams become realities. The first important task ahead is the formulation of a prioritized instrumentation plan with realistic goals. The astronomical programs call for instruments covering the whole wavelength range from 0.5 to 20 μm; some require simultaneous operation in two widely separated wavelength bands for fringe tracking and data taking. Monitoring programs take a lot of observing time and thus need the auxiliary telescopes. To image complex objects, the auxiliary telescopes have to be combined with the four 8 m UTs, which is possible only if a sufficient number of delay lines is available. Off-source fringe tracking calls for dual star feeds and additional differential delay lines. Narrow-angle astrometry needs extremely precise metrology on top of that. The sensitivity in the visible and near-IR depends strongly on the availability and quality of adaptive optics.

It is clear that all these aspects will not be realized from the beginning. Making the right choices about a sufficiently capable, but at the same time affordable and technically viable, "first light" configuration will be crucial for the health of the VLTI program. At the same time, a plan has to be drawn up on how to phase in additional instruments and capabilities. In this process the balance has to be kept between relatively simple low cost instruments, more capable and versatile (but also more complex and more expensive) solutions, and specialized techniques optimized for a single key application. These workshop proceedings will be of invaluable help for this purpose, since they will give us a good indication of the technical requirements for each scientific project. This mapping from the space of possible instrumental configurations into the space of astronomical problems that can be tackled should form the basis of discussions between astronomers and engineers about the detailed implementation plan for the VLTI. Where the mapping is still incomplete – for example in the area of Solar System research – further feasibility studies are highly desirable.

In parallel with the technical planning, observing strategies for short observations and for long-term programs have to be prepared. This involves an understanding of operational aspects like calibration procedures and scheduling requirements. These issues are more complex for the VLTI than for a single telescope, since visibility measurements are strongly influenced by "dirty" atmospheric and instrumental effects, and because of the peculiarities of aperture synthesis. Some observing programs will require special preparatory work. An example would be the systematic search for random associations of high-redshift galaxies and quasars with suitable foreground stars that can be used as references for fringe tracking.

While planning for the VLTI is a complex process in itself, we must not forget that it is strongly influenced by outside events. Financial aspects and unexpected astronomical discoveries aside, the progress made by other interferometer projects on the ground and in space, and developments in the area of adaptive optics are the two most important factors that could change the boundary conditions for the VLTI. A number of interferometers with comparatively small apertures are already producing data or will come into operation very soon. In addition, instruments with sensitivity comparable to the VLTI (the Keck, LBT, and Magellan interferometers) may see first light soon after the VLTI. Finally, plans for interferometric space missions have been on the drawing board for quite a while. With so many contenders in the starting blocks, the competitiveness of the VLTI will depend critically on timing during all stages of its implementation.

The availability of adaptive optics is a perfect illustration of this point. When the VLTI will be used at wavelengths for which its telescopes are limited by the seeing, it will not do better in terms of sensitivity than any of the already existing arrays with much smaller apertures. It is only by active correction of the atmospheric seeing through adaptive optics that we will be able to fully exploit the huge collecting area of the VLTI and dive into the regime where much fainter objects can be studied. With this in mind, it is clear that equipping the telescopes of the VLTI with adaptive optics systems should be very high on the list of the desirable upgrades.

6 The VLTI and Beyond

It will take a few years until the VLTI becomes operational, and even longer before its full potential gets realized with the combination of the four 8 m telescopes using high-order adaptive optics. Nevertheless, a look ahead into the more distant future is useful. It appears that the VLTI will be close to the limit of what can be done sensibly from the ground. Other instruments will have somewhat longer baselines, or better *uv* coverage, but it is difficult to envisage an affordable way of obtaining significantly more sensitivity, accuracy, wavelength flexibility or sky coverage than what will be provided by the VLTI.

On the other hand, interferometry from above the atmosphere looks very promising. Space is quiet, empty, and cold. In the absence of atmospheric fluctuations, the coherent integration time on any point in the sky is only limited by the observing time available, giving an enormous boost to sensitivity even with moderate (1 m class) apertures. At $\lambda \gtrsim 2.5\,\mu$m, the advantage of space is even larger, because the whole telescope can be radiatively cooled to a temperature of 40 K or so. The undistorted wavefront facilitates the calibration of the visiblities and makes advanced techniques for high-contrast observations feasible. Interferometer concepts based on "free flyers" – where each telescope is mounted on an independent small satellite – make very long baselines possible and provide superb imaging quality because the *uv* plane can be covered efficiently.

Bearing all this in mind, it seems plausible that the next big step after the VLTI will be the leap into space. Should ESO play a role in this adventure? There are very good arguments for an ESO involvement in space interferometry. Firstly, the VLTI offers unique opportunities for the development of instrumental concepts and techniques, which have to be tested on the ground before they can be incorporated into space missions. Secondly, the VLTI will play an important role in defining the key scientific topics that can be addressed with space-based interferometry. Thirdly, the efficient use of a space facility will require a user community well-trained in interferometric observations. It thus appears that some coordination between ESO and ESA could be beneficial for both sides.

A panel discussion with representatives of both agencies that took place during the workshop was devoted to the question of how a joint effort in the area of interferometry could work. Differences between ESO and ESA in their member nations, organizational structures, and scientific communities to be served will have to be addressed in order to define the scope and terms of a fruitful collaboration. With ESO's VLTI back on the track to fast implementation, and an interferometric cornerstone in ESA's Horizon 2000+ program, there is an enormous potential for a strong European role in the development of interferometry. Generating synergism between these programs can only make them stronger. We can thus look forward to exciting opportunities, and to a rich harvest of astronomical results in the near future and in the decades to come. George Miley boldly predicted that the IAU will hold its third (!) symposium on "Optical Interferometry of AGN" in November 2010. We look forward to seeing his paper at that conference, and to many other results at the meetings that will be held on imaging of T Tauri disks, stellar surface features, spectroscopy of extrasolar planets ...

References

ESO (1986), VLT report no. 49
ESO (1989), VLT report no. 59
ESO (1992), VLT report no. 65
Paresce, F., et al. (1996), The Messenger 83, 14
Walsh, J.R., and Danziger, I.J. (1995), Science with the VLT, Springer-Verlag
Whitney, A.R., et al. (1971), Science 173, 225

POSTER PAPERS

Simulations of VLTI/VISA Imaging Observations of Young Stellar Objects at 2.2μm

Nancy Ageorges[1], Oskar Von der Lühe[1]

ESO, Karl Schwarzschild Strasse 2, D-85748 Garching b. München, Germany

Star and planet formation is one of the most exciting science goals of the VLT Interferometer. Circumstellar disks related to star and planet formation are believed to be a common feature of both young and main sequence stars in different stages of development.

We have studied the potential of the VISA mode of VLTI to do disk's morphology observations by computer simulations. Our main concern is the synthesis capability of various VLTI configurations. We have concentrated here on the study of a limited sample of disk morphologies at high angular resolution. Disk parameters were derived from HST observations (Mc Caughrean et al., 1996). The simulated object is located at 440pc at a declination of -30°; its FWHM of 26 AU has an apparent extent of 60 mas. Adopting a star magnitude m_K of 9.49 and a disk luminosity of 0.5 L_*, we constructed different kind of models: *a/* a featureless disk with an exponential intensity profile and a central unresolved source, viewed with an inclination angle of 75° ("star + disk" model), *b/* a star + disk model as above with a gap of 10 AU between the star and the inner boundary of the disk, *c/* a star + disk system with a void between 12 and 20 AU from the star.

We used the CalTech VLBI software package (Pearson, 1991) for our simulations using only elliptical gaussians and uniform ellipsoids. To simulate VLTI observations we used the actual layout for 8m Unit and 1.8m Auxiliary telescopes of the VLT Observatory on Cerro Paranal. We used six configurations of three Auxiliary telescopes at any given time, the measurements from which were merged into a single data set which was subject to hybrid mapping and model fitting.

The wavelength is 2.25μm. The integration time on source is 5 min., observations are repeated after 10 min. and are carried out between 03:00:00 and 21:00:00, with the source culminating at 12:00:00. The hour angle is limited by the zenith distance limits of the telescopes (60°). The only noise sources are photon noise and a low level of detector readout noise (5e- /pixel).

Table 1 presents the parameters for the initial model (*b*), the input model as derived from the inverted and CLEANed maps and the iterated parameters after the non-linear least-squares model fitting process.

The CLEAN maps reveal that the simple approach using Fourier inversion and CLEANing is inappropriate for studying extended objects like the disk since CLEAN tries to decompose any source into a collection of point sources. At best, these maps can serve as an input for constructing initial models.

Initial model parameters can deviate substantially from the "true" values.

		FWHM (mas)		
Simulated model				
central star	0.0085	1.0000	1.0000	0.0000
gaussian disk	0.0042	60.0000	0.2500	20.0000
Definition of the	-0.00112	45.0000	0.2500	20.0000
planetesimal ring	0.00105	30.0000	0.2500	20.0000
Input model for model fitting				
central star	0.0100	1.0000	1.0000	0.0000
gaussian disk	0.0005	74.0000	0.2000	12.7000
Definition of the	-0.0010	39.9000	0.2000	15.2500
planetesimal ring	0.0050	28.0000	0.2000	17.15000
Iterated parameters				
central star	0.0085	1.0000	1.0000	0.0000
gaussian disk	0.0043	66.3553	0.25155	20.0408
Definition of the	-0.0005	44.2581	0.27248	19.9609
planetesimal ring	0.0008	27.8947	0.24074	19.6639

Table 1. Parameters of the star + disk model with the trace of a planetesimal between 12 and 20 AU. The last two boxes represents the parameters of the input model (for model fitting) and the results for the iterated parameters.

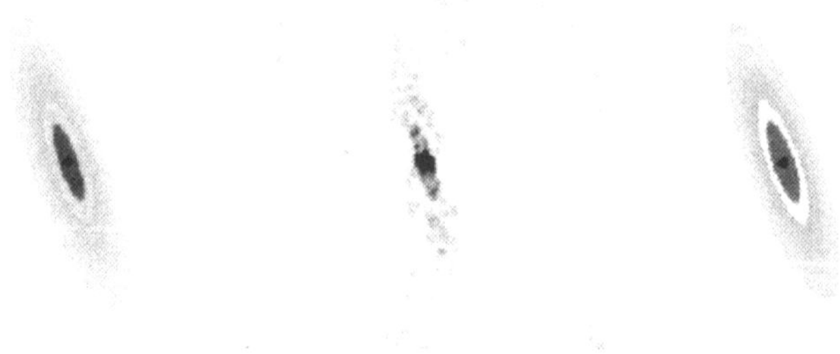

Fig. 1. Simulated model (left), CLEANed image and model derived from model fitting (right).

In most cases studied, a reasonable input model would result in a final model which describes the source quite well after 40 iterations. Parameters related to the features on the disks are quite sensitive while the initial flux considered for the large disk component would always converge close to its truth value.

We tried reducing the number of configurations, with little encouraging results. We therefore conclude that it takes quite some investment in observing time to find structures in circumstellar disks.

References

Mc Caughrean, M. J.; O'Dell, C. R. (1996): Astron. Journ. 111, p. 1977ff.

Pearson, T. J. (1991): Introduction to the Caltech VLBI Programs, California Institute of Technology.

Speckle Observations of Post-Hipparcos Late Type Double Stars

Eric Aristidi[1], Bruno Lopez[2], Marcel Carbillet[1], Pierre Cruzalèbes[2], François Mignard[2], Yves Bresson[2], and Georges Helmer[2]

[1] Département d'Astrophysique de l'Université de Nice-Sophia Antipolis URA 709 du CNRS, Parc Valrose, F-06108 Nice cedex 02, France
[2] Observatoire de la Côte d'Azur, B.P. 229, F-06304 Nice Cedex 4

1 Introduction

Late type stars are surrounded by an envelope of gas and dust. Several observations have shown the asymmetry of the envelope when the central star is a post-Asymptotic Giant Branch star. Everyone has in mind the spectacular image of Egg-Nebula made by the Space Telescope. The recent speckle-masking image reconstructions of the AGB Mira star R Cas (Hoffman et al, 1995) show a clearly elongated disc. These stars lose a fraction of their mass with a geometry that is still not well understood (bipolar jets, equatorial discs). Among the different models proposed to explain the geometry of the dust shell and of the mass loss, Morris (1987) referred to the presence of a double star system.

This assumption is supported by several observations. Since the discovery of the companion of Mira (Joy, 1926), about 20 binary systems involving a late-type star have been discovered using various techniques (spectroscopy, astrometry, speckle,...) from both ground-based telescopes and spatial observations (see Castelaz *et al.*, 1995 and references therein). A hundred of other systems have been discovered by the satellite Hipparcos during its observing run, with separations as close as 100 mas and up to a magnitude difference of 2.5 (Mignard *et al.*, 1995).

The study of such systems is of high interest for the knowledge of the stellar evolution and the understanding of the mass loss process of late type stars. Interacting mass exchange may exist in binary systems. Photometric variations of the companion could then result from this mass exchange. These photometric fluctuations have been observed for the prototype of Mira stars, Omicron Ceti, by the satellite IUE in the ultaviolet (Reimers *et al.*, 1985). Paul Couteau has observed Mira with the 50 cm refractor of Nice and has noticed some color variation of the companion (a B-type star).

2 The Project

Our present project is to make systematic observations of these stellar systems in order to study the light curves of the companions. Due to the variability of the late -type stars (for example *o* Ceti has a V magnitude ranging from 3 to 9 with a period of 311 days, its companion is of 11th magnitude) the observations must be foreseen at the main star's minimum. The close separations of these systems require long baselines and /or speckle interferometric techniques. The separation between Mira A and B is about 1 arcsec, corresponding to $\simeq$ 100 A.U. at the assumed distance of Mira (114 pc). With an interferometer of 1 mas resolution, one could observe such systems up to a distance of 100 kpc.

The stars discovered by Hipparcos have separations larger than 100 mas. They can be observed with a 1.50 m telescope. A speckle experiment has been developed at the "Laser Lune" telescope (Observatoire de la Côte d'Azur). It is currently under realisation and has given its first light on 05/30/1996. It is a simplified version of the "Tavélographe Masqué" (see below).

3 First Observations

Preliminary observations were carried out in December 1995 at the Pic du Midi 2.0 m Bernard Lyot telescope using the specklegraph developed by the Aperture Synthesis group of Observatoire Midi-Pyrénées (France) and an ICCD video camera. Observations were made in green light (λ = 550 nm, $\Delta\lambda$ = 70 nm). Among the 15 post-Hipparcos observed stars, five have been processed and two have been detected as double. Results are presented in figure 1.

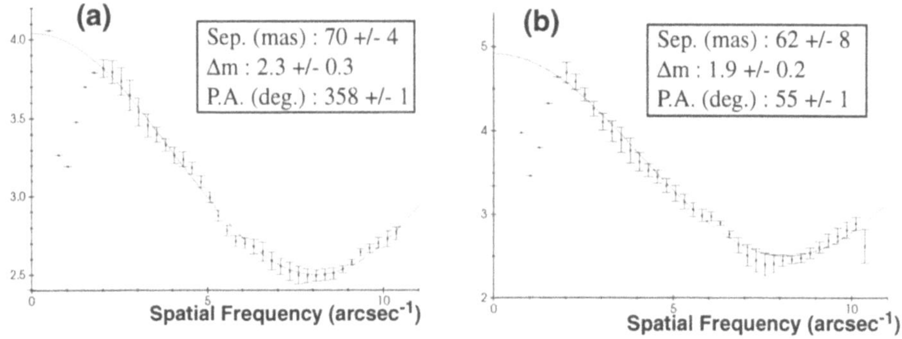

Fig. 1. Visibility functions for HIP 21046 (a) and HIP 21059 (b). The computation was made on $\simeq$ 1000 images (20 ms exposure) of the double stars and their reference star.

References

Hoffman K. *et al.*, *Bull. Spec. Astroph. Obs.* **39**, 59, (1995)
Morris, *P.A.S.P.* **99**, 105, (1987)
Joy A.H., *ApJ* **63**, 333, (1926)
Castelaz M.W., Mccollum B., *Astron. J.* **109**, 341, (1995)
Mignard F. *et al. Astron. Astrophys.*, **304**, 94 (1995)
Reimers D., Cassatella A., *ApJ* **297**, 275, (1985)

Wavefront Characterization for High Angular Resolution Techniques

R. Avila[1,2], A. Ziad[1], J. Borgnino[1], F. Martin[1], A. Agabi[1], A. Tokovinin[3]

[1] Département d'Astrophysique, Université de Nice-Sophia Antipolis, Parc Valrose, 06108 Nice, France
[2] *On leave from* Instituto de Astronomía, UNAM, México
[3] Sternberg Astronomical Institute, Universitetsky prop. 13, 119899, Moscow, Russia

Abstract. We present the theoretical 2-D angle of arrival covariance and temporal cross-spectrum, for the wavefront characterization with the Grating Scale Monitor.

1 Introduction

In response to the need of systematic measurements of the atmospheric turbulence optical effects at long baselines, the Grating Scale Monitor (GSM)[1,2] is beeing developed. It consists in a Shack-Hartmann-kind technique made of synchronized independent modules measuring the angle of arrival (AA) fluctuations in a given direction, at different points of the wavefront (3 at the moment, 4 in a near future), with baselines varying from 14 cm to several tens of meters in any direction. The Fried parameter r_0 and the *spatial coherence outer scale* $\mathcal{L}_0$ are derived from the spatial variance and covariances of the AA measured with several pairs of modules. Their temporal cross-spectrum contains spatio-temporal information. The new version of GSM can track a star, particularly needed in the southern hemisphere, which implies the extension of the theoretical models to any baseline direction. Here we present the theoretical 2-dimensional AA covariance and temporal cross-spectrum.

2 Angle of Arrival 2-D Covariances

Until now theoretical expressions of AA covariances existed only for baselines either parallel or perpendicular to the AA measurement direction[3]. Following a similar calculation as in Borgnino et al. (1992)[3], in the case of the Von Karman model the covariance for a baseline of modulus B and making an angle θ with the AA measurement direction, can be written as

$$C_\alpha(B,\theta) = A \int_0^\infty f^3 (f^2 + \frac{1}{\mathcal{L}_0^2})^{-11/6} \left[2 \frac{J_1(\pi D f)}{\pi D f} \right]^2 [J_0(2\pi f B) - \cos(2\theta)\, J_2(2\pi f B)]\, df \ , \quad (1)$$

where f is the modulus of the spatial frequency, $A = 0.0228\pi\lambda^2 r_0^{-5/3}$ and D is the diameter of both pupils. J_0, J_1 and J_2 are bessel functions.

Using this expression a mean outer scale of 40 m was derived from the data obtained at Plateau de Calern site of the Observatoire de la Côte d'Azur during the nights of July the 11th and the 12th 1995.

3 Angle of Arrival Temporal Cross-Spectrum

The temporal cross-spectrum (TCS) has been calculated under the Taylor hypothesis, and considering a single effective turbulent layer. It is a function of the temporal frequency ν, the pupils diameter D, r_0, $\mathcal{L}_0$, the baseline vector (B,θ),and the wavefront propagation velocity vector (v,δ), both in polar coordinates, θ and δ beeing angles with respect to the AA measurement direction. Figure 1 shows some examples of the TCS modulus for three different baseline directions. We notice two cutt-off frequencies $\nu_1 \simeq v/\mathcal{L}_0$ and $\nu_2 \simeq v/(B\sin(\delta-\theta))$. If $\theta = \delta$, ν_2 goes to infinity, and the noticeable high cutt-off frequency is $\nu_c \simeq 0.5v/D$. In principle, measuring the TCS modulus for 2 different baseline directions θ, one can retrieve v and δ from the expression of ν_2. Nevertheless, a more detailed analysis including several turbulent layers has to be undertaken in order to estimate the measurability of the cutt-off frequencies.

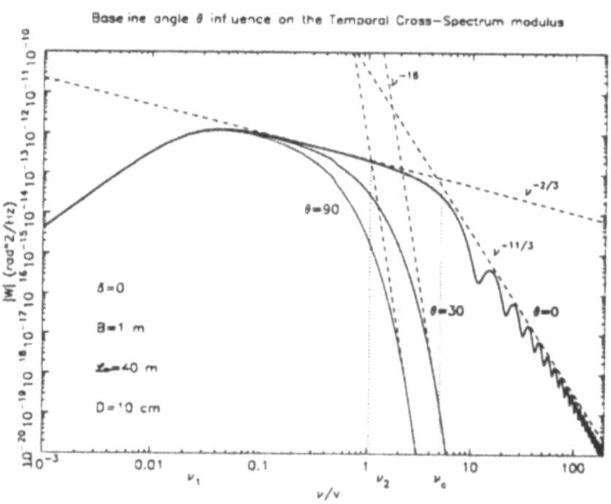

Fig. 1. Temporal cross-spectrum modulus for 3 different baseline directions θ. The cutt-off frequencies are $\nu_1 \simeq v/\mathcal{L}_0$, $\nu_2 \simeq v/(B\sin(\delta-\theta))$ and $\nu_c \simeq 0.5v/D$

References

1 Martin, F., Tokovinine, A., Agabi, A., Borgnino, J., Ziad, A.(1994): A&A Suppl. Ser. **108**, 173-180
2 Agabi, A., Borgnino, J., Martin, F., Tokovinine, A., Ziad, A.(1995): A&A Suppl. Ser. **109**, 557-562
3 Borgnino, J., Martin, F., Ziad, A.(1992): Opt. Com. **91**, 267

Interfero Coronography at O.C.A.

Pierre Baudoz[1], Yves Rabbia[2], and Jean Gay[1]

[1] Observatoire de la Côte d'Azur (O.C.A.), Département Fresnel, URA CNRS 1361, BP 04229, 06304 Nice Cedex 4, France
[2] O.C.A., Dpt. Fresnel, URA CNRS 1361, avenue Copernic, 06130 Grasse, France

Abstract. We report on an interfero-coronograph under developement at Observatoire de la Côte d'Azur (O.C.A.) which should allow, when coupled with adaptive optics on large telescope, to detect faint companion and/or circumstellar features as close as the first Airy's ring with magnitude difference roughly up to 10.

Key Words : coronography, interferometry, double stars, circumstellar matter

Besides angular and spectral resolution, high dynamical range is needed in the study of stellar environment, and a coronographic approach is required. The one based on destructive interferences is expected to yield higher angular resolution than the ones using a mask in image plane (Lyot, 1930 ; Malbet, 1996). In this framework, an interfero-coronograph has been devised and put into development at O.C.A.

This device essentially is a Michelson (beam splitting) interferometer modified by inserting on one arm a 180 degrees pupil rotation and an achromatic π dephasing of the lightwave (see figure 1, left). Both rotation and dephasing are performed by means of a focusing cat's eye. Our coronograph is presently pertaining to a single aperture, but its adaption to diluted aperture is conceivable (Gay & Rabbia, 1996). The essential condition to satisfy is that the complex amplitude entering the coronograph must remain centro-symetric up to the recombinaison step. Departures from this situation decrease the rate of extinction. This makes necessary to use adaptive optics when observing from the ground. Hence, our device is intended to work in the near infrared domain.

Achievable angular resolution leads to directly observe surrounding features as close as the first angular Airy ring for the used aperture. Another specific interest is that the extinction results from weighting to zero the light from the unresolved on-axis source so that removal concerns the whole image plane instead of darkening a part of it, as done in other approaches. Also achromaticity allows operation with wide spectralwidths. A laboratory prototype has been build to demonstrate feasibility (Gay & al., 1996) and an improved set-up is under development.

The assumed origin of achromatic dephasing is an additional focusing (Gouy, 1890 ; Boyd, 1980) however we found it safer to check that point. By comparing interferograms (integrated signal in image plane versus optical path difference, using a pointlike on-axis source) respectively recorded without and with additional focusing the doubt has been ruled out (see figure 1, right).

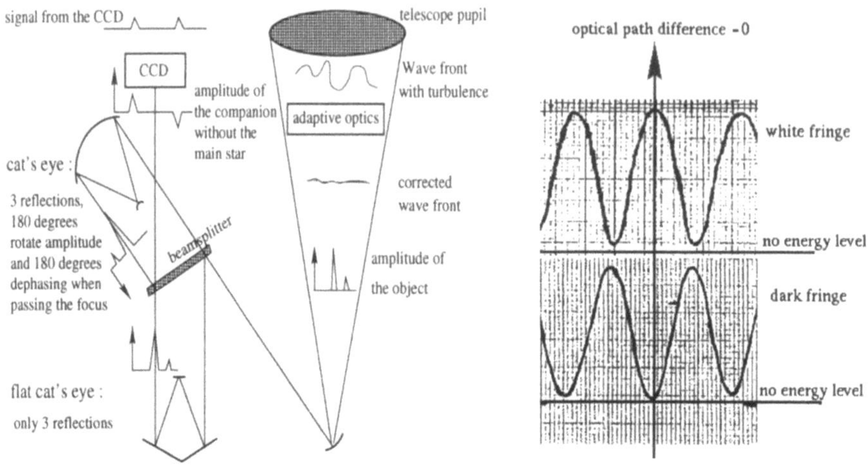

Fig. 1. left : schematic of set-up and principle, right : interferograms showing reversal of fringe around zero-OPD when replacing "flat"reflector by cat's eye

An extinction of 98 % of the light coming from the on-axis source has been readily obtained with the first and quick-made set-up, in spite of rather poor optical quality and unmatched coatings of the components. In parallel preliminary rough estimates yield an extinction better than 10 magnitudes when using adaptive optics correcting up to the ninth order.

Eventhough the present estimates fall short to detect exoplanets as the ones recently discovered (Mayor & Queloz, 1995) it is conceivable that coupling our device with the "dark speckle" approach (Labeyrie, 1995) will reduce the gap, and also much favourable star-planet configurations might occur. Anyhow, expected capabilities of our coronograph gives reasonable hopes for imaging faint features around stars.

References

Boyd R. W., 1980, J.O.S.A., vol.70, 7, p.877
Gay J. & Rabbia Y., 1996, C.R.Acad.Sci.Paris, vol 322, serie IIb, p.625
Gay J., Rabbia Y. & Manghini C., 1996, poster presented at the "IR Space Interferometry Workshop", Toledo, Spain, March 1996
Gouy L., 1890, C.R.Acad. Sci. Paris, vol 110,p.1251
Labeyrie, 1995, A & A, 298, p.544
Lyot B., 1930, C. R. Acad. Sci. Paris, 191, p.834
Malbet F., 1996, A & A Suppl, 115,p. 161
Mayor M. & Queloz D., 1995, Nature 378, 355-359

Simulation of Speckle Images for a Field Wider than the Isoplanatism Angle

H. Beaumont, C. Aime, L. J. Sánchez, H. Lantéri

Département d'Astrophysique, Université de Nice-Sophia Antipolis, Parc Valrose, 06108 Nice, France

We present in this communication a simulation of speckle images for a field of view wider than the isoplanatic angle. The model assumes a single layer of turbulence located at a distance h above the telescope pupil; h is a parameter of the model. For different points of the object field, the telescope aperture projects itself onto different regions of the remote layer of turbulence; for each of them, a different point spread function (PSF) is computed and added in the focal plane of the telescope with an intensity proportional to the object source. For the simulation, a wavefront is synthesized with a fractal algorithm, and the diffraction effect over the distance h is neglected. The PSF thus obtained are almost fully decorrelated for an angle θ wider than D/h for a telescope of diameter D. This very simple single layer model leads, however, to unrealistic values of h and prevent it to be interpreted in terms of a real physical parameter.

The general form for the object-image relationship in absence of isoplanatism is that of a Fredholm integral. Denoting O(s) the brightness intensity of the extended object and S(r, s) the instantaneous point spread function at the focus of the telescope, a point of the object produces in the focal plane of the telescope a brightness intensity of the form O(s)S(r, s). The image I(r) at the focal plane is the addition in intensity of all these contributions:

$$\mathrm{i}(r) = \int O(s)\ S(r,s)\, ds \qquad (1)$$

This expression, reduces to the well known relation of convolution in the limiting case where S(r, s) does not depend on s and can be written as S(r-s). It is important to emphasize that, in relation (1), S(r, s) may be considered as a function of r depending of the position s of the point source. In a real experiment, if we could observe point sources with different separations s, close by PSF's would be very similar, and PSF's corresponding to far away positions would appear as uncorrelated. The effect of gradual decorrelation of S(r, s) for different s values is the result of the three dimensional aspect of the atmospheric turbulence. A complete simulation of these effects would therefore include the modelling of multiple layers of atmosphere. The implementation of such a 3D model would be very time-consuming. The procedure we used to simulate these effects consists in using a single layer of turbulence that we assume to be located at a distance h above the telescope aperture. Depending upon the direction of observation s, different regions of the wavefront are used to contribute to the formation of the PSF S(r, s) as illustrated in figure 1.

Considering simple geometrical considerations, for a telescope of diameter D, an almost complete decorrelation of S(r, s) occurs for a difference of position h.s larger than D. The drawback of thus model is that an isoplanatic angle of 10" is obtained with a 1 meter telescope with a h value of 20km, which is not realistic.

Figure 2 shows the result of simulations for the same seeing conditions but reproducing a decrease of the isoplanatic angle value.

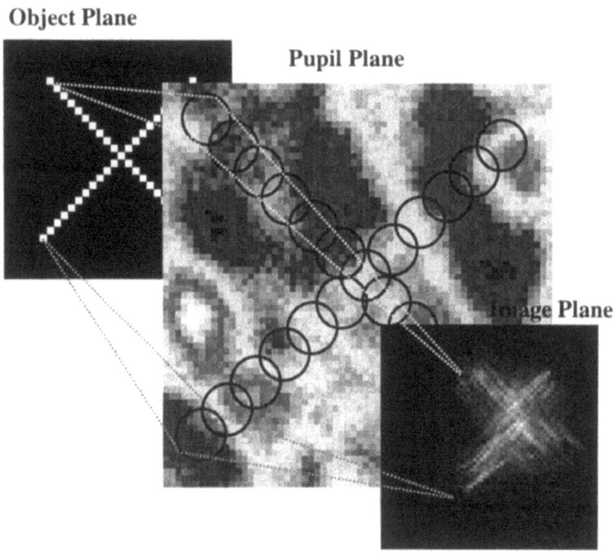

Fig. 1. For each point of the object O(s), s being a position angle, a point spread function S(r, s) is computed using a different region of the wavefront ψ(u)P(u-h.s), where P(u) is the aperture of the telescope and h a parameter that can be associated with the altitude of a single layer of turbulence. The quantity O(s)S(r, s) represents the contribution in intensity in the image plane of the point source at the position s in the object. The overall image is obtained as the sum of all intensity contributions of the sources. If the object is defined over N points, N Fourier Transform of the ψ(u)P(u-h.s) are necessary to synthesize the observed image.

Fig. 2. Simulations for identical seeing conditions and different h values. The isoplanatic angle decreases from left to right.

Visibility Losses Due to Atmospherical Spectral Decorrelation in Michelson Interferometry

Philippe Berio[1], Denis Mourard[1] and Farrokh Vakili[1]

Observatoire de la Côte d'Azur, URA 1361 du CNRS Caussols, F-06460 St. Vallier-de-Thiey, France

Abstract. This paper aims at revealing the effects of the atmospherical spectral decorrelation on the visibility measurements and at showing a method to overcome this problem.

1 Introduction

In the GI2T experiment, fringes are spectrally dispersed. The measurements method selects two different spectral bands in order to compute the cross correlation $\langle I(\boldsymbol{\theta}, \lambda_1, \Delta\lambda_1) \otimes I(\boldsymbol{\theta}, \lambda_2, \Delta\lambda_2)\rangle$ and then the polychromatic cross-spectrum (PCS) $\langle \tilde{I}(\mathbf{f}, \lambda_1, \Delta\lambda_1)\tilde{I}^*(\mathbf{f}, \lambda_2, \Delta\lambda_2)\rangle$. Our analysis consists of a theoretical computation of this spectrum, based on a numerical model for the atmospheric turbulence. From these theoretical computations, we are going to reveal the effect of the increase of the separation between the spectral bands.

2 Theoretical Computation

From the splitting of the spectral bands in elementary bands $\delta\lambda$, the PCS computation is reduced to several dichromatic cross-spectrum (DCS) computations.

$$\langle \tilde{S}(\mathbf{f}, \Delta\lambda_1)\tilde{S}^*(\mathbf{f}, \Delta\lambda_2)\rangle = \frac{1}{NM}\sum_{i=1}^{N}\sum_{j=1}^{M}\langle \tilde{S}(\mathbf{f}, \lambda_i)\tilde{S}^*(\mathbf{f}, \lambda'_j)\rangle, \tag{1}$$

The DCS at λ_1 and λ_2 can be written as (Borgino and al. 1989):

$$\langle \tilde{S}(\mathbf{f}, \lambda_1)\tilde{S}^*(\mathbf{f}, \lambda_2)\rangle = \frac{1}{s^2}\int M_{1,2}(\mathbf{f}, \boldsymbol{\rho})A_{1,2}(\mathbf{f}, \boldsymbol{\rho})d\boldsymbol{\rho}, \tag{2}$$

where $M_{1,2}(\mathbf{f}, \boldsymbol{\rho})$ is the fourth-order momentum of the phase fluctuations and $A_{1,2}(\mathbf{f}, \boldsymbol{\rho})$ is the overlap area between four pupils.
The Von Karman model is able to compute $M_{1,2}(\mathbf{f}, \boldsymbol{\rho})$ as a function of the turbulent parameter r_0 and L_0, and then the PCS. So, the visibility is determined by the ratio between the energy in the high and low frequency peaks of the spectrum (Roddier and al. 1976).

3 Theory-Experiment Comparison

This section aims at checking the robustness of the model by comparing predicted visibility losses and actual measurements. To do this, we need to know the values of the turbulent parameters for the night of observation. The Fried parameter is determined by fitting the low frequency peak (seeing peak) of the spectrum with the theoretical expression given by Korff. The outer scale is adjusted to reproduce correctly the visibility losses.

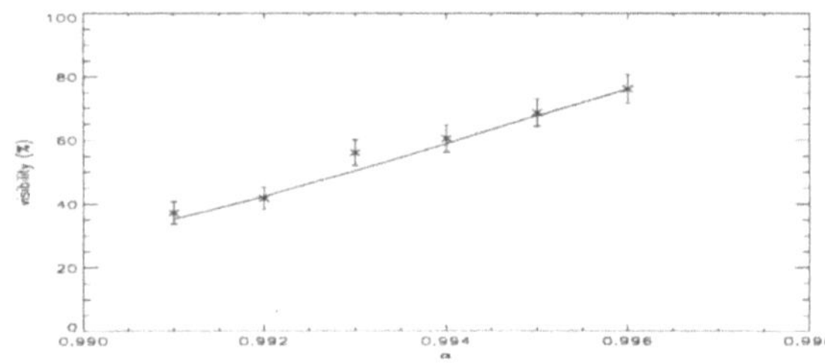

Fig. 1. Theoretical visibilities versus $\alpha = \lambda_1/\lambda_2$, for $\lambda_1 = 6730\mathring{A}$, $\Delta\lambda_1 = 30\mathring{A}$, $\Delta\lambda_2 = 6\mathring{A}$, $B = 29m$, $r_{0_1} = 14cm$, and the estimated value $L_0 = 5m$ corresponding to the best fit. And the visibility measured by GI2T for the night of 17.10.94.

4 Conclusion

From the comparisons between the experimental and theoretical visibilities, we have shown that the model is able to reproduce the losses of visibility due to spectral decorrelation. The same analysis could be realized on the effect of the spectral bandwidth increasing (Berio and al. 1996). So, this is the first model simulating the atmospheric turbulence effects on the spectral decorrelation of fringes obtained from a multi-speckle long baseline optical interferometer. This analysis could be used to improve the visibility calibration of interferometers like GI2T or VLTI.

References

J. Borgnino (1989): *Polychromatic transfer functions in stellar interferometry* J. Opt. Soc. Am. A 6, 244-251.

C. Roddier and F. Roddier (1976): *On the fringe visibility in a Michelson interferometer* J. Opt. Soc. Am. 66, 580-584.

P. Berio (1996): *Effects of the atmospherical spectral decorrelation on the visibility measurements in Michelson interferometry.* submitted to J. Opt. Soc. Am.

High Resolution Measurements of Close Binary Stars and the Probability Imaging Technique

Marcel Carbillet, Claude Aime, Eric Aristidi, and Gilbert Ricort

Département d'Astrophysique de l'Université de Nice-Sophia Antipolis
URA 709 du CNRS, Parc Valrose, F-06108 Nice cedex 02, France

Abstract. We report in this communication the first experimental results in the visible range obtained by the Probability Imaging technique applied to close binary stars. Speckle data of the well known binary β Del as well as the newly discovered one Moai 1 (Carbillet et al. 1996) are processed. An analysis of the probability density functions, combined with the classical visibility, is used to reconstruct the binary systems.

1 Introduction

An alternative to the computation of moments usualy done in the different speckle interferometry techniques is the analysis of the probability density functions (pdf) at several points in space in the speckle pattern. This is what Probability Imaging (PI) does – see Aime (1993). The two-fold pdf $P^{(2)}(\Omega_1, \Omega_2; \boldsymbol{\rho})$ is a function of two random variables of intensity Ω_1 and Ω_2, and a space-lag $\boldsymbol{\rho}$. Whatever the value of $\boldsymbol{\rho}$, the two-fold pdf for a point source has a symmetrical structure in Ω_1 and Ω_2; when for $\boldsymbol{\rho}$ close to the separation vector $\mathbf{d}$, the two-fold pdf for a binary star has an arrow-head shape with a trend towards a direction Ω_2 of the order of $\alpha\Omega_1$, where α is the oriented intensity ratio between the two components of the binary system, with no ambiguity about their relative position.

2 Gaussian Model

Assuming that the complex amplitude of the wave at the focus of a large telescope is a circular Gaussian process, the two-fold pdf of a binary star can be expressed as the product between the two-fold pdf of a point source and a function $Q(\Omega_1, \Omega_2)$ containing all the information about α. The value of α can be then easily retrieved doing a simple radial integration of Q.

3 Application

In practice, the Q function is obtained by dividing the two-fold pdf of the binary star by the point-source one. If the oriented intensity ratio α is not close to 1, one can compute $Q - Q^T$, where Q^T is the transpose quantity of Q. The radial integration gives then α as well as $\frac{1}{\alpha}$. Otherwise, if α is close to 1, one only has to compute the radial integration of Q, since $Q - Q^T$ becomes irrelevant.

The data processed here were obtained in the red (λ=650 nm) using a speckle-graph developed by the Aperture Synthesis group of the Observatoire Midi-Pyrénées, and coupled with an intensified CCD video camera and an acquisition system. This instrumentation was set at the Cassegrain focus of the 2 m Bernard Lyot telescope of the Pic du Midi, France. The complete procedure to obtain the relevant parameters d, α and PA (position angle) uses both the classical visibility function and the Q function, as shown in Figure 1.

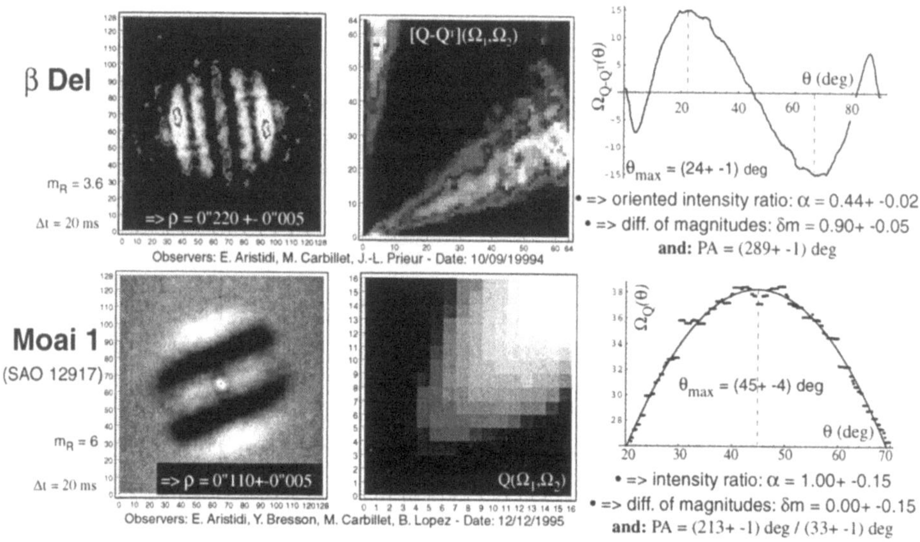

Fig. 1. Square visibility function, $Q - Q^T$ (top) or Q (bottom) function and the radial integration of it, for β Del (top) and Moai 1 (bottom).

4 Perspectives

The possibility to avoid the use of a reference star is under examination, using the two-fold pdf of the double star computed for $\boldsymbol{\rho} \perp \mathbf{d}$. We also plan to apply this technique as a routine for our current speckle observations, and especially for Post-Hipparcos late-type binary star ones – see Aristidi et al. (1996).

References

Aime C. (1993): Trends in Opt. Eng. **1**, 15
Aristidi E., Lopez B., Carbillet M., et al. (1996): this conference
Carbillet M., Lopez B., Aristidi E. et al. (1996): to appear in Astron. Astrophys.
All references available on `http://boulega.unice.fr/imager/g2-pub.html`

Late Type Stars at High Angular Resolution: Observations and Modelling

Pierre Cruzalèbes[1], Bruno Lopez[2], and Manfred Bester[3]

[1] Observatoire de la Côte d'Azur, URA CNRS 1361, Av. Copernic, 06130 Grasse, France
[2] Observatoire de la Côte d'Azur, Département Fresnel, BP 4229, Bd de l'Observatoire, 06304 Nice Cedex 4, France
[3] Space Sciences Laboratory, University of California at Berkeley, Berkeley, CA 94720, USA

Abstract. We present first images of some late type stars obtained with different high angular resolution techniques. To interpret them, we use a non spherical radiative transfer modelling.

1 Infrared Speckle Observations

We have reconstructed near-infrared diffraction-limited images of the Red Rectangle (CRL 915) by using the bispectral analysis method (Cruzalèbes et al. 1996). Fig. 1 shows an image at 2.2 μm obtained at the KPNO 3.80m telescope with 360 frames of 1/30s integration time each (courtesy of J.M. Beckers). The effective angular resolution is about $0.2''$, the field of view is $3.7'' \times 3.7''$, north is up and west is left. In order to show the details of the nebula, the highest level has been thresholded to 20% of the maximum of intensity. The lowest level is 0.2%, corresponding roughly to the level of the noise. Fig. 2 and 3 show images at 3.87 μm and 4.8 μm obtained at the CFHT 3.60m telescope with 5000 short exposure frames of 0.16s integration time (courtesy of E. Tessier). In these 2 images the resolution is $0.3''$ and the field of view $3.7'' \times 3.7''$ (north up, west left). The highest level is 20% of the maximum of intensity. The lowest is 0.2% for the 3.87 μm image and 0.5% for the 4.8 μm image.

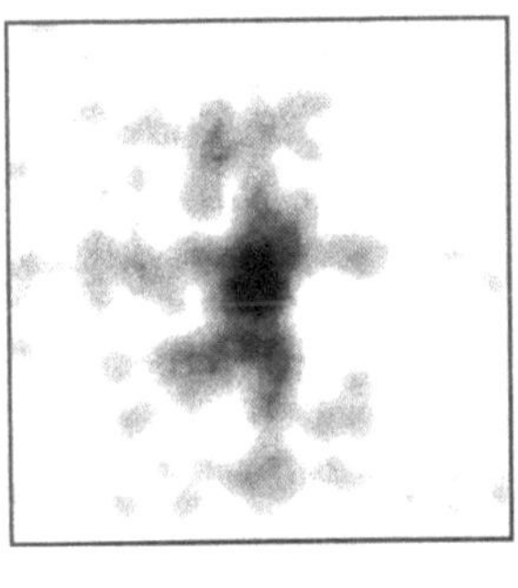

Fig. 1. CRL 915 in the K-band

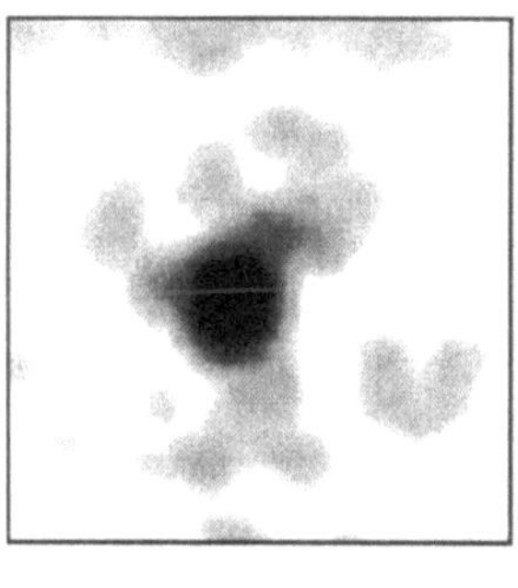

Fig. 2. CRL 915 in the L-band

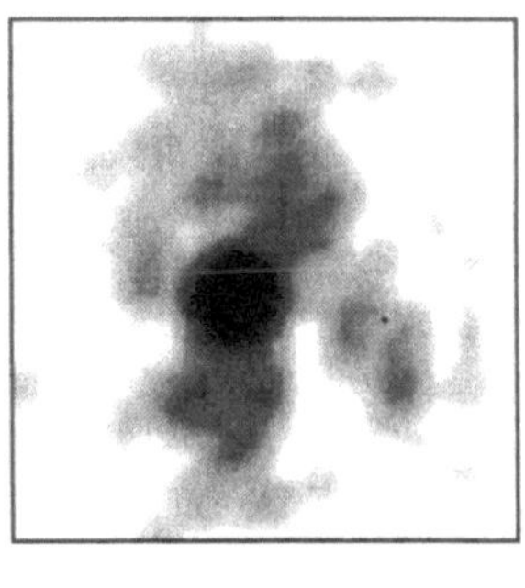

Fig. 3. CRL 915 in the M-band

The general features of the nebula are the same in each of these near-IR images: the overall shape of the nebula is bipolar. The shematic geometry of the nebula shows a central part 0.7″ in size lying between two side lobes, 1.2″ in size for the north-eastern lobe and 1″ for the south-western one. The opening angle of the outflow indicated by the NIR extensions is of the order of 20° (from north to east). These images, especially that in the K-band, have been used as observational constraints for the Red Rectangle modelling (see section 3).

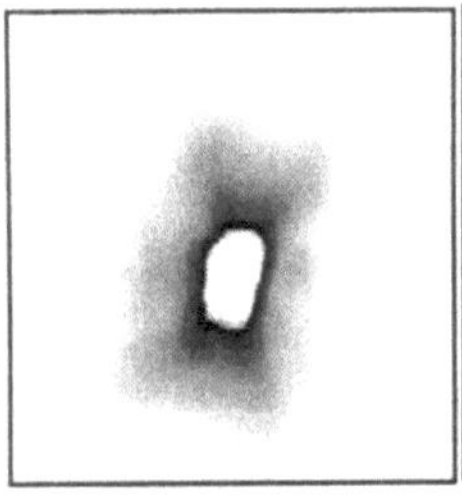

Fig. 4. CRL 915 in K

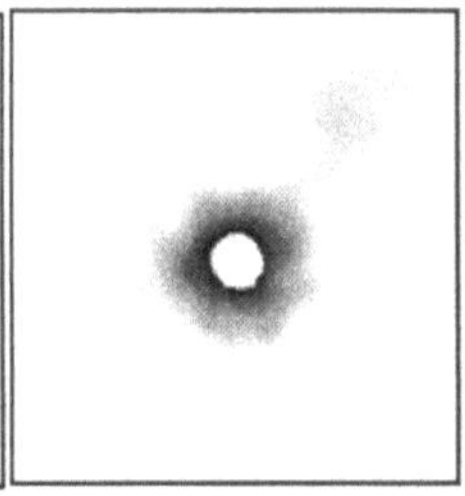

Fig. 5. VY CMa in K

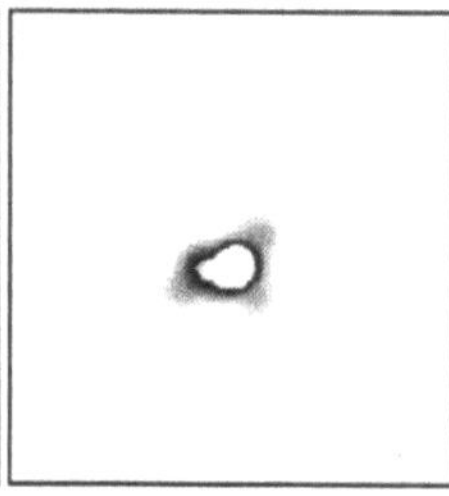

Fig. 6. α Sco in K

Fig. 7. TX Psc in K

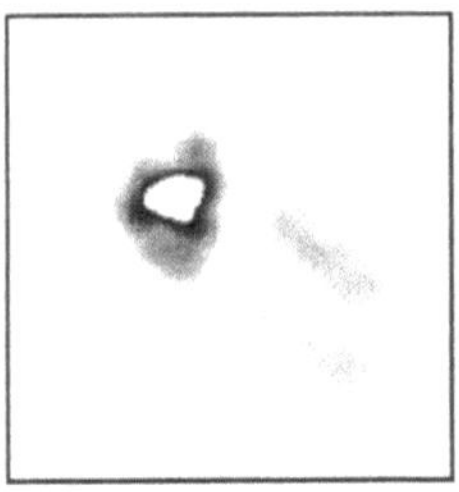

Fig. 8. o Ceti in J

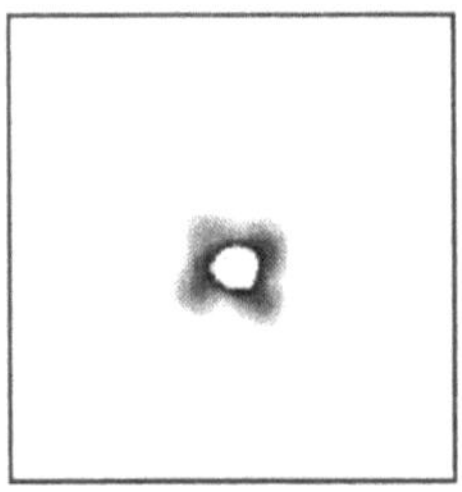

Fig. 9. o Ceti in H

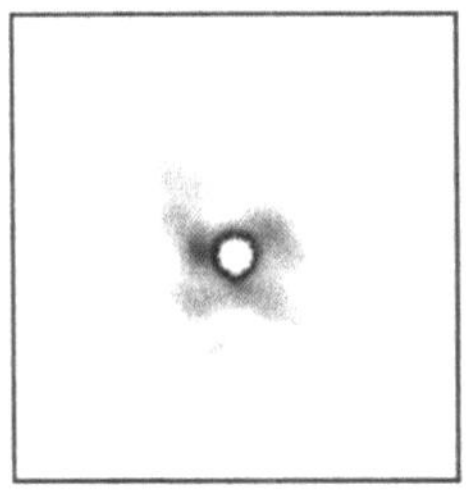

Fig. 10. o Ceti in K

2 Adaptive Optics Observations

Five late type objects have been observed at ESO in 1994 with the Come-On+ adaptive optics system. One night has been obtained on the 3.60m telescope to observe the Red Rectangle, VY Cma, TX Psc and α Sco (all in the K-band), and o Ceti in the J-, H- and K-bands. The camera used was the SHARPII camera. The data were off-line processed with the MPE-IRSI software using the Shift and Add algorithm (for co-adding the short exposure frames) and the Lucy one (for the deconvolution). Each deconvolved image presents a noise level of about 0.1% and an effective final angular resolution of 0.15″, the field of view being 3″ × 3″ (north up, west left). Fig. 4 shows the Red Rectangle nebula (exposure time of 2000s), Fig. 5 VY Cma (260s), Fig.6 α Sco (660s), Fig. 7 TX Psc (640s), all at 2.2 μm. Fig. 8, 9 and 10 show o Ceti at 1.25 (600s exposure time), 1.65 (1400s) and 2.2 μm (600s). In order to show the details of the envelopes surrounding the stars, the highest level has been thresholded to 15% of the maximum of intensity (in Fig. 4 to 10). There still exists an ambiguity on the certain detection of the envelopes surrounding TX Psc in K, α Sco in K and o Ceti in H, with these exposure times. According to our images, their sizes could be smaller than one arcsecond. In the opposite, the images of VY Cma in K and o Ceti in J clearly show non symmetrical accumulations of matter at a distance nearly greater than one arcsecond. These features could be explained by the presence of a companion star, the accumulations beeing elongated towards its expected position. We also confirm the detection in K of the envelope of the Red Rectangle (non spherical, $2'' \times 1.5''$), and of that of o Ceti (roughly spherical, $1.5''$ of diameter).

3 Modelling the Red Rectangle Nebula at 2.2 μm

Fig. 11 shows non spherical simulated maps of the Red Rectangle nebula at 2.2 μm obtained with the radiative transfer model (Lopez et al. 1995). These must be compared to Fig. 1 and 4. Although our simulated images are not as elongated as observed, our model allows to extract the optical depth, constraints on the size of the grains, as well as geometrical parameters such as the dust density law (Lopez et al. subm. to A&A). In choosing a right set of parameters we succeeded in obtaining a good fit for the photometric spectrum from 0.4 to 100 microns and a reasonable ratio between the flux measured through a $2.8''$ aperture at the center of the nebula and the flux measured at $10''$ distance in the northern lobe. We also obtained square shapes of the nebula at 2.2 μm.

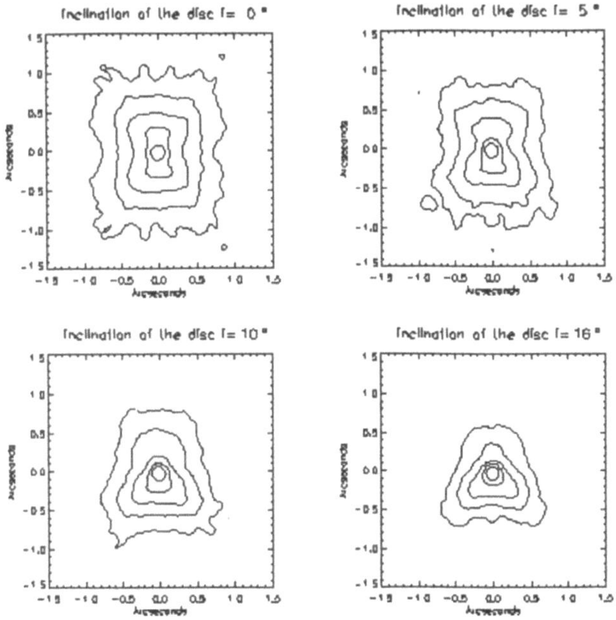

Fig. 11. Non spherical simulated maps of CRL 915 for different view angles (contour levels = 0.45, 0.2, 0.09, 0.04, 0.02)

4 Modelling Omicron Ceti at 1.25 μm

Fig. 12 shows a non spherical simulated image of *o* Ceti at 1.25 μm obtained with the radiative transfer model (field of view $2'' \times 2''$). This must be compared to Fig. 8. Fig. 13 shows a cut of the simulated image along the direction of the clump. This model is mainly based on mid-infrared long baseline interferometric observations carried out in 1993-94 with the ISI interferometer (Lopez et al. subm. to ApJ). Substantial variations in the 11 μm visibility (performed from 1989) point out that a large concentration of dust has occured close to the photosphere in 1992. This tends to indicate that dust is not formed periodically following the luminosity cycle of the Mira variable, as generally thought. Our observations confirm that the distribution of the circumstellar matter around *o* Ceti is asymmetrical, and suggest that a clump of dust may be present at the position of the companion (a white dwarf surrounded by an accretion disk).

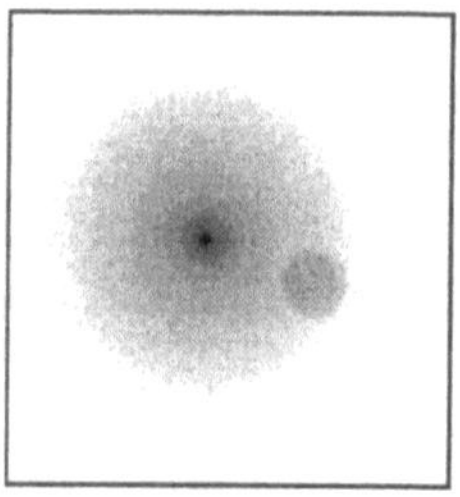

Fig. 12. Model of o Ceti in J

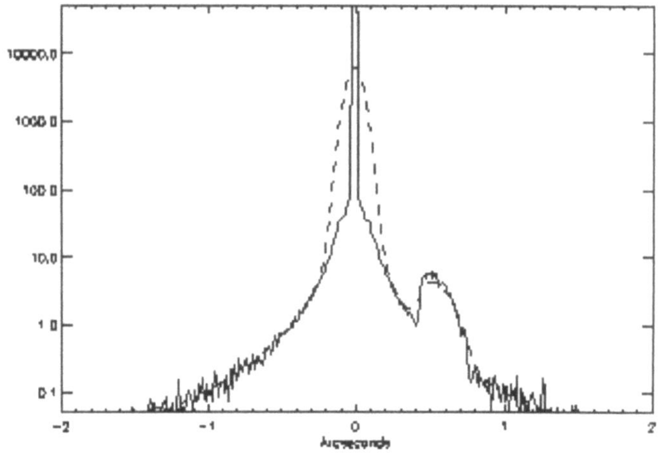

Fig. 13. Cut of the model of o Ceti in J (photometric level in arbitrary units)

References

Cruzalèbes, P., Tessier, E., Lopez, B., Eckart, A., Diphène, D., 1996, AAS 116, 597

Lopez, B., Mékarnia, D., Lefèvre, J., 1995, A&A 296, 752

Lopez, B., Tessier, E., Cruzalèbes, P., Lefèvre, J., "Dust Envelope Modelling of the Red Rectangle Nebula", submitted to A&A

Lopez, B., Danchi, W.C., Bester, M., et al., "Non-Spherical Structures and Temporal Variations in the Dust Shell of *o* Ceti Observed with a Long Baseline Interferometer at 11 μm", submitted to ApJ

Interferometric Observations of Mira Stars

Karl-Heinz Hofmann[1], Yuri Balega[2], Michael Scholz[3] and Gerd Weigelt[1]

[1] Max-Planck-Institut für Radioastronomie, Auf dem Hügel 69, D-53121 Bonn, Germany
[2] Special Astrophysical Observatory, Nizhnij Arkhyz, Zelenchuk region, Karachai-Cherkesia, 357147, Russia
[3] Institut für Theoretische Astrophysik der Universität Heidelberg, Tiergartenstrasse 15, D-69121 Heidelberg, Germany

1 Observations with the Russian 6 m SAO Telescope

We present the first diffraction-limited speckle masking (Weigelt 1977; Lohmann et al. 1983) observations of a Mira star (R Cas) with the Russian 6 m SAO telescope (Hofmann et al. 1995, Weigelt et al. 1996). At 700 and 714 nm wavelength a resolution of $1.22\lambda/D = 30$ mas was achieved. The reconstructed *true images* (Fig. 1) are the images with the highest resolution obtained up to now in this wavelength region. The speckle interferograms of R Cas were recorded through interference filters with center wavelength/bandwidth of 671 nm/6 nm and 714 nm/6 nm (strong TiO band absorption), 700 nm/6 nm (moderate TiO absorption), and 1043 nm/23 nm (continuum). The images of the R Cas disk are non-uniform and elongated. For example, the size of the 714 nm TiO absorption band image is 42×56 mas (axial ratio 0.76; see Fig. 2 and Table 1). The goal of our project is to provide the basic observations for modeling the atmospheric structure of Mira variables (e.g. Bessel et al. 1996).

2 Observations with the VLTI

Our goal of future VLTI observations is the determination of continuum diameters and intensity profiles (limb darkening) of stellar disks, and measurements of shape (asymmetries) and stellar surface structures of Mira variables. The continuum diameters of the nearest Mira stars are ≤ 30 mas. But the resolution of a single VLT 8 m telescope (UT) at λ=1043 nm (continuum) is only 34 mas. Precise diameters can only be obtained if at least the first minimum of the visibility (i.e. baselines ≥ 9 m) can be measured. Therefore, such measurements require at least one UT plus one AT. Determinations of limb darkening of stellar disks require measurements up to about the second minimum of the visibility function, i.e. baselines of about 18 m. Measurements of star disk shapes and stellar surface structures require a good uv-coverage (see Reinheimer et al. "Infrared interferometric imaging with the VLTI in the multi-speckle mode" and "Multi-speckle interferometric imaging with the VLTI and the LBT at optical wavelengths" in this proceedings volume).

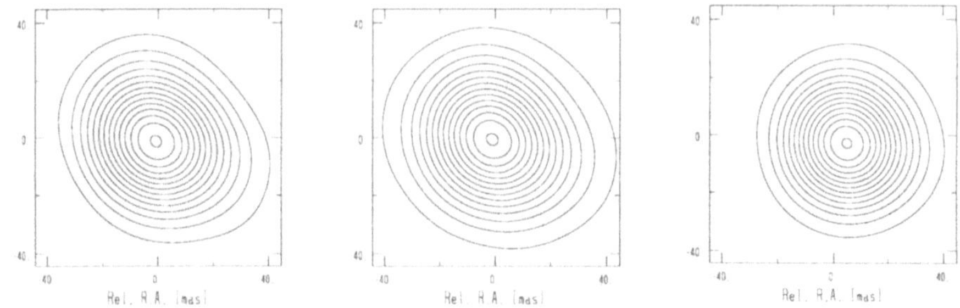

Fig. 1. Diffraction-limited speckle masking reconstructions of R Cas: (left) 671 nm, (middle) 714 nm , and (right) 700 nm. In each panel the contour levels are plotted from 7-98% of peak intensity in steps of 7%. North is at the top and east to the left.

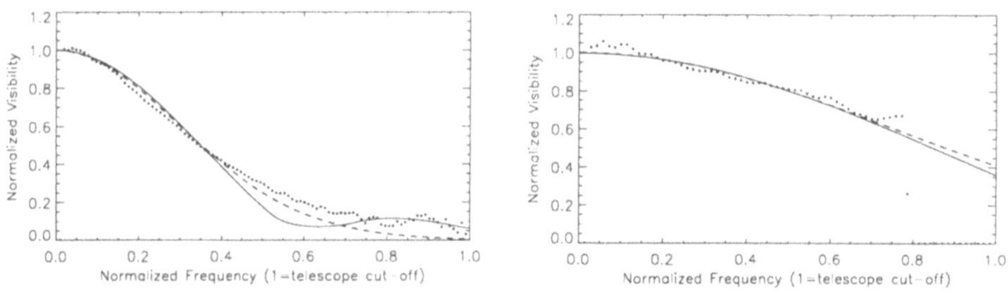

Fig. 2. Azimuthally averaged visibilities of R Cas (dots): (left) 714 nm and (right) 1043 nm plus UD (solid line) and Gaussian model visibility (dashed line).

References

Bessell M.S., Scholz M., Wood P.R. 1996, A&A 307, 481
Hofmann K.-H., Balega Y., Scholz M., Weigelt G. 1995, in: Stellar surface structure, IAU Symp.176, Strassmeier K.G. (ed.), Kluwer, Poster Proceedings p.45
Lohmann A.W., Weigelt G., Wirnitzer B. 1983, Appl. Opt. 22, 4028
Weigelt G.P. 1977, Opt. Commun. 21, 55
Weigelt G., Balega Y., Hofmann K.-H., Scholz M.. 1996, A&A submitted

Table 1. Disk parameters derived from elliptical UD fits to the R Cas visibilities.

Data set	Av. diameter (mas)	Major axis (mas)	Minor axis (mas)	PA (°)	Axis ratio
671/6	43.6 ±2.0	51.0 ±2.4	35.6 ±2.4	52 ±7	0.70 ±0.07
700/6	36.2 ±2.0	38.8 ±2.4	33.6 ±2.4	52 ±7	0.86 ±0.10
714/6	49.2 ±2.0	55.6 ±2.4	42.2 ±2.4	58 ±7	0.76 ±0.07
1043/23	29.5 ±4.0	–	–	–	> 0.9

Results of Studies on the Environmental Factors Affecting the VLTI Performance

B. Koehler, S. Lévêque, and O. von der Lühe

European Southern Observatory, Karl-Schwarzschild-Strasse 2, D–85748 Garching

Abstract. We have undertaken a number of studies and experiments in the last few years to assess the impact of various environmental factors on the performance of the ESO Very Large Telescope Interferometer (VLTI). The topics include atmospheric turbulence, wind loads on the telescope structure, vibrations created by equipment, thermal environment along the beam paths, natural and man-made seismic noise, as well as acoustic noise.

This paper gives an overview of the final results obtained. A more detailed report of these activities can be found in a recent paper (Koehler, 1996).

The main result is the very good confidence that the VLT 8m telescope and the infrastructure design is adequate for interferometric use at optical wavelengths down to the visible.
Keywords: Interferometry, optical aperture synthesis, performance, environment, seeing, thermal effects, vibrations, wind load, seismic, acoustic.

1 Introduction

The technical challenge of coherently combining for the first time large multi-mode telescopes implies a thorough system analysis. A major element of it is the top-down generation and bottom-up validation of error budgets for each major performance characteristic of the interferometer which is required by the science objectives. Figure 1 illustrates this process. It constitutes the framework of the activities reported here which focus on the impact of environmental factors with the goal (i) to better understand the effects of the environment on the interferometer performance, (ii) to check that these effects can be maintained at a level compatible with the top-level requirements, and (iii) to specify, when necessary, various sub-system requirements and operational constraints aiming at reducing them.

2 Structural Deformation Inside the UT

When considering the use of the VLT 8m Unit Telescope (UT) in an interferometric mode, one of the first question is: will the UT be stable enough to ensure the very drastic requirement imposed by interferometry on the dynamic stability of the Optical Path Difference (OPD)? To answer this question, a complete and detailed Finite Element Model of the telescope (figure 2) has be used to compute its dynamic behaviour under a number of load cases. These include

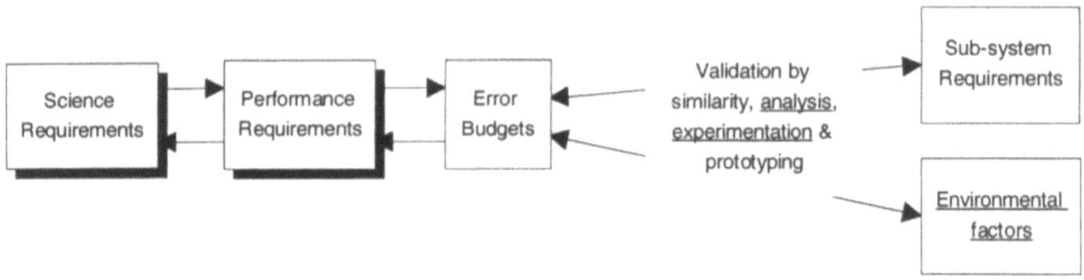

Fig. 1. The requirement flow chart illustrating the need to address the effects of environmental factors as part of the validation of the top-level error budgets.

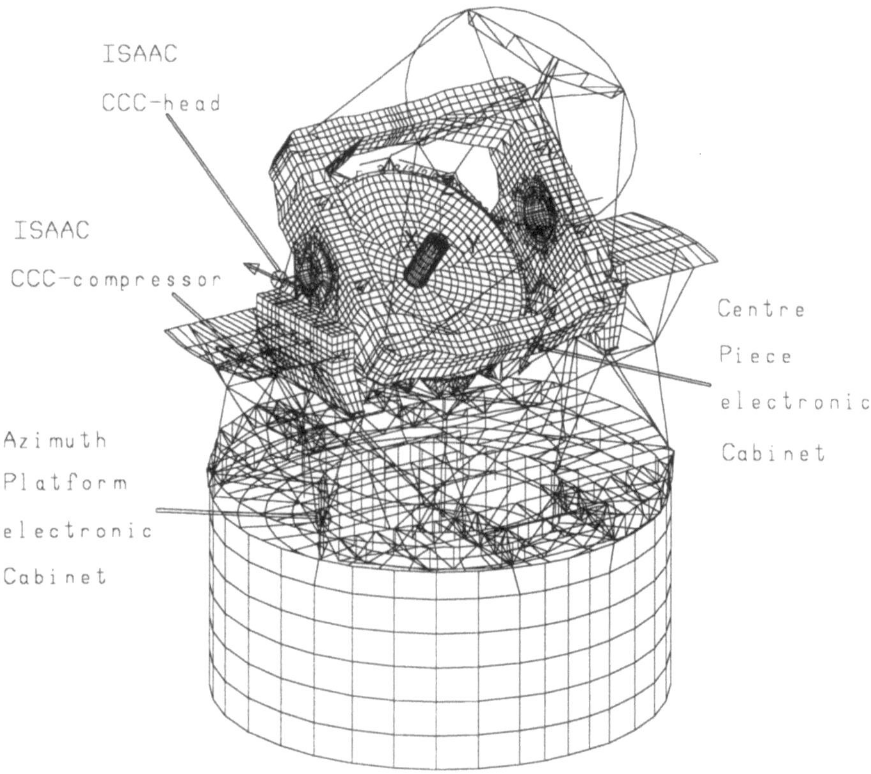

Fig. 2. FEM of the VLT 8-meter Telescope used to evaluate the Optical Path Length Stability under dynamic wind load, seismic motion, and vibrating equipment.

the wind load on the telescope structure, the ground motion due to natural and man-made micro seismic activity, as well as vibrations generated by equipment located on the telescope.

The static wind load on each structural element was computed using results from VLT wind tunnel tests and applying standard wind load computation methods. A Von Karman wind spectrum model with parameters derived from Paranal wind measurement statistics was used to define the load dynamic properties. The OPD response was then computed from the harmonic responses of each mirror of the optical train down to the Coudé focus. Thanks to the outstanding stiffness of the telescope structure and in spite of the relatively large loading, the results (Koehler, 1996) obtained are well within the stringent requirement imposed by observation in the visible (OPD fluctuation of 9 nm rms over an exposure time of 10 msec for a specification of 14 nm).

A more severe impact was found for the response to micro seismic loads (Koehler, 1996, 1994, 1995). A dedicated measurement campaign at Paranal provided the magnitude and statistics of micro earthquakes. It was found that the specified OPD stability limits is reached, on average, every 40 minutes and every 1 hour 40 minutes for observation in the visible and in the near-infrared respectively. This led us to implement a continuous seismic monitoring system as part of the VLT Astronomical Weather Station in order to flag observations disturbed by seismic events. This seemingly severe operational constraint and observation efficiency loss imposed by the site is actually offset by the outstanding atmospheric conditions at Paranal. The effect of several man-made seismic noise sources was also investigated. As an example, the disturbance from a car driving in the vicinity of the observatory top was measured. Its effect was found to remain small as long as the traffic stays off the Observatory Platform. This indicates the merits of having the Control Building away from the Observatory Platform.

Several types of vibration-generating equipment which are attached to the telescope structure and could degrade the OPD stability during interferometric observation have been identified. Among the most significant ones are the compressor and the cold heads of the closed cycle coolers for IR instruments. Their disturbance force spectrum was measured on the ISAAC lab prototype and applied to the telescope structure through de-coupling devices with variable characteristic frequency and damping property. Upper limits (5 Hz) of the characteristic frequencies were found necessary to ensure a negligible contribution to the OPD stability error budget of the telescope. These results have been implemented in the final design of the VLT IR instruments.

In addition to the above load cases, the vibration of the primary mirror due to acoustic loading has been investigated using a 3D elasto-acoustic model of the telescope placed inside the enclosure acoustic cavity (figure 3). The acoustic noise generated by the wind interacting with the dome entrance slit was measured on representative La Silla telescopes and acoustic noise from the closed cycle compressor and cold heads was measured in the lab. Results showed a negligible contribution (< 2 nm) for the latter sources and a significant, but acceptable,

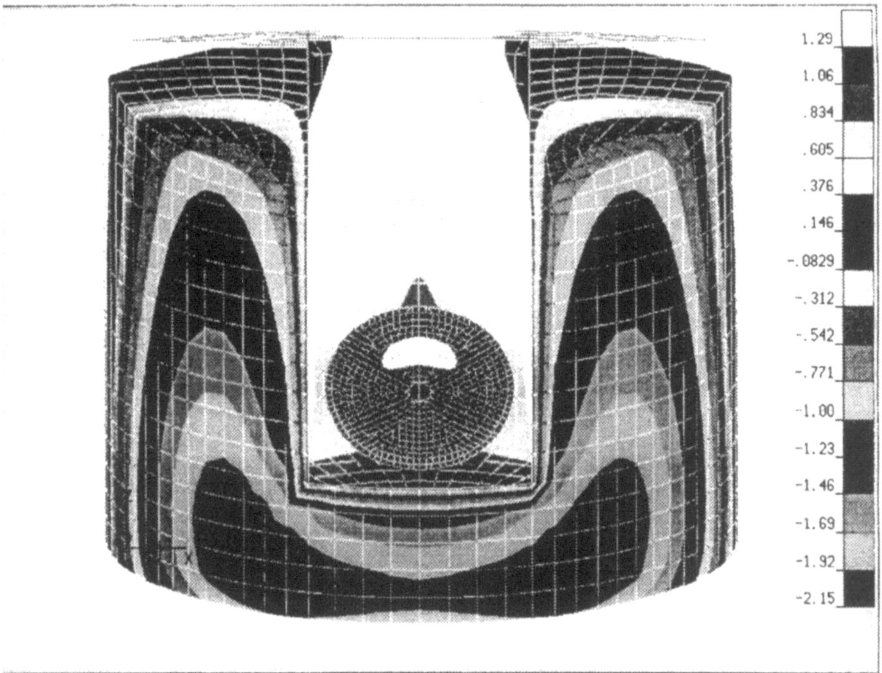

Fig. 3. Dynamic acoustic loading (at 13 Hz) on the VLT enclosure and primary mirror due to acoustic noise generated by the interaction of the wind with the entrance slit.

effect due to wind-generated acoustic noise (10 nm rms over a 10 msec exposure for a 10 m/sec wind speed).

3 Atmospheric and Internal Seeing

A number of VLTI-specific studies related to *external (atmospheric) seeing* and *internal seeing* effects have been carried out. These include the use of data from balloon-borne probes launched from Paranal during the PARSCA campaign in order to derive the spectrum of the atmospheric tilt and OPD disturbances both on-axis and off-axis (anisoplanatic effects). The results have been used to assess the performance of tilt and OPD active compensation as well as their related field limitations. Internal seeing is a concern for the VLTI since the delay lines are placed in air. Experiments have been undertaken in a long underground tunnel representative of the delay line tunnel at Paranal. Both the lateral beam wander and the OPD fluctuations due to air turbulence along long propagation paths were measured. Results of all these seeing-related studies can be found elsewhere (Koehler, 1996, Lévêque 1995). The main outcome is the good confidence that the level of turbulence strength inside the delay line tunnel imposed by the error budget ($Cn^2 < 10^{-15}$ $m^{-2/3}$) can be achieved.

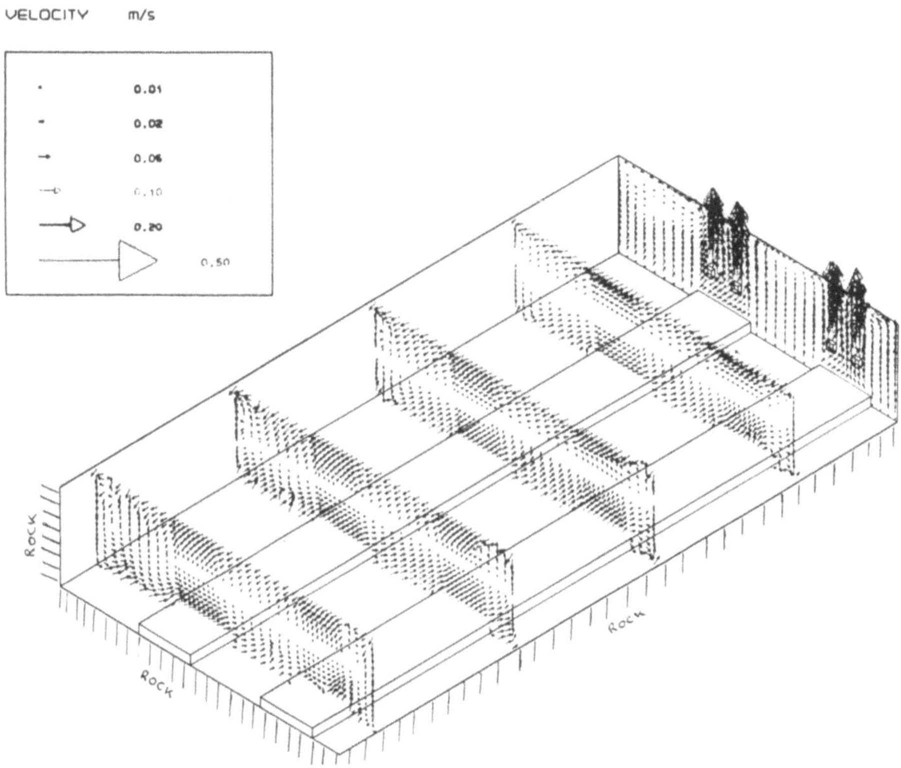

Fig. 4. Flow pattern inside the Delay Line Tunnel created by four pieces of equipment dissipating 50 W each. The arrows represent the mean air velocity.

4 Thermal Environment

Because of their impact on the wave front quality and beam geometry, the thermal conditions encountered along the internal beam path have been carefully investigated. An extensive thermal study has been undertaken with the goal to (i) assess the thermal environment under the *natural* outside conditions (daily & seasonal variations), (ii) evaluate the thermal influence of *artificial* heat sources (electronic equipment & heat dissipation during maintenance periods). Various 2D and 3D Finite Difference thermal models were built to investigate the situation in various parts of the VLTI infrastructure (auxiliary telescope station, light beam ducts, delay line tunnel, interferometric laboratory). After optimisation of the building insulation a satisfactory thermal stability and homogeneity all over the infrastructure could be achieved (vertical gradient within $\pm 0.2\,^\circ$C/m, longitudinal homogeneity 0.2 C, daily variation $< \pm 0.02\,^\circ$C, seasonal variation $\pm 2.4\,^\circ$C). A last part of the study focused on the simulation of air flow pattern (figure 4) and evaluation of Cn^2 under various conditions. A more detailed description of the investigations and results can be found elsewhere (Koehler, 1996).

5 Conclusion

A number of studies related to environmental factors have allowed us to *define* or *refine specifications* for several VLT/VLTI subsystems which are particularly critical for the VLTI operation. This has concerned, among others, the tunnel insulation, the heat dissipation inside the tunnel, the first axial eigenfrequency of the M2 unit, the stiffness of the Coudé mirror supports, the vibration decoupling devices of the instrument close cycle coolers, and the need for seismic monitoring.

The results obtained have *confirmed that the performance requirements set for the VLTI are achievable.* They have shown that the design of the VLT Observatory and, in particular, the 8m Unit Telescopes is compatible with the stringent requirements imposed by visible interferometric observations.

The decisive confirmation of this conclusion on the real hardware has already started and will continue as hardware becomes available and commissioned on the site.

References

B. Koehler, S. Lévêque, O. von der Lühe, "Results of studies on the environmental factors affecting the VLTI performance", Optical Telescopes of Today and Tomorrow, being published in Proc. SPIE Vol. 2871, Landskrona 1996.

B. Koehler, "Hunting the Bad Vibes at Paranal!", The Messenger N76 (June 1994), pp 4-9, 1994.

B. Koehler et al., "Impact of the Microseismic Activity on the VLT Interferometer", The Messenger N79 (March 1995), pp 4-9, 1995.

S. Lévêque et al., "Measurement of optical path fluctuations due to internal seeing for the VLTI", Advanced Imaging Technologies and Commercial Applications, N. Clark (Ed.), in Proc. SPIE Vol. 2566, pp 156-163, San Diego 1995.

Internal Optical Path Calibration System for VLTI

Samuel Lévêque, Bertrand Koehler, Oskar von der Lühe

European Southern Observatory, Karl Schwarzschild Str. 2,
D-85748 Garching bei München

Abstract. A metrology system is needed to calibrate the internal Optical Path Difference (OPD) within the interferometric array. The objective is to determine the differential optical path between any pair of telescopes of the array with an accuracy of some 100μm. This OPD will be monitored from the Interferometric Laboratory to a retroreflector located on the telescope's secondary mirror or to a flat mirror located at the folded Nasmyth focus. It will be performed "off-line" every time a change in the interferometer configuration has happened or when thermal disturbances have occurred. A concept has been derived which uses an external light source located in the pupil plane area. In order to verify its applicability, a test bench has been procured and integrated at ESO. This test bench includes a mock-up of two delay lines with translation stages and control units, and metrology units. First results are presented.

1 Need for Internal Calibration

In order to obtain fringes from an astronomical source with an interferometer which consists of several independent telescopes, the optical paths as seen from the object while propagating through individual telescopes towards a common beam combiner must be equalised.

The OPD to be created by the delay line is

$$OPD = \mathbf{S}.\mathbf{B} + OPD_{int} \tag{1}$$

where **S** is the 3D unit vector pointing towards the object under study at a given time.
B is the vector Baseline defined in 3 dimensional space.
OPD_{int} is the internal OPD obtained when the delay lines are at their mechanical reference positions.

The measurement of OPD_{int} is needed to rapidly acquire fringes even after relocation of Auxiliary telescopes and to trace internal optical path changes within the array and update the OPD model (Lévêque et al, 1996). The desired accuracy is 100 μm.

2 Concept for an Internal OPD Calibration

The internal OPD calibration is based on white light interferometry since the goal is to detect the white light fringe corresponding to the Zero Path Difference

(ZPD) position (Lévêque, 1995). Laser diodes are used instead of conventional white light sources. Indeed, laser diodes offer a better collimation efficiency which is important when long distances are involved, a lower power dissipation and are less expensive. The emitted light propagates in both interferometric arms, is retroreflected at each telescope's secondary mirror and recombined. One optical path is modulated for fringe detection. The Zero Path Difference (ZPD) is obtained by moving a delay line until the superposition of the 2 returning beams creates an interferometric signal with a maximum contrast. The delay line position is then recorded by its own metrology system. Figure 2 presents the calculated fringe visibility as a function of optical path difference for a source whose spectral distribution is shown on Figure 1.

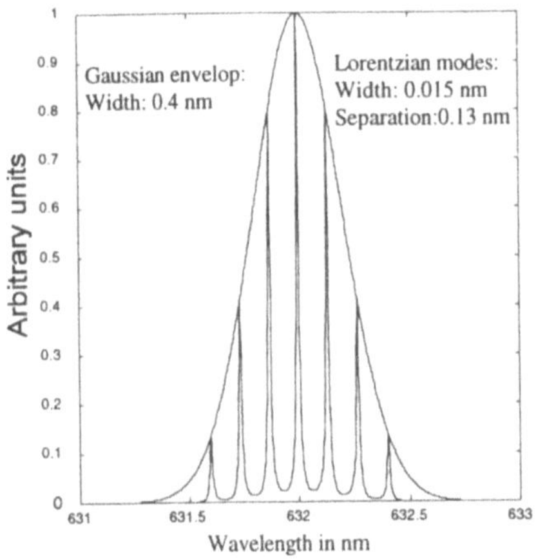

Fig. 1. Source spectral distribution

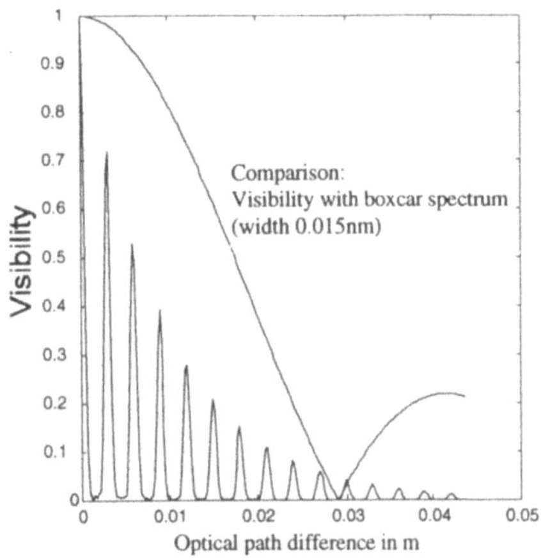

Fig. 2. Corresponding fringe visibility

3 Test Bench

A test bench has been integrated at ESO in order to demonstrate the feasibility of the concept applied to VLTI using multimode laser diodes, identify source of errors and determine the measurement accuracy. Two translation stages with linear motors and digital controllers are used. The displacement range is 150 mm, the resolution is 80 nm and the positioning accuracy is 4 μm. Multimode laser diodes have a peak power of 15mW and wavelength are centred in the red. Beams are spatially filtered with a monomode optical fiber whereas an optical path modulation of 4λ at 70Hz is obtained by a corner cube mounted on a piezo. Fringe detection is performed by a silicon photodiode and the acquisition system is developed under Labview.

4 Results

Figure 3 shows an example of a series of fringe visibility local maxima obtained by scanning one delay line around the ZPD position using a multimode laser diode centred on 680 nm. The number of local maxima depends directly on the number of modes emitted. However, the maximum visibility can be easily discriminated from other local maxima which are at least 10% smaller in amplitude for a separation of about 3mm from the ZPD position. Figure 4 presents a finer visibility scan on one side of the visibility maxima. The accuracy of the ZPD position is better than 20 μm.

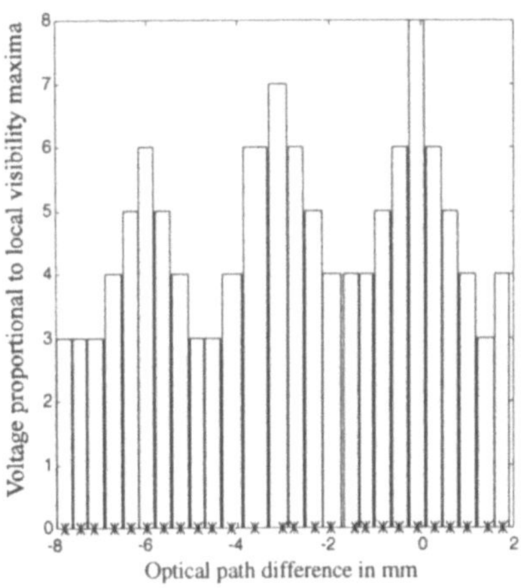

Fig. 3. Position of local visibility maxima

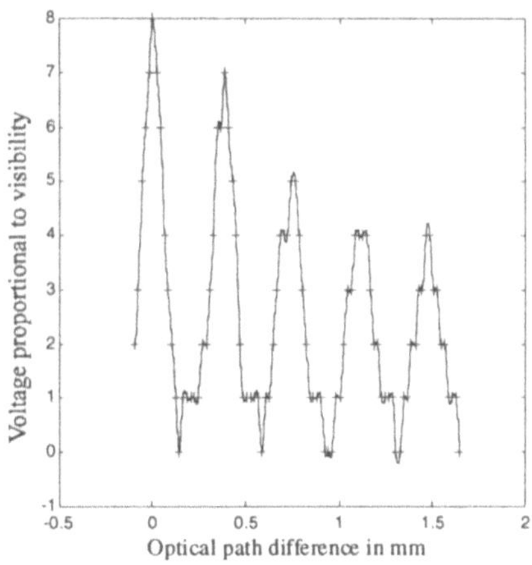

Fig. 4. Visibility scan close to ZPD position

5 Conclusion

The need for an internal OPD calibration system has been presented. A concept has been proposed to determine with high accuracy the zero path difference position between a pair of telescopes. A test bench has demonstrated that the zero path difference position could be found with a 20 μm accuracy.

References

S. Lévêque et al. "Optical Path difference Model for VLTI", Optical telescopes of today and tomorrow, Landskrona, 1996.

S. Lévêque, "Overview of optical path measurement techniques for the VLTI metrology systems", ESO report VLT-TRE-ESO-15000-0765, March 1995.

M. Dobosz, "Polychromatic light interferometer for high accuracy positioning", Optics and laser engineering 24, 1995, p43–56.
K. Petermann, "Laser diode modulation and noise", ADOP advances in Optoelectronics, Kluwer academic publishers.
Y. Ning, "Characterisctics of laser diode for interferometric use", Applied optics, 1989, vol.28, No.17, p3657.

Pupil Mask Interferometry of M8E-IR

M. Löwe[1], B. Stecklum[2], O. von der Lühe[3] and A. Quirrenbach[4]

[1] Max-Planck-Gesellschaft, Arbeitsgruppe "Staub in Sternentstehungsgebieten", Schillergäßchen 2-3, 07745 Jena, Germany
[2] Thüringer Landessternwarte Tautenburg, Sternwarte 5, 07778 Tautenburg, Germany
[3] European Southern Observatory, Karl-Schwarzschild-Str. 2, 85748 Garching, Germany
[4] Max-Planck-Institut für Extraterrestrische Physik, Giessenbachstraße, 85740 Garching, Germany

Abstract. We present NIR pupil mask observations of the young stellar object M8E-IR with the NTT interferometer. For these observations the entrance pupil of ESO's NTT was covered with a non-redundant 7-hole mask to suppress the low spatial frequencies. We compare our results with previous high-angular resolution observations.

1 Introduction

A mask with holes placed in a pupil plane conjugate of a ground based telescope acts as an interferometer and may help attaining diffraction-limited resolution. The holes should be configured as a non-redundant array and they must be smaller than the atmospheric coherence diameter (Fried parameter) r_0. Imaging without a mask uses all baselines between r_0-sized patches which fit into the primary mirror. High resolution imaging of a bright, compact, and simply structured source requires mainly the long baselines.
In comparison to conventional speckle interferometry, pupil mask interferometry is substantially less sensitive to changes of atmospheric conditions and therefore can be much easier calibrated. The advantages of the pupil masks were extensively discussed by Haniff et al.(1989) and Buscher et al.(1993).

2 Observations

In June 1995, we observed M8E-IR with the MPE infrared camera SHARP (Eckart et al. 1991) on the ESO-NTT in the K band ($\lambda = 2.2\mu$m). The entrance pupil of the telescope was covered by a non-redundant 7-hole mask mounted on the baffle in front of mirror 3. The mask layout is a modified Cornwell design (Bedding et al. 1993).
We obtained 3000 0.2s-exposures of M8E-IR, giving a total integration time of 10 minutes. To calibrate the wavefront distortion introduced by the seeing, the unresolved single star HR 6746 was observed as point spread function (PSF) reference. Due to the celestial position, the parallactic angle of M8E-IR changed during the observation by 20°. This improves the uv-coverage significantly and allows aperture synthesis to gain a better image quality. Additionally, we took data of the binary star η Ophiuchi to prove the image orientation and scaling.

3 Results and Conclusion

M8E-IR is a source of intense IR and radio line emission (Walker et al. 1975, Wright et al. 1977) presumably harbouring a B0 star (Simon et al. 1984) and sharing similarities with the BN-object in Orion. Simon et al. (1985) decomposed the extended structure detected by the lunar occultation observations into a small (compact) and a large component. At 3.8μm wavelength, the projected sizes of these components are 6 mas (FWHM) and 100 mas, respectively, each contributing to the total flux with 94 and 6 per cent. The similarity of their results at 10μm wavelength leads to the suggestion that the extent of these components is not strongly dependent on wavelength. Thus, the numbers quoted above should be representative for the K band as well. From their results for three different position angles, Simon et al. (1985) concluded that the small component is a circumstellar disk seen nearly edge-on with the major axis at the position angle of about 150°. The large component should have spherical symmetry. The length of the major axis of the disk was derived to be 72 mas in order to deliver the observed flux.

From our observations, we conclude that M8E-IR is resolved in north-south direction with an extended component to the south. While this seems to be consistent with the results of the occultation observations, it poses some problems for modelling the source when CO radio line measurements are taken into account. The position angle of the lobes of the molecular outflow is about 145° (Mitchell et al. 1992). The disk which drives this flow should be perpendicular to its orientation. From this point of view, it seems to be reasonable that the structure claimed to be a disk by Simon et al. is actually due to scattered light in the innermost regions of the outflow lobes. The appearance of similar sources (e.g. GGD27-IRS2) is dominated by light scattered off dust particles at wavelengths shorter than 4μm with the central object emerging only at longer wavelengths. While the resolution of our observations was not sufficient to fully resolve these lobes, a trace of this scattered light presumably has been found.

For further information please contact B.Stecklum (stecklum@tls-tautenburg.de).

References

Bedding,T. R., von der Lühe,O., Zijlstra,A.A., Eckart,A., Tacconi-Garman,L.E.: 1993, The Messenger, **74**, 2

Buscher,D.F., Haniff,C.A.: 1993, J. Opt. Soc. Am. A., **10**, 1882

Cornwell, T.J.: 1988, IEEE Trans. Antennas and Propagation, **36**, 1165

Eckart,A., Hofmann,R., Dehoux,P., Genzel,R., Drapatz,S.: 1991, The Messenger,**65**, 1

Haniff,C.A., Buscher,D.F., Christou,J.C., Ridgway,S.T.: 1989, MNRAS, **241**, 51p

Mitchell,G.F., Hasegawa,T.I., Schella,J.: 1992, ApJ, **386**, 604

Simon,M., Cassar,L., Felli,M., Fischer,J., Massi,M., Sanders,D.: 1984, ApJ, **278**, 170

Simon,M., Peterson,D.M., Longmore,A.J., Storey,J.W.V., Tokunaga,A.T.: 1985, ApJ, **298**, 328

Walker,R.G., Price,S.D.: 1975, AFCRL Infrared Sky Survey, AFCRL-TR-75-0373

Wright,E.L, Lada,C.J., Fazio,G.G., Kleinmann,D.E: 1977, AJ, **82**, 132

Observations of the Close Binary System β Lyrae with the GI2T Interferometer

Frédéric Morand[1], Daniel Bonneau[1], Denis Mourard[1], and Petr Harmanec[2]

[1] Observatoire de la Côte d'Azur, Equipe GI2T, Station de Calern, 2130 route de l'Observatoire, F-06460 Caussols, France
[2] Astronomical Institute, Academy of Sciences of the Czech Republic, Fričova 1, CR-25165 Ondřejov, Czech Republic

Abstract. We present the results of the 1994 international multitechniques campaign of observations of the well-known binary system β Lyrae, emphasizing on the interferometric observations and results obtained with the GI2T interferometer.

1 First Observations of β Lyrae with GI2T

The GI2T is located on Plateau de Calern (France). It is a Michelson stellar interferometer with two large apertures (1.52m) continuously movable on a North-South baseline, combining a high angular resolution (close to 1 mas) and a medium spectral resolution (R = 4500), operating at visible wavelengths (Mourard et al. 1994a). Data recording and reduction have been described in Mourard et al. (1994b).

β Lyrae is an eclipsing and spectroscopic binary, with an orbital period of $12\overset{d}{.}93$, increasing at a rate of 19 sec/yr. This star exhibits very strong and variable Hα and He emission lines. β Lyrae is presently described as a close binary system, in which the more massive component is a Be star surrounded by an accretion disk fead by the matter lost by a B giant filling its Roche lobe.

Preliminary observations of this star were carried out in summer 1991. Spectral dependence of the visibility amplitude across the Hα line was evidenced, indicating that the region emitting this line is partially resolved but no visibility phase dependence was detected (Mourard et al. 1992).

2 The 1994 International Multitechniques Campaign

An international campaign, involving spectroscopic, photometric, polarimetric and interferometric observations of β Lyrae has been organized during summer 1994 (Harmanec et al. 1996).

The initial goals of interferometric observations were to resolve the binary system in continuum light and to characterize the morphology of the regions emitting the Hα and HeI λ6678 lines. To achieve that, the maximum angular resolution has been used and observations of γ Lyr as a reference star have been carried out before and/or after the β Lyr ones.

Knowing the spectrophotometric diameter of γ Lyr and thus its theoretical visibility, calibrated visibilities of β Lyr were calculated for each night.
Four spectral bandpasses per spectral line (continuum, whole line, violet peak, red peak) have been defined on GI2T spectra for the analysis of the visibility function. The boundary wavelengths have been determined for each orbital phase by using the RVs given for each emission feature by the high resolution Ondřejov and DAO spectra.
Interferometric analysis leads to the following conclusions :

- – the continuum source appears unresolved in the N-S direction at any orbital phase, indicating that the binary must be oriented close to the E-W direction.
- – the Hα and HeI sources are resolved at any phase, but without any orbital phase dependence, showing that theses sources are more extended than the binary and probably located, at least partly, outside the orbital plane. Furthermore, Hα violet and red peaks have lower visibilities than the whole line, meaning that Hα source is at least a triple and not centrosymmetric object.

Observations with the other techniques have permitted an improvement of the ephemeris accuracy, detection of non orbital photometric variations (P = $282\overset{d}{.}4$) and rapid spectral variations (P = $4\overset{d}{.}73$). Spectroscopy and polarimetry confirm the presence of material outside the orbital plane.
The main conclusion of this campaign is then the detection of **bipolar jet structure** in β Lyr.We do not need to postulate any matter outflow via Lagrangian point L2 and do not have any observational evidence of it (Harmanec et al. 1996).

3 Perspectives

Close binary systems offer a unique opportunity to study the processes involved in mass transfer and mass loss and how they are related to stellar evolution. As demonstrated with β Lyr, interferometric observations provide new insights into the nature of these objects.
The bi-dimensional uv coverage of VLTI should be used at visible wavelengths to derive information of prime importance about the morphology of circumbinary and circumstellar regions.
Combined with spectroscopic observations and improved models of close binaries, high angular resolution observations will allow to set tighter constraints on physics of semi-detached systems.

References

Harmanec, P., Morand, F., Bonneau, D., et al. (1996): A& A **312**, 879
Mourard, D., Bonneau, D., Blazit, A., et al. (1992): *Complementary Approaches to Double and Multiple Star Research*, IAU Col. 135, ASP Conferences Series, **32**, 510
Mourard, D., Tallon-Bosc, I., Blazit, A., et al. (1994a): A& A, **283**, 705
Mourard, D., Tallon-Bosc, I., Rigal, F., et al. (1994b): A& A, **288**, 675

Stellar Winds of T Tauri Stars

Gernot Paatz and Max Camenzind

Landessternwarte Heidelberg, Königstuhl, D-69117 Heidelberg, Germany

Abstract. We present calculations of the structure of magnetized winds of classical T Tauri stars (CTTSs) in an axisymmetric, stationary, polytropic approximation in a given magnetosphere (i.e. the wind dynamics is in general not in force equilibrium with the magnetic flux tubes). According to the widely accepted model of CTTSs (see Camenzind 1990) the magnetosphere is dominated by a strong stellar magnetic field ($B_* \simeq 1$ kG) which possibly is of dipolar type. We obtain the structure of the underlying magnetosphere from a semi-quantitative treatment, in which a dipolar stellar field of kilo-Gauss strength is squeezed by a surrounding diamagnetic accretion disk (Riffert 1980). This approximation allows to derive an analytic expression for the resulting magnetic field. After decreasing according to a dipole in the innermost region the magnetic flux-tube function shows an almost flat behaviour until the collimation region is reached. This picture should yield a reasonably good description for stellar winds, however it does not take into account disk winds. The shape of the resulting flux function poses constraints on the wind parameters and thus on its initial conditions and could give some hints to the origin of the outflows (star and/or accretion disk).

1 Introduction

Classical T Tauri stars are low-mass protostellar objects (Appenzeller & Mundt 1989). They often show bipolar outflows which are collimated on scales of 100 or even several 10 AU. Even though it is clear that magnetic forces are involved in the collimation process the details are only poorly understood. The VLTI will for the first time allow to zoom into the closest stellar surroundings and directly resolve the collimation region and even in some detail the uncollimated wind.

2 Magnetized Winds

A magnetohydrodynamic (MHD) wind-magnetosphere structure is mathematically determined by the simultaneous solution of wind equation and the Grad-Schlüter-Shafranov (GSS) equation for the magnetic flux tubes Ψ which is an unsolved problem up to now. Instead we solve the polytropic wind equation for a stellar wind in a magnetosphere as described above. The flat behaviour of the flux tubes in the region beyond a few stellar radii constraints the possible wind solutions. A wind can only exist at large distances from the star if

$$\frac{4\pi\eta}{\Omega^{F^2}} \sqrt{\left(\frac{2E}{3}\right)^3} > \Psi_\infty \,, \tag{1}$$

where Ψ_∞ represents the magnetic flux in that region. η decxribes the mass flux per magnetic flux tube, Ω^F the angular velocity of the flux tube and E the total energy of the flow. Relation (1) follows from the energy equation of a polytropic wind at large stellar distances (Paatz & Camenzind 1996). On the one hand a high energy and a high mass flux favours the existance of wind solutions at large distances from the star whereas low-mass winds break down before the collimation region (Fig. 1). On the other hand the value of the energy is fixed by the requirement that the flow must smoothly pass through the fast magnetosonic point. Our treatment is however limited by the fact that our flux tubes are fixed irrespective of the force equilibrium perpendicular to the flux tubes. As a next step we will construct wind solutions which approximately obey this equilibrium (Paatz & Camenzind 1997). These solutions could then serve as a starting source term of the GSS equation.

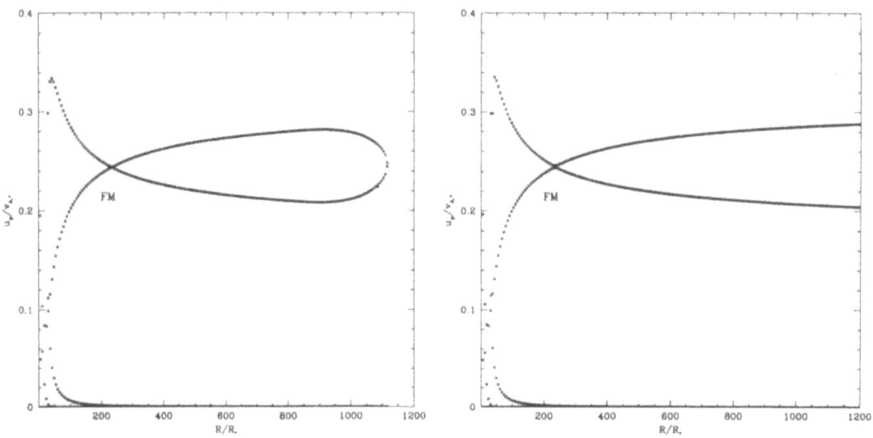

Fig. 1. Poloidal velocity u_p (normalized to the Alfvén velocity on the star) of stellar winds as a function of the radius R (normalized to the foot point cylindrical radius). If the mass flux per magnetic flux tube η is too low (i.e. almost force-free solutions) the wind solution breaks down and does not reach the jet regime (Fig. 1.a). Higher values of η allow the existence of asymptotic wind solutions (Fig. 1.b). FM denotes the fast magnetosonic point

Acknowledgement: This work is supported by the Deutsche Forschungsgemeinschaft.

References

Appenzeller, I., Mundt, R. 1989, A&A Review 1, 291
Camenzind, M., 1990 in *Rev. Mod. Astron.* 3, Ed. G. Klare, Springer (Heidelberg), 234
Paatz, G., Camenzind, M. 1996, A&A, 308, 77
Paatz, G., Camenzind, M. 1997, A&A, in prep.

Operating the VLTI in the Thermal Infrared

Andreas Quirrenbach

Max-Planck-Institut für Extraterrestrische Physik, Postfach 1603, D-85740 Garching

1 Science with the VLTI at 10 and 20 μm

Many of the most interesting scientific programs for the VLTI require operation of the interferometer in the thermal infrared. Observations of small bodies in the Solar System will benefit from mid-IR capabilities, and 10 μm is the best wavelength for attempting direct detections of giant extrasolar planets in orbits of $\sim$ 1 AU around their parent stars. The characterization of exo-zodiacal light and the study of dust disks around main-sequence stars (e.g., β Pic, ε Eri, α PsA) will also require observations in the thermal IR. Dust shells around cool stars and dust formation in the later phases of classical novae can be studied in great detail. The surface structure of cool stars can be imaged in the visible, while the fringes are being tracked at 10 μm. Mid-IR observations of disks around pre-main-sequence stars, and of pre-main-sequence binaries and infrared companions, will vastly improve our understanding of the early phases of stellar evolution. Looking towards the Galactic Center, we can map the distribution of warm dust and search for an infrared counterpart to Sgr $A^\star$. VLTI observations of bright AGN at 10 μm will reveal the structure and spatial scale of the obscuring tori that are at the heart of unified theories, and high-resolution maps of some strong fine structure lines (e.g., [S IV] 10.511 μm and [Ne II] 12.814 μm) can be used to probe the physical conditions of the ionized gas. At high redshift, studies of galaxies at wavelengths corresponding to the near-IR in the rest frame will allow us to investigate their morphology and measure their star-forming activity without much disturbance from dust obscuration.

2 Expected Performance of the VLTI in the Thermal IR

With a fringe spacing of 16 mas at 10 μm, and 32 mas at 20 μm for the longest baseline between the 8 m telescopes, the resolution of the VLTI will by far surpass all existing instruments operating at these wavelengths. It is this unique capability that will enable us to carry out the projects listed above. In the mid-IR, the expected sensitivity depends critically on good baffling, in order to minimize the thermal background. Assuming an efficiency of the warm part of the interferometer (telescopes, delay lines) of 0.4, and of the cold part (beam combiner, detector) of 0.3, and a diameter of $1''$ of the cold field stop, the *fringe tracking* sensitivity of the VLTI is estimated to be 0.4 Jy at 10 μm and 0.8 Jy at 20 μm. When off-source fringe tracking is possible, the sensitivity for a 3σ detection in 10 minutes should be 3.5 mJy at 10 μm and 10 mJy at 20 μm. The sensitivity scales linearly with the diameter of the field size. For the 1.8 m auxiliary telescopes, 4.4 times stronger sources are needed with the same field size.

3 "First Fringes" at 10 μm?

While the thermal background makes astronomy in the mid-IR quite difficult in general, the $\lambda^{6/5}$ scaling of the atmospheric coherence length and time, and the looser instrumental tolerances at longer wavelengths, favor the 10 μm region for interferometry. In good seeing, even an 8 m aperture is fully coherent at this wavelength with only tip-tilt correction; no high-order adaptive optics is required. In the thermal-background-limited regime, there is no sensitivity loss for dispersed fringe detection. This means that a relatively simple instrument providing both fringe tracking and spectrally resolved observations with R up to a few hundred can work near the theoretical sensitivity limit. Such an instrument would provide a fairly large interferometric field-of-view (of order 1-2$''$). It would work both with the 8 m UTs and the 1.8 m telescopes of VISA; a "hybrid" combination of the 8 m and 1.8 m telescopes would also be possible. Since an interferometer acts as an efficient filter for the uncorrelated sky noise, chopping and nodding would not be necessary. All in all, excellent science, little competition, and efficient use of the large apertures make the thermal IR an attractive wavelength region for a "first light" instrument for the VLTI.

4 MIMI: Concept for a Mid-IR Instrument

MIMI (Mid-Infrared Michelson Instrument) is a conceptual idea for a first generation instrument for the VLTI working in the N and Q bands (10 and 20 μm). The beams from the individual telescopes would be combined pairwise, using 50% beam splitters (see Buscher 1988, Mozurkewich 1994). Each output beam would be spectrally dispersed in a low-resolution spectrograph ($R \approx 200$, or possibly even higher). The outputs of all spectrographs could be directed to a single array detector, similar to the ones used in present mid-IR cameras. The optical path difference would be varied with a triangular signal at ≈ 20 Hz using piezo mirrors. The detector readout has to be synchronized with the delay dither, enabling detection of the dispersed fringes, and providing filtering of the thermal background noise. A variable cold field stop is needed, since the smallest field size allowed by the science target and by the current seeing should always be chosen in order to minimize the thermal background. MIMI should reach the sensitivity quoted in the previous paragraph; it should thus be suited for addressing many of the astronomical topics mentioned above. A detailed study of MIMI is needed to ascertain its performance, and to compare it to possible alternatives.

References

Buscher, D.F. (1988), MNRAS 235, 1203
Mozurkewich, D. (1994), Proc. IAU Symp. 158, p. 177

Multi-Speckle Interferometric Imaging with the VLTI and the LBT at Optical Wavelengths

Th. Reinheimer, K.-H. Hofmann, M. Schöller, and G. Weigelt

Max-Planck-Institut für Radioastronomie, Auf dem Hügel 69, 53121 Bonn, FRG

Abstract. We present a method for interferometric imaging with arrays of large optical telescopes in the multi-speckle mode. The raw data for computer experiments were produced by simulating light propagation in the atmosphere, earth rotation, and photon noise. Since a Fried parameter r_0 smaller than the telescope diameter was chosen, multi-speckle long-baseline interferograms were obtained. From the various data sets, diffraction-limited images were reconstructed by speckle masking interferometry (Weigelt 1977, Lohmann et al. 1983, Reinheimer et al. 1993, 1996) and the iterative building block method (Hofmann and Weigelt 1993).

1 Laboratory Simulation and Computer Experiments

Fig. 1 shows one of our laboratory experiments (see Reinheimer et al. 1996 for more details). The figures show (a) the pupil function of the LBT, (b) the uv-coverage, (c) the object, (d) one of the 200 simulated speckle interferograms of a single star, (e) one of the 400 simulated speckle interferograms of the object, and (f) the diffraction-limited image reconstructed by speckle masking and the building-block method (1000 iterations).

Fig.2 shows one of our computer experiments (see Reinheimer et al. 1993). The figures show (a) the VLTI pupil function, (b) the uv-coverage, (c) the object (a galaxy with diameter ≈ 9 milli-arcsec for $\lambda = 7000$ Å and 100 m diameter of the array), (d) a point source multi-speckle interferogram (Fried parameter $r_0 = 2$ m, without photon noise), (e) one of the 48 000 simulated speckle interferograms of the object with photon noise corresponding to 8000 photoevents per interferogram, and (f) the diffraction-limited image reconstructed by speckle masking and the building-block method (6000 iterations). Figures 2c and f are enlarged by a factor of 12 compared to d and e.

References

Hofmann, K.-H., Weigelt, G. 1993, A&A 278, 328

Lohmann, A.W., Weigelt, G., Wirnitzer, B. 1983, Appl. Opt. 22, 4028

Reinheimer, T., Hofmann, K.-H., Weigelt, G. 1993, A&A 279, 322

Reinheimer, T., Hofmann, K.-H., Schöller, M., Weigelt, G. 1996, *Speckle masking interferometry with the Large Binocular Telescope*, A&AS *accepted*

Weigelt, G. 1977, Optics Commun. 21, 55

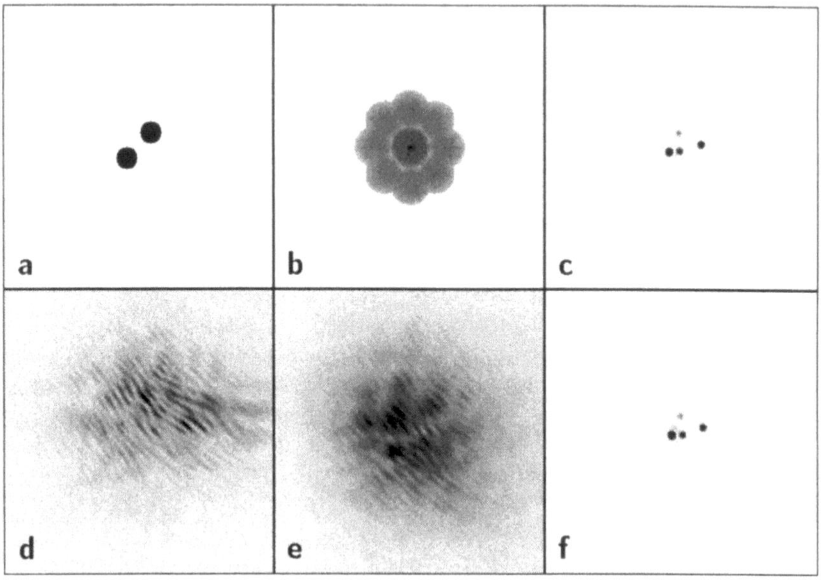

Fig. 1. Laboratory LBT simulation.

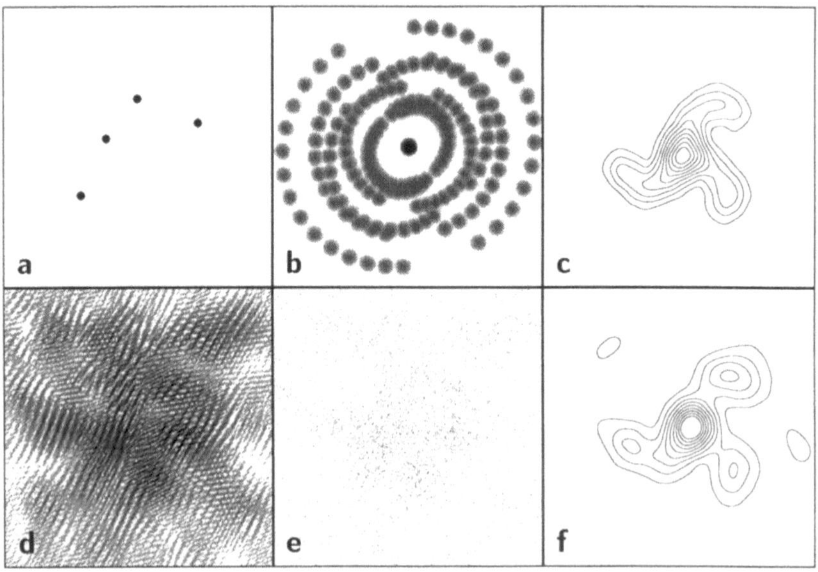

Fig. 2. Computer simulation of multi-speckle long-baseline interferometry.

Infrared Interferometric Imaging with the VLTI in the Multi-Speckle Mode with a Combination of the 8 m UTs and the 2 m ATs

Th. Reinheimer and G. Weigelt

Max-Planck-Institut für Radioastronomie, Auf dem Hügel 69, 53121 Bonn, FRG

Abstract. We present computer simulations of interferometric imaging with the VLT Interferometer at infrared wavelengths. Various observing modes, 4 UTs plus 3 ATs and 1 UT plus 3 ATs, were studied. Image reconstruction was performed by speckle masking (Weigelt 1977, Lohmann et al. 1983, Reinheimer and Weigelt 1987, Reinheimer et al. 1993) and the iterative building block method (Hofmann and Weigelt 1993).

1 Results for a Stellar Surface and a Star Cluster

Figure 1 shows one of our VLTI image reconstruction experiments. In this experiment the object is an extended stellar surface with limb darkening and two dark spots on the surface. The figures 1a to 1f show (a) the VLTI pupil function (full operation hybrid mode; 4 UTs plus 3 ATs at positions J1, J4, and E0), (b) the uv-coverage, (c) the object, (d) a point source multi-speckle interferogram (Fried parameter $r_0 = 2\,\mathrm{m}$; infrared wavelengths), (e) one of the 2100 simulated speckle interferograms of the object (additive detector noise with an amplitude of about 10% of the maximum of the noise-free speckle interferograms was simulated), and (f) the diffraction-limited reconstruction by speckle masking and the building-block method (5000 iterations; contours are plotted from 0 to 0.9 of the peak brightness in steps of 0.1).

Figure 2 shows a second computer experiment. In this experiment the object is a star cluster consisting of 60 stars with random positions and intensities from 1 to 0.078. The figures 2a to 2f show (a) the pupil function (UT1 plus 3 ATs at positions A1, B0, E0), (b) the uv-coverage, (c) the object, (d) one of the 2100 simulated speckle interferograms of the object (Fried parameter $r_0 = 2\,\mathrm{m}$) with additive detector noise (10% of noise-free maximum intensity), (e) the diffraction-limited reconstruction of the stellar field for an observation with only the 8-m telescope, and (f) diffraction-limited reconstruction of the stellar field for the whole array (3000 iterations).

References

Hofmann, K.-H., Weigelt, G. 1993, A&A 278, 328
Lohmann, A.W., G. Weigelt, Wirnitzer, B. 1983, Appl. Opt. 22, 4028
Reinheimer, T., Weigelt, G. 1987, A&A 176, L17
Reinheimer, T., Hofmann, K.-H., Weigelt, G. 1993, A&A 279, 322
Weigelt, G. 1977, Optics Commun. 21, 55

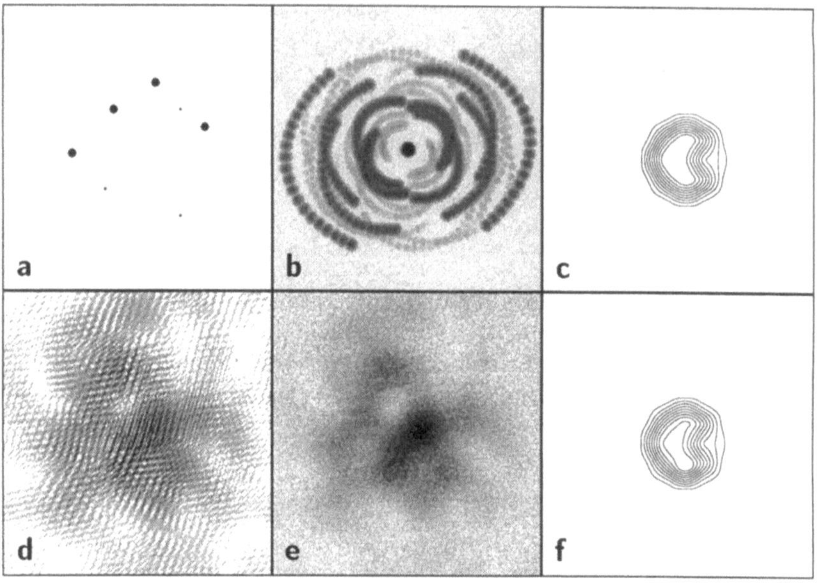

Fig. 1. Computer simulation of multi-speckle interferometric imaging with the VLT Interferometer (hybrid mode: 4 UTs plus 3 ATs at positions J1, J4, and E0; see text). Figures c and f are zoomed by a factor of 8 compared to Figs. d and e.

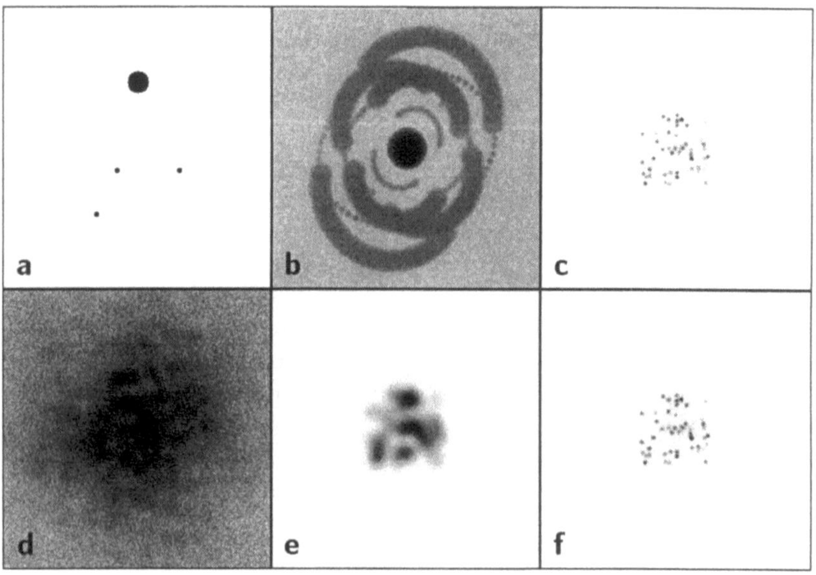

Fig. 2. Computer simulation of multi-speckle interferometric imaging with the VLT Interferometer (UT1 plus 3 ATs at positions A1, B0, and E0; see text).

Thermal IR Imaging with MAX: Pushing the Limit of Single–Dish Ground Based Observations

M. Robberto[1,2], T.M. Herbst[1], S.V.W. Beckwith[1], C. Birk[1] and P. Bizenberger[1]

[1] Max-Planck-Institute für Astronomie, Königstuhl 17, D-69117 Heidelberg, Germany
[2] Osservatorio Astronomico di Torino, Str. Ossevatorio 20, I-10025 Pino Torinese, Italy

Abstract. Diffraction limited images (Strehl ratio > 0.3) are routinely taken at 10-20μm with the new MPIA thermal IR camera MAX at UKIRT. The high read-out rate of the MAX electronics allows us to study the spatial and temporal variations of the telescope point spread function introduced by atmosphere, telescope and mirror flexure, etc. Comparing data taken before and after the recent (August 1996) upgrade of the UKIRT top-end, we can analyze these variations and obtain information useful for effective interferometry at large telescopes.

1 Seeing Effects

Standard seeing conditions ($\simeq 1''$) at visible wavelengths roughly correspond to $0''\!.5$ seeing in the N band, a value comparable to the size of the Airy disk at UKIRT at 10μm. Therefore, depending on the seeing quality, MAX can provide almost diffraction limited images. Strehl ratios as high as 0.53 were reached in April 1996 under excellent seeing conditions (Fig.1). As the major contribution to the seeing comes from image motion, shift and add techniques produce a significant increase in the Strehl ratio. Starting from a Strehl ratio $\simeq 0.5$, the gain at 10μm provided by software processing is about 20%.
Before the top-end upgrade, it was known that the major source of seeing degradation at UKIRT was the wind shake on the telescope structure. During windy nights, we had seeing as bad as $1''\!.3$ at 10 microns. The new top-end (Glindemann 1996, *in preparation*) presents a structure much less sensitive to wind shake and even in 40 Mph wind, we can now easily attain Strehl ratios $\simeq 0.8$ (seeing $\simeq 0''\!.65$ at K). Working routinely with superb seeing conditions, the intrinsic optical quality of the telescope becomes the limiting factor. We find that at UKIRT, in the 10-20μm bands, the adaptive optics correction –also available with the new top-end as tip-tilt at the secondary mirror– does not provide any significant improvement.

2 Mechanical Effects

In the best seeing conditions, changes of the PSF due to variations of the telescope mechanical and thermal status can be easily observed in real time. The structure of the first Airy ring turns out to be extremely sensitive to these effects. Here we present a typical case of sudden variation from an almost perfect

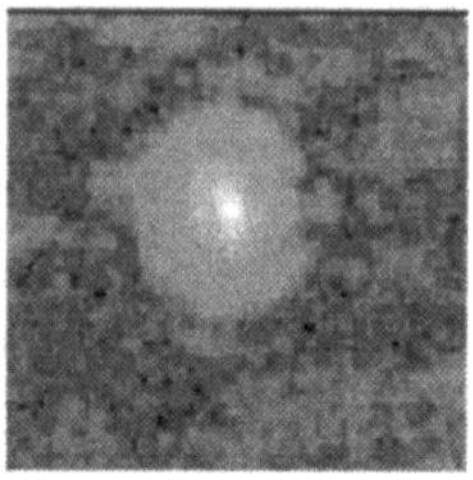

Fig. 1. Image of BS5340. Integration time is 2 s at 11.6 μm. Strehl ratio is 0.53

ring to a "trilobate" profile (Fig.2). The pattern of the lobes reflects the position of the main supports of the telescope primary mirror.
The distorted profiles tend to remain stable for times of the order of several seconds. Frequent adjustements of the telescope intrinsic optical quality through the active control of the primary mirror can eliminate this effect.
We note that the diffraction pattern of UKIRT is strongly affected by the size ($\simeq$ 1.2 m) of the central hole of the primary mirror. This has two major effects:

1. a reduction of the encircled energy in the central peak of the Airy profile from 83% to 68%. Working in diffraction limited conditions, this is a major disadvantage.
2. a modulation of the intensity of the outer rings. Together with the first ring, the fourth and the seventh rings are the most prominent. Our images clearly show this effect.

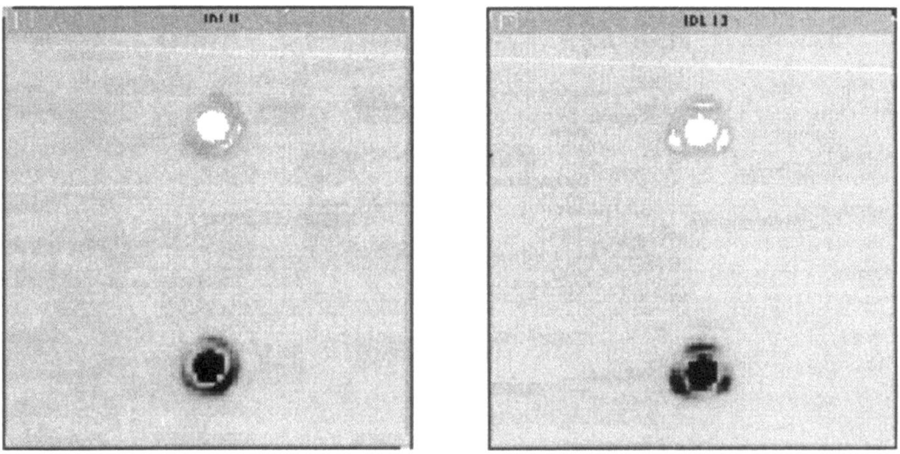

Fig. 2. Real-time PSF variations. Chopping-mode observations of BS 8775 at 11.6 μm. The integration time for each frame is 98ms (coadd of 10 frames 9.8 ms each). The two frames are separated by 2.04s.

Differential Interferometry and the Non-Axisymmetric Envelope of γ Cas

L. J. Sánchez[1,2], R. G. Petrov[1], V. Vasyuk[3], S. Lagarde[1], A. Blazit[4], S. Cuevas[2] and R. Foy[5]

[1] Dépt. d'`Astrophysique, Université de Nice, Parc Valrose, 06108 Nice, France
[2] Instituto de Astronomía, UNAM, Apartado Postal 70-264, 04510 México D.F.
[3] Special Astrophysical Observatory, Nizhnij Arkhyz 357147, Karachaj-Cherkassija, Russia
[4] Observatoire de la Côte d'Azur, 06460 Saint-Vallier de Thiey, France
[5] Observatoire de Lyon, 69561 Saint-Genis-Laval, France

Abstract. We report on Differential Speckle Interferometry observations of the Be star γ Cassiopeiae. The photodisplacement $\vec{\varepsilon}(\lambda)$ measured in the Hα line is strongly incompatible with an axisymmetric envelope.

Differential Interferometry measures the variation $\vec{\varepsilon}(\lambda)$ of the photocentre of a *non-resolved source* with wavelength (see Petrov et al. in these proceedings). In the case of the Be star γ Cas, long-baseline interferometry seemed to support an axisymmetric model for the envelope. Quirrenbach et al. (1993) observations are well fitted by an oblate inclined Gaussian disk with major axis (3.2 mas FWHM) perpendicular to the intrinsic measured polarisation (P-axis in **Fig. 1.b**). Stee et al. (1995) described an axisymmetric radiative wind model giving acceptable representations of emission line profiles and variations of visibility with wavelength observed at the GI2T interferometer. Within this model, the rotation velocity law is much better constrained by $\vec{\varepsilon}(\lambda)$ than by the line profile or the wavelength dependent visibilities (Chalabaev 1991, Stee 1996). Other effects, such as the X-ray emission or the temporal variations of the lines must be explained by additional features such as a density wave in the envelope, a low mass companion or stellar surface activity. Each of these phenomena has a specific $\vec{\varepsilon}(\lambda)$ signature.

We obtained Differential Speckle Interferometric data on γ Cas at the Russian SAO 6 m telescope with the Franco-Mexican Differential Speckle Interferometer (Cuevas et al. 1994) whose x-λ images were recorded by the CP40 photon counting camera. After careful gain and distortion corrections, the average cross-spectrum between the λ and the reference channel yields the photodisplacement $\vec{\varepsilon}(\lambda)$ and its uncertainty $\sigma_\varepsilon(\lambda)$. The orientation of the slit on the sky was inverted every minute in order to remove any residual instrumental bias. The differential atmospheric refraction (up to 3.6 mas for 40 Å) was measured together with the $\vec{\varepsilon}(\lambda)$ of γ Cas and found in good agreement with prediction. The measures in four different directions were combined in the vectorial representation of $\vec{\varepsilon}(\lambda)$ shown in **Fig. 1.b**.

The total excursion of $\vec{\varepsilon}(\lambda)$ is of about 3 mas, which is compatible with the estimated equivalent width of the envelope from long-baseline interferometry. The uncertainty $\sigma_\varepsilon(\lambda)$ ranges from 0.16 mas to 0.5 mas per 1Å channel. The

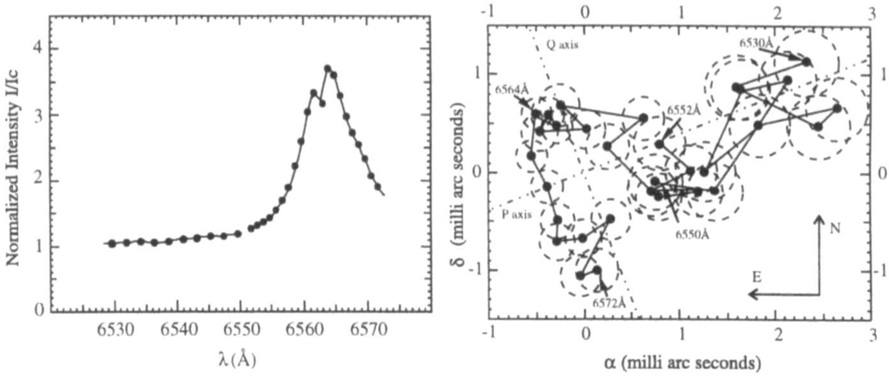

Fig. 1. **(a)** Hα emission line of γ Cas on Sept 5th 1993. **(b)** Vectorial representation of the photodisplacement $\vec{\varepsilon}(\lambda)$. Each dot represents the difference between the photocentre in one of the 32 channels shown in **(a)** and the photocentre in the reference channel (full Hα line) set arbitrarily at the origin of the coordinate system. The λ for each dot can be found if one follows the curve from 6530 Å in the upper right to 6572 Å in the lower left corners. Circles represents the measure uncertainty (radius is σ).

photocentre of the "Hα dominated image" and the photocentre of the "continuum dominated image" are separated by 1.9 ± 0.18 mas. This corresponds to an actual distance between the B star and the photocentre of the envelope of about 4 mas ($\sim 20\,R\star$). It is quite tempting to say that the 3.2 mas (or less) cloud emitting the Hα line does not "envelop" the B star. In the Hα channel the photocentre of the zone of the envelope with a small relative radial velocity is clearly not a centre of symmetry for the $\vec{\varepsilon}(\lambda)$ curve. This "horseshoe" shape is compatible with material on a very elliptic orbit around a companion as described by Harmanec (1987). If such a companion exists, it should have a mass close to $2\,M\odot$ to have a Roche lobe large enough to contain a ~ 2 mas envelope at 4 mas of a $16\,M\odot$ star. But if its mass is larger than $2\,M\odot$, it should have been detected by radial velocity measurements (except if the orbit is very excentric). Other explanations are currently under examination.

References

Chalabaev, A. (1991) Proc. ESO Conf. **39** 403
Cuevas, S., Petrov, R.G., Chelli, A. et al. (1994) Proc. SPIE **2200** 501
Harmanec, P. (1987) Proc. Coll. IAU **92** 339
Petrov, R. G., Sánchez, L. J. and Lagarde, S. in these proceedings
Quirrenbach, A., Hummel, D. F., Buscher, J. T. et al. (1993) ApJ **416** L25
Stee, Ph. (1996) A&A **311** 945
Stee, Ph., Araujo, F.X., Vakili, F. et al. (1995) A&A **300** 219

HST Imaging of GM Aurigae: A Circumstellar Disk Seen in Scattered Light

Karl Stapelfeldt and The WFPC2 Science Team

Mail Stop 183-900, Jet Propulsion Laboratory, 4800 Oak Grove Drive, Pasadena, CA 91109, USA

1 Introduction

GM Aurigae (HBC 77, IRAS 04519+3017) is a K3V classical T Tauri star located at a distance of 140 pc. It is a young (1-2x10^6 yrs), low-mass (0.7 $M_\odot$), low-luminosity (0.8 $L_\odot$) object with far-infrared and submillimeter excesses indicating substantial circumstellar dust. The absence of optical jets or high velocity CO indicates that the source lacks any bipolar outflow. The excess emission has been well-modeled by assuming the source possesses a circumstellar disk of diameter 300 AU and mass 0.1 $M_\odot$ (Beckwith et al. 1990 AJ 99 924). The IRAS 12 μm flux is below disk model predictions, however, which has led to speculation that the disk has an inner hole of radius 0.4 AU (Marsh and Mahoney 1993 Ap.J. 405 L71).

Millimeter interferometry of GM Aur by Koerner, Sargent, and Beckwith (1993 Icarus 106 2) finds a compact (diameter $<$ 300 AU) continuum core at the center of an extended bar of ^{13}CO at PA $\approx 50°$. The molecular gas bar has a diameter of 1000 AU and a velocity structure that strongly resembles models of a disk in Keplerian rotation. Recent observations at higher spatial resolution (Dutrey et al. 1996, Koerner 1996, both in preparation) confirm this general picture.

The optical starlight is polarized by 1-2% at PA 155°, indicating that (1) significant reflection nebulosity is associated with the star and (2) for a symmetric nebula the reflected light should be extended along PA 65°. We observed GM Aurigae with HST in order to determine the spatial and geometric relationship between this nebulosity and the disk inferred by the infrared and millimeter observations.

2 Observations & Results

GM Aurigae was observed on the Planetary Camera with the following filters and exposure times: F555W, 30 sec; F675W, 16 sec and 400 sec; and F814W, 10 sec and 230 sec. PSF subtraction was accomplished using PC1 images of the star HDE283572 as a reference.

Nebulosity is detected in all filters and exposures. It extends symmetrically about the star to a diameter of 900 AU, with a minor axis $\approx$250 AU. The reflected light extends along virtually the same position angle as the molecular as disk seen in ^{13}CO, and thus we conclude that the nebulosity arises via reflection off dust grains within the molecular disk. The reflected light contributes about 8% of the total light from the GM Aur system, and has a surface brightness of about 14 mag/sq″ at a distance of 40 AU. The radial brightness profile at the midplane follows a power law of $r^{-2.3}$, consistent with models of an optically thick flared disk.

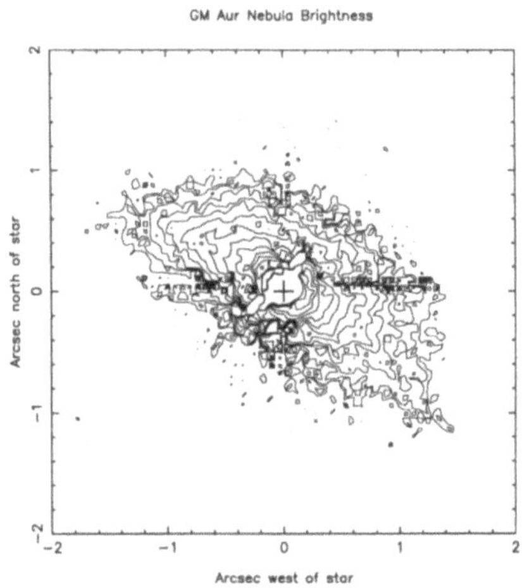

Fig. 1. Contour map of PSF-subtracted F555W Planetary Camera image

3 Discussion

Infrared SED modelling of GM Aurigae clearly indicates the disk must be optically thick at visual wavelengths. Thus starlight does not penetrate the interior of the disk, it only illuminates the disk's upper and lower surfaces. We have calculated models of the reflection nebulosity of such disks for our work on HH 30. A disk viewed about 25° from edge-on reproduces the morphology of the GM Aur nebulosity, allows the star to be directly seen, and is consistent with the nebular surface brightness. It should be possible to retrieve the flaring function of the disk from the radial brightness profile; this work is in progress.

4 Prospects with the VLTI

The circumstellar nebulosity of GM Aurigae and other T Tauri stars are well-suited to studies with the VLTI or Keck interferometers. The bright central star provides an ideal wavefront reference (V= 12, K= 7), and the VLTI/Keck interferometers should be able to image radial structures in these disks on scales of a few AU, if they exist. A measurement of the source diameter at $\lambda = 10\mu m$ would provide a sanity check to disk emission models. For GM Aurigae, it would be particularly interesting to see if the there actually is an inner hole in the central few milliarcsec of the disk, as suggested by the infrared spectral energy distribution.

Software Concept for Four-Dimensional Bispectral Analysis of Stellar Interferograms

Hans Weghorn

Sternwarte der Universität Bonn, Germany
AEG Mobile Communication, Ulm, Germany

Abstract. During the past decades interferometric observation techniques were able to establish themselves in astronomy. This was possible due to a rapid development of astronomical instruments, fast computer hardware, and fast sensitive detectors directly producing digital images, e. g., CCD cameras. On the other hand, the interferometric processing methods are not yet fully elaborated, and the currently most complex method, i. e. four-dimensional bispectral analysis, can not be performed in real-time in detail. The concept for implementing bispectral processing can be considered an as complicated topic as the instrumental hardware. With complexity theory a three-staged software concept is motivated here, which allows pipelined processing of stellar interferograms.

1 Introduction

As further development of the first astronomical speckle method – A. Labyrie's stellar speckle interferometry[1] – bispectral analysis was proposed as general, two-dimensional mathematical concept on one-dimensional input data including astronomical examples in 1983[2]. Later, bispectral analysis was recognized not only as being helpful for interferometric observations with single-dish telescopes but also for evaluating data from multiple-aperture telescopes like optical interferometers with long baselines.

In true astronomical observations the input interferograms are two-dimensional requiring a four-dimensional bispectral processing. For the latter, a first mathematical concept was shown in 1988[4]. The mathematical concept is one aspect of bispectral analysis; since the technical realization has to be made with computers, the software implementation must also be taken into account.

2 Complexity of Bispectral Processing

Assuming astronomical interferograms with a size of N^2 pixels as input, the complexity of the bispectrum calculation obviously is dominated by $O(N) = N^4$ because the bispectrum is four-dimensional (note: the Fourier transformation yields only a part of $O(N) = N \log N$ during the processing of each interferogram).

This complexity is meet exactly, when the program, which averages the bispectrum, is provided with a list containing all bispectral coordinates to be processed. The direct approach of determining the required elements during calculation of the bispectrum would increase the complexity to $O(N) = N^8$, because each element would have to be checked for re-occurrence resp. **u**,**v**,**w**-exchange symmetry by a double-search (search for same **w**, and search for same **u**).

3 Software Design for Bispectral Analysis

Therefore, the software concept for bispectral processing conveniently is divided into three parts (Fig. in[7]):

1. Generation of the look-up-table (LUT) which lists all relevant bispectral vector coordinates; i. e. omitting all redundant elements from $\{(\mathbf{u}, \mathbf{v}, \mathbf{w} = \mathbf{u} + \mathbf{v}) \in Z^6\}$ or elements carrying no information (diluted **u**,**v**-coverage for interferometers). The LUT is sorted with increasing $|\mathbf{w}|$ up to the cut-off frequency, and contains all applicable **u**,**v**-combinations satisfying $\mathbf{u} + \mathbf{v} \equiv \mathbf{w}$, $|\mathbf{u}| \leq |\mathbf{w}|$, and $|\mathbf{v}| \leq |\mathbf{w}|$. This sorting is required for minimal computing amount in the third processing step.

2. Calculation of the average 4-d bispectrum: in a loop all K input interferograms $i_k(\mathbf{x})$ are Fourier transformed, and their bispectrum $I^{(3)}(\mathbf{u}, \mathbf{v}) = \sum_{k=1}^{K} I_k(\mathbf{u}) I_k(\mathbf{v}) I_k^*(\mathbf{w} \equiv \mathbf{u} + \mathbf{v})$ is accumulated according to the LUT. For fast processing, the LUT is vectorized before the averaging is performed.

3. Extraction of the object Fourier spectrum from the bispectrum: this cannot be performed directly but has to be done by recursive[6] or iterative techniques[4]. With certain starting points at the frequency origin, each Fourier element is generated from the bispectrum and from lower Fourier spectrum elements which are already known. In this manner, the whole frequency information can be computed from the inner to the outer parts of the object spectrum.

The high-resolution object information is preserved in this averaging process of third order. After a correction of the wrong frequency weighting produced by the influence of the earth atmosphere, and after Fourier transformation, often a diffraction-limited image of the observed astronomical object is obtained.

For performing the scientific work in[4] and [6], this bispectral scheme was under development as part of an interferometric image processing software package since 1985. The software is used in Unix environments since several years; the first[3] and recently obtained astronomical results are reported in[7].

References

1. Labyrie, A. (1970), A&A **6**, pp85
2. Lohmann, A. W., et al. (1983), Appl. Opt. **22**, pp4028
3. Baier, G., et al. (1988), ESO Messenger **52**, pp11
4. Weghorn, H. (1988), diploma thesis, University of Erlangen
5. Weghorn, H. (1993), SPIE Proc., Vol. 1983, pp448
6. Weghorn, H. (1994), Ph.D. thesis, Hänsel-Hohenhausen, DHS 869, Egelsbach
7. Weghorn, H., et al. (1995), Abstract Series No. 11 of the Astronomische Gesellschaft, K. Weber, Leimen, p111 & p158

Real-Time Image Compression Method for Astronomical Interferograms

Hans Weghorn[1,2], Jörg Colberg[3], and Thilo Bauer[4]

[1] Sternwarte der Universität Bonn, Germany
[2] AEG Mobile Communication, Ulm, Germany
[3] Max-Planck-Institut für Astrophysik, Garching, Germany
[4] Observatorium Hoher List der Sternwarte der Universität Bonn, Daun, Germany

Abstract. For astronomical interferometric observations, the life-time of the disturbing turbulences in the earth atmosphere leads to short exposure times ($\leq \frac{1}{10}$s), and consequently to the desire of high frame rates ($\geq 8\text{s}^{-1}$) for a good exploitation of telescope time. Additionally, a high spatial sampling demands detector arrays with at least 2k×2k pixels when using large optical interferometers like, e. g., the ESO VLTI. In total, such detectors produce output data rates for the astronomical images of at least 32 MByte per second. Obviously, these high data amounts can only be recorded with a reasonable technical effort when data compression methods are applied.

1 Digital Images & Standard Compression Techniques

The analog input signal of astronomical exposures is usually converted to digital images using a linear mapping. E. g., for each pixel (**pic**ture c**el**l) of a CCD sensor 4096 ($\equiv 2^{12}$) equidistant intensity levels are distinguished. These values appear as integer numbers in binary notation, which is clearly required when mathematical operations have to be applied to the astronomical images using digital computers. Shannon's theory of information teaches us that this is not a space-efficient coding scheme for storing such images, because the probability distribution $N(i)$ of the pixel intensities i is heavily non-uniform for astronomical exposures as especially for astronomical interferograms. The minimum number of bits required for storing an image consisting of N pixels with a bit depth of B, can be evaluated as the total entropy to

$$E_{\text{tot}} = \sum_{i=0}^{2^B-1} -N(i)\log_2 N(i)/N \quad [\text{bit}]$$

The optimization of storing large amounts of data was investigated early in computer science. MPEG defines the currently most efficient image compression scheme achieving rates of up to 100. Unfortunately, this method is not useful for interferometric data in astronomy because it restores high frequency information with low and unpredictable quality. For scientific imaging applications, the classical non-destructive compression schemes should be considered first: **i)** Run-length encoding requires many pixel sequences with identical intensities, and is not applicable to astronomical data due to noise; **ii)** Delta-coding takes advantage of slowly varying signals and is not directly useful for astronomical interferograms because of high fringe contrast and photon noise; **iii)** Huffman coding (HC) stores rare intensity values with long, and frequent intensity values with short codes; **iv)** The Lempel-Ziv (LZ) scheme detects repeating pixel sequences and stores only a back-index on the second occurrence. The latter two methods yield sufficient compression rates for astronomical interferograms.

2 A New Scheme: Signed Huffman Coding

In the new compression scheme, which is based on Huffman coding, the differences between neighbour pixels rather than the pixel values themselves are stored. Due to a symmetric probability distribution of these pixel-to-pixel variations for astronomical interferograms, the identical Huffman coding sequence can be used for a negative or a positive variation by the same number of counts. The up/down information is coded by an extra sign bit at the beginning of each code for all non-zero values.

pixel difference	classical Huffman	signed Huffman
0	0	0
±1	10 110	1S0
±2	1110 11110	1S10
±3	111110 1111110	1S110
⋮	⋮	⋮

Assuming an image with K_0 occurrences of 0, K_1 occ. of ±1, ... , the total image size E_H for HC and E_{SH} for signed Huffman (SHC) can be calculated: $E_H = K_0 + \sum_{i=1}^{2^B-1}((2i) + (2i+1)) \cdot K_i$ $\quad E_{SH} = K_0 + \sum_{i=1}^{2^B-1} 2 \cdot (2+i) \cdot K_i$
Obviously, SHC requires less bits for coding the high-order terms:

Δ intensity	0	±1	±2	±3	±4	...
classical Huffman	$1 \cdot K_0$	$(2+3) \cdot K_1$	$(4+5) \cdot K_2$	$(6+7) \cdot K_3$	$(8+9) \cdot K_4$	...
signed Huffman	$1 \cdot K_0$	$(3+3) \cdot K_1$	$(4+4) \cdot K_2$	$(5+5) \cdot K_3$	$(6+6) \cdot K_4$	...
difference	–	$+K_1$	$-K_2$	$-3K_3$	$-5K_4$	...

SHC yields compression rates of about 4.8 on astronomical interferograms which is better than for HC and LZ (~4.6). These values were measured using interferometric data which was taken with a special CCD camera at the ESO Danish 1.5 at La Silla, Chile. De-biasing the raw interferograms with the fixed-pattern offset of the CCD sensor, further improves the compression rate up to 10. Since the computing complexity of SHC and HC is only linear resp. to the number of image pixels – the processing consists of two loops: building a histogram, and mapping the pixels to their code –, they run about five-times faster than LZ, which has a higher complexity due to the involved back-search.

3 Conclusion

Reducing the storage space of astronomical images is not only relevant for interferometric observation techniques, but for the latter it may become essential for being able to record the extremely high data rates. Generally, the proposed signed Huffman coding, which achieves slightly better performance than other methods, is applicable to any astronomical image. Totally, there is a demand for a new standard format for astronomical exposures containing compressed pixel data, which will be directly supported by SW packages like IRAF and MIDAS.

Synthetic Images from Dynamic Model Atmospheres of C-Rich Long-Period Variables*

W. Windsteig, E. A. Dorfi, S. Höfner, J. Hron and F. Kerschbaum

Institut für Astronomie der Universität Wien, Türkenschanzstr. 17, A-1180 Wien, Austria

Abstract. Based on self-consistent models of dust driven winds from C-rich long-period variables (LPVs) we have performed detailed frequency-dependent radiative transfer calculations. We present spatial brightness profiles as well as synthetic VLTI images of the circumstellar dust shells.

1 Modelling Method

The dynamical models are obtained by solving the coupled system of grey radiation hydrodynamics and time-dependent dust formation as described in Höfner & Dorfi (1995), employing an implicit numerical method and an adaptive grid. Considering C-rich stars we assume the formation of amorphous carbon grains (Gail & Sedlmayr 1988). The effects of stellar pulsation are included in our calculations by applying a piston and a variable luminosity at the inner boundary of the models which is located below the stellar photosphere.

We solve the frequency-dependent radiative transfer equation for given density and temperature structures as well as dust grain properties resulting from the dynamical model calculations. The spatial brightness profiles are obtained using dust opacities of amorphous carbon grains according to Maron (1990) and SiC data by Pégourié (1988) with an abundance ratio of 4:1.

To compute synthetic images shown in Fig. 1, we convolved the intensity profiles of a LPV model ($10^4\,L_\odot$, $1.0\,M_\odot$, $\varepsilon_C/\varepsilon_O = 1.80$, $\dot{M} = 1.6 \cdot 10^{-5}\,M_\odot\,\mathrm{yr}^{-1}$, $\langle u \rangle = 25$ km/s) and a stationary model ($10^4\,L_\odot$, $1.0\,M_\odot$, $\varepsilon_C/\varepsilon_O = 2.30$, $\dot{M} = 2.9 \cdot 10^{-7}\,M_\odot\,\mathrm{yr}^{-1}$, $\langle u \rangle = 12$ km/s) with an Airy diffraction pattern of a 100 m telescope at the wavelengths of 3.78 μm (L'-band), 9 μm (continuum) and 11.3 μm (peak of the SiC feature). The Airy pattern was truncated at the second minimum. We have assumed that the VISA and VIMA/Hybrid modes can give a good sky and u,v coverage for a maximum baseline of 100 m.

2 Synthetic VLTI Images

The wavelength dependence of the shell-like structure of the LPV model and the telescope resolution give the most pronounced structure at 3.78 μm and 11.3μm.

* This work is supported by the Österreichischer Fonds zur Förderung der wissenschaftlichen Forschung under project numbers S7305-AST and S7308-AST.

On the other hand, the source size is comparable at 3.78 μm and 9 μm while the longest wavelength shows the most extended emission. At all studied wavelengths and out to distances of about 500 pc, the images from the dynamic models are markedly different from those of the stationary case.

It seems save to conclude that out to distances of about 500 pc the (multiple) shells should be resolvable with a 100 m class telescope at all three wavelengths. Taking into account the luminosities and colours of C-stars (e.g. Kerschbaum et al. 1996) the VISA and VIMA sensitivity should give a sufficient S/N for all stars in this volume. From the number densities of carbon stars with mass loss rates comparable to our models (Groenewegen et al. 1992), about 100 carbon stars should be accessible with our chosen VLTI parameters.

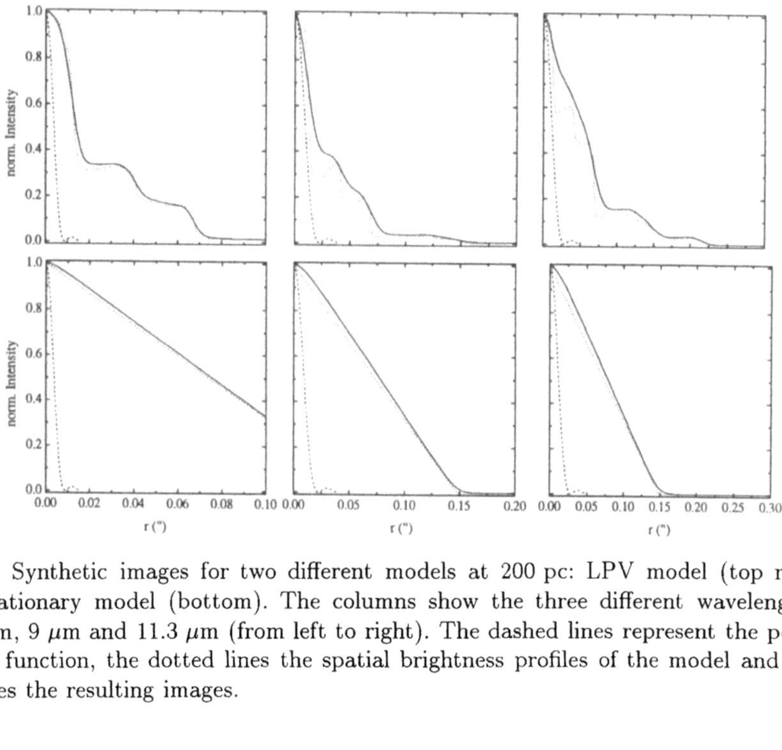

Fig. 1. Synthetic images for two different models at 200 pc: LPV model (top row) and stationary model (bottom). The columns show the three different wavelengths 3.78 μm, 9 μm and 11.3 μm (from left to right). The dashed lines represent the point spread function, the dotted lines the spatial brightness profiles of the model and the full lines the resulting images.

References

Gail H.-P., Sedlmayr E. 1988, A&A 206, 153

Groenewegen M.A.T., de Jong T., van der Bliek N.S., Slijkhuis S., Willems F.J.: 1992, A&A 253, 150

Höfner S., Feuchtinger M.U., Dorfi E.A.: 1995, A&A 297, 815

Kerschbaum F., Habison P., Loidl R., Olofsson H., Hron J.: 1996, *The Carbon Star Phenomenon*, IAU Symp. 177, ed. Wing R.F.; Reidel, Dordrecht, in press

Maron, M.: 1990, Astrophys. Space Sci. **172,** 21

Pégourié, B.: 1988, A&A **194**, 335

Brightness Profiles and Spatial Spectra of Dynamical Models for C-Rich Circumstellar Dust Shells Around Long-Period Variables

Jan Martin Winters, Axel J. Fleischer, Erwin Sedlmayr

Inst. f. Astron. & Astroph., TU Berlin, Hardenbergstr. 36, D-10623 Berlin, Germany

Poster copy including the figures is available via http://export.physik.tu-berlin.de/Publikationen/

1 Introduction

Based on dynamical models of circumstellar dust shells (CDS) around long-period variables (LPVs), which include time-dependent hydrodynamics and a detailed treatment of the processes of formation, growth and evaporation of carbon grains, angle- and frequency-dependent radiative transfer calculations have been carried out. The models are completely determined by 6 parameters, comprising the 4 fundamental stellar parameters (M_*, T_*, L_*, $\epsilon_{\rm i}$) and the pulsation period and velocity amplitude at the base of the atmosphere (P, $\Delta u_{\rm P}$). It turns out that these models result in a discrete onion-like structure of the CDS (cf. Fleischer et al. 1992) which decisively influences the shape of the synthetic lightcurves (Winters et al. 1994) as well as the calculated spatial intensity profiles and the corresponding spatial spectra (Winters et al. 1995).

We present consistent theoretical models for the CDS of the extreme carbon stars IRC +10216 and AFGL 3068. The synthetic light curves, the spectral energy distributions at different phases, the terminal wind velocities etc., are in excellent agreement with the observed properties of these objects. Therefore, the values of the fundamental stellar parameters and the distance to the objects, which are not directly accessible by observations, can be derived from the models.

2 IRC +10216

The model for IRC +10216 is determined by: stellar mass $M_* = 1 M_\odot$, stellar luminosity $L_* = 2.4\,10^4 L_\odot$, stellar temperature $T_* = 2500$ K, carbon-to-oxygen ratio $\epsilon_{\rm C}/\epsilon_{\rm O} = 1.20$, pulsation period $P = 650$ d, and a piston amplitude $\Delta u_{\rm P}$ $= 6$ km/s, which results in a calculated distance to IRC +10216 d = 170 pc, final outflow velocity $v_\infty = 14.3$ km s^{-1}, mass loss rate $\dot{M} = 1.2\,10^{-4}$ M$_\odot$ yr^{-1}, dust-to-gas mass ratio $\rho^{\rm d}/\rho^{\rm g} = 8.1\,10^{-4}$.

Figure 1 shows the spectral energy distribution for IRC +10216, which gives a distance of 170 pc. **Fig. 2** shows observed (Ridgway & Keady 1988) and calculated spatial spectra for IRC +10216 at different wavelengths. Note the internal structures in the visibility functions at $\lambda = 12\mu$ and 25μ ! **Fig. 3** shows the theoretical brightness profiles, which exhibit a clear step-like structure on a scale of a few stellar radii at wavelengths $\lambda \geq 5\mu$. In **Fig. 4**, the 1.5μ brightness profile is compared to the result of a lunar occultation observation (Richichi 1993, private communication). A step-like structure in the observed profile is clearly visible already at this short wavelength. The radial scale, on which the steps occur in the model as well as in the observations is of the same order of only a few stellar radii.

3 AFGL 3068

The model for AFGL 3068 is determined by: stellar mass $M_* = 0.9 M_\odot$, stellar luminosity $L_* = 1.3\,10^4 L_\odot$, stellar temperature $T_* = 2100$ K, carbon-to-oxygen ratio $\epsilon_C/\epsilon_O = 1.33$, pulsation period $P = 696$ d, and a piston amplitude $\Delta u_P = 8$ km/s, which results in a calculated distance to AFGL 3068 d = 1.3 kpc, final outflow velocity $v_\infty = 15.7$ km s^{-1}, mass loss rate $\dot{M} = 1.7\,10^{-4}$ M$_\odot$ yr^{-1}, dust-to-gas mass ratio $\rho^d/\rho^g = 1.6\,10^{-3}$.

Figure 5 shows the spectral energy distribution for AFGL 3068, which gives a distance of 1300 pc. **Fig. 6** shows calculated spatial spectra for AFGL 3068 at different wavelengths. Note that internal structures in the visibility functions are almost absent in this case. **Fig. 7** shows the theoretical brightness profiles, which again exhibit a step-like structure on a scale of a few stellar radii. These steps only occur at even longer wavelengths than in the case of the IRC +10216 model, caused by the larger optical depth in the AFGL 3068 model. In **Fig. 8**, the 1.5μ and 12μ brightness profiles are shown. In the considerably more extended 12μ profile a step-like structure is visible.

4 Concluding Remarks

Time resolved, high angular resolution observations of circumstellar dust shells around long-period variable stars would provide very specific information about the physical processes relevant for the mass loss mechanism in dusty AGB stars. These observations could be directly used to test and improve the reliability of current theoretical models. Finally, this would lead to a more thorough understanding of the nature of circumstellar dust and its role for the hydrodynamic and thermodynamic structure and for the temporal behaviour of the circumstellar shell. **Acknowledgements:** This work was supported by the BMBF (grant 05 3BT13A 6) and by the DFG (grant Se 420/8-1).

References

Fleischer A. J., Gauger A., Sedlmayr E. (1992): A&A 266, 321–339
Ridgway S. T., Keady J. J. (1988): ApJ 326, 843–858
Winters J. M., Fleischer A. J., Gauger A., Sedlmayr E. (1994): A&A 290, 623–633
Winters J. M., Fleischer A. J., Gauger A., Sedlmayr E. (1995): A&A 302, 483–496

Author Index

Agabi, A. 353
Ageorges, N. **349**
Aime, C. 357, 361
Angel, J.R.P. 80
Aretxaga, I. 112
Aristidi, E. **351**, 361
Avila, R. **353**
Bagnuolo, W.G. 133
Balega, Y. 206, 367
Baraffe, I. **72**
Baudoz, P. **355**
Bauer, T. 399
Beaumont, H. **357**
Beckwith, S.V.W. 391
Bedding, T.R. **3**
Berger, J.-P. 248
Berio, P. **359**
Berruyer, N. 227
Bessell, M. 332
Bester, M. 200, 363
Birk, C. 391
Bizenberger, P. 391
Blazit, A. 393
Bonneau, D. 381
Booth, A.J. **332**
Borgnino, J. 353
Bresson, Y. 351
Camenzind, M. 383
Carbillet, M. 351, **361**
Carleton, N.P. 80, 318
Chabrier, G. 72
Colberg, J. 399
Coudé du Foresto, V. **86**, 318
Cruzalèbes, P. 200, 351, **363**
Cuevas, S. 393
Danchi, W.C. 200
Davis, J. **143**, **179**, 332
Dorfi, E.A. 401
Dougados, C. 240
Drapatz, S. 259
Dutrey, A. 248
Duvert, G. 248
Eckart, A. **259**
Eislöffel, J. **240**
Encrenaz, T. **57**
Fleischer, A.J. 192, 403
Foy, R. 393
Freytag, B. **316**
Gay, J. 355
Gendron, E. 119
Genzel, R. 259, 267
Guilloteau, S. 248
Habing, H. **183**
Haniff, C.A. **212**
Harmanec, P. **171**, 381
Hartkopf, W.I. 133
Hatzes, A.P. **293**
Helmer, G. 351
Henri, G. 102
Herbst, T.M. 391
Hofmann, K.-H. 206, **367**, 387
Hofmann, R. 259
Höfner, S. 401
Holweger, H. 316
Hron, J. 401
Katterloher, R. 259
Käufl, H.-U. 153
Kerschbaum, F. 401
Koehler, B. **369**, 375
Krabbe, A. 267
Kudritzki, R.P. 267
Lafon, J.-P.J. **227**
Lagarde, S. 393
Lai, O. **119**
Langer, N. 206
Lantéri, H. 357
Lévêque, S. 369, **375**
Lopez, B. **200**, 351, 363
Löwe, M. **379**
Ludwig, H.-G. 316
Lutz, D. 267
Malbet, F. **102**, 248
Mariotti, J.-M. 86, 318, 339
Martin, F. 353

McAlister, H.A. 133
Melnick, J. 112
Menard, F. **248**
Meylan, G. **274**
Mignard, F. 351
Miley, G.K. **91**
Moffat, A.F.J. **179**
Monin, J.-L. 248
Morand, F. **381**
Mourard, D. **334**, 359, 381
Najarro, F. **267**
Natta, A. **237**
Osterbart, R. 206
Paatz, G. **383**
Perrin, G. 86, **318**
Petitjean, P. 100, 102
Petrov, R.G. 393
Quirrenbach, A. **35**, **163**, 259, **339**, 379, **385**
Rabbia, Y. 355
Reinheimer, Th. **387**, **389**
Richichi, A. 153, **326**
Ricort, G. 361
Ridgway, S.T. 133, 318
Rigaut, F. 119
Robberto, M. 391
Rouan, D. 119
Sánchez, L.J. 357, **393**
Schöller, M. 387
Scholz, M. 367
Sedlmayr, E. **192**, 403
Shao, M. **48**
Smith, L.F. 179
Stapelfeldt, K. **395**
Stecklum, B. **153**, 379
Steffen, M. 316
Tacconi-Garman, L. 259
ten Brummelaar, T.A. **133**
Terlevich, R. 112
Théodore, B. **100**
Tokovinin, A. 353
Townes, C.H. 200
Traub, W.A. **80**, 318
Turner, N. 133
Tytler, D. **121**
Vakili, F. 359
Van Winckel, H. 220
Vasyuk, V. 393
Voit, G.M. **104**
von der Lühe, O. **13**, **303**, 349, 369, 375, 379
Waelkens, C. **220**
Waters, R. 220
Weghorn, H. **397**, **399**
Weigelt, G. **206**, 367, 387, 389
Windsteig, W. **401**
Winters, J.M. 192, **403**
Wuchterl, G. **64**
Ziad, A. 353

ESO ASTROPHYSICS SYMPOSIA
European Southern Observatory

Series Editor: Philippe Crane

P. Crane (Ed.), **The Light Element Abundances**
Proceedings, 1994. XVI, 432 pages. 1995.

J. R. Walsh, I. J. Danziger (Eds.), **Science with the VLT**
Proceedings, 1994. XXV, 477 pages. 1995.

C. G. Tinney (Ed.), **The Bottom of the Main Sequence – And Beyond**
Proceedings, 1994. XVII, 309 pages. 1995.

G. Meylan (Ed.), **QSO Absorption Lines**
Proceedings, 1994. XXIII, 471 pages. 1995.

D. Minniti, H.-W. Rix (Eds.), **Spiral Galaxies in the Near-IR**
Proceedings, 1995. X, 350 pages. 1996.

H. U. Käufl, R. Siebenmorgen (Eds.), **The Role of Dust in the Formation of Stars**
Proceedings, 1995. XXII, 461 pages. 1996.

P. A. Shaver (Ed.), **Science with Large Millimetre Arrays**
Proceedings, 1995. XVII, 408 pages. 1996.

J. Bergeron (Ed.), **The Early Universe with the VLT**
Proceedings, 1996. XXII, 438 pages. 1997.

F. Paresce (Ed.), **Science with the VLT Interferometer**
Proceedings, 1996. XXII, 406 pages. 1997.

MIX
Papier aus verantwortungsvollen Quellen
Paper from responsible sources
FSC® C105338

If you have any concerns about our products, you can contact us on
ProductSafety@springernature.com

In case Publisher is established outside the EU, the EU authorized representative is:
Springer Nature Customer Service Center GmbH
Europaplatz 3, 69115 Heidelberg, Germany

Printed by Libri Plureos GmbH
in Hamburg, Germany